On Sunspots

on sunspots

GALILEO GALILEI & CHRISTOPH SCHEINER

Translated & introduced by
Eileen Reeves & Albert Van Helden

The University of Chicago Press :: Chicago & London

Eileen Reeves is professor of comparative literature at Princeton University. Albert Van Helden is professor of the history of science at the University of Utrecht and the translator of Galileo's *Sidereus Nuncius*, also published by the University of Chicago Press.

The University of Chicago Press, Chicago 60637
The University of Chicago Press, Ltd., London

Printed in the United States of America

19 18 17 16 15 14 13 12 11 10   1 2 3 4 5

ISBN-13: 978-0-226-70715-0 (cloth)
ISBN-10: 0-226-70715-6 (cloth)
ISBN-13: 978-0-226-70716-7 (paper)
ISBN-10: 0-226-70716-4 (paper)

Library of Congress Cataloging-in-Publication Data

Galilei, Galileo, 1564–1642.
On sunspots / Galileo Galilei and Christoph Scheiner ; translated and introduced by Eileen Reeves and Albert Van Helden.
p. cm.
Includes bibliographical references and index.
Summary: "A history of the controversy over sunspots with translations of the letters contained in Galileo's Istoria e dimostrazioni intorno alle macchie solari e loro accidenti. The material that was added during the printing of the Istoria, the dedication, preface, laudatory poems, and the note from the printer to introduce Scheiner's tracts, have been added in an appendix." Provided by publisher.
ISBN-13: 978-0-226-70715-0 (cloth : alk. paper)
ISBN-10: 0-226-70715-6 (cloth : alk. paper)
ISBN-13: 978-0-226-70716-7 (pbk. : alk. paper)
ISBN-10: 0-226-70716-4 (pbk. : alk. paper) 1. Sunspots—Early works to 1800. I. Reeves, Eileen Adair. II. Van Helden, Albert. III. Scheiner, Christoph, 1575–1650. Tres epistolae. English. IV. Scheiner, Christoph, 1575–1650. Accuratior disquisitio. English. V. Galilei, Galileo, 1564–1642. Istoria e dimostrazioni intorno alle macchie solari e loro accidenti. English. VI. Title.
QB525.G22 2010
523.7'4—dc22

2009042191

∞ The paper used in this publication meets the minimum requirements of the American National Standard for Information Sciences—Permanence of Paper for Printed Library Materials, ANSI Z39.48-1992.

Contents

Abbreviations

AD	Christoph Scheiner S.J., *De Maculis solaribus et de stellis circum Jovem errantibus accuratior disquisitio* (Augsburg, 1612).
JKGW	Johannes Kepler, *Johannes Kepler Gesammelte Werke*, 20 vols (Munich: C. H. Beck, 1937–).
OGG	Galileo Galilei, *Le Opere di Galileo Galilei*, 20 vols, ed. Antonio Favaro. Edizione Nazionale (Florence: G. Barbèra, 1890–1909, reprinted 1929–1939, 1964–1968).
RU	Christoph Scheiner, S.J., *Rosa Ursina* (Bracciano: Apud Andream Phaeum, 1626–1630). (Note that the pagination of *Rosa Ursina* is highly irregular and at times has to be indicated by recto and verso additions.)
SN	Galileo Galilei, *Sidereus Nuncius, or, The Sidereal Messenger*, translated with introduction, conclusion, and notes by Albert Van Helden (Chicago: University of Chicago Press, 1989).

Preface

This book project has a long history. It began with discussions between Mario Biagioli and Albert Van Helden in the early 1990s about the several reasons for providing an English translation of Galileo's controversy with Christoph Scheiner over the nature of sunspots. Biagioli's *Galileo Courtier* (1993) was then in press, and Van Helden's translation of Galileo's *Sidereus Nuncius* (1989) had recently appeared. Three developments in the historiography of science came together in their conversations: scientific controversies, Jesuit science, and the "visual language" of science. Steven Shapin and Simon Schaffer's *Leviathan and the Air-Pump* (1985) had shown the importance of scientific controversies in epistemological shifts, and the debate between Galileo and Christoph Scheiner about the nature of sunspots was an obvious target. This was all the more timely, because Jesuit science in the early modern period was becoming an area of attention. To this must be added the increasing interest in the role of pictorial representations in the early modern period.

The sources for the controversy between Scheiner and Galileo were available in Italian and Latin, but problematic in other languages. Scheiner's two tracts in Latin had never been translated into other languages, and only certain portions of Galileo's three sunspot letters were available in English in Stillman Drake's *Discoveries and Opinions of Galileo* (1957). Drake's abridged translation, moreover, did not include the long visual appendix of the second letter, which showed Galileo's observations and was a crucial part of his argument.

An English edition of *Istoria e dimostrazioni intorno alle macchie solari e loro accidenti* (1613) seemed therefore to be a worthwhile project, and the National Endowment for the Humanities agreed to the point of granting funds to make possible a cooperative project between Biagioli and Van Helden. Because of other projects, however, Biagioli had to withdraw. Van Helden then joined up with Eileen Reeves, who shared the original intellectual motivations for the project, but who also brought to it a broad knowledge of sources usually outside the scope of historians of science. The focus of the work shifted accordingly.

The manuscript was submitted to the University of Chicago Press in 2005, and was originally an annotated edition of *Istoria e dimostrazioni intorno alle macchie solari e loro accidenti*. The referees and several of the commentators pointed out an important structural problem in that configuration of Galileo's and Scheiner's texts. Galileo had read only Scheiner's first tract, *Tres Epistolae de Maculis Solaribus* (1612), when he wrote his first and second letters in response. In the meantime, Scheiner had already finished his second work, *De Maculis Solaribus et Stellis circa Iovem Errantibus Accuratior Disquisitio* (1612), and upon reading this tract, Galileo wrote his third and final letter. Half of the print run of *Istoria e dimostrazioni intorno alle macchie solari e loro accidenti*, also included, after Galileo's letters, Scheiner's two Latin tracts, an addition that altered the original sequence of the exchange. In writing the introduction to the edition, we had discussed the texts in the order in which they were written, but this arrangement somewhat obscured the sequence of the translated texts. We were strongly advised to rearrange the letters in chronological order, and we have accepted that advice. The book is therefore no longer an English edition of the 1613 publication, but a history of the controversy with translations of the letters. The material that was added during the printing of *Istoria*, the dedication, preface, laudatory poems, and the note from the printer to introduce Scheiner's tracts, have been added in an appendix.

Over the years, we have had aid, advice, and feedback from a number of colleagues, to whom we are very grateful. The initial translation of Scheiner's *Tres Epistolae* and *Accuratior Disquisitio* were corrected and improved by James Houlihan; the entire manuscript was commented on by Robert S. Westman, Noel Swerdlow, Michele Camerota, Ugo Baldini, William R. Shea, Lodewijk Palm, Isabelle Pantin, Maurice A. Finocchiaro, Owen Gingerich, Peter Dear, Sven Dupré, Christoph Lüthy, Paolo Galluzzi, and the anonymous referees.

Steven Harris, Rita Haub, and Michael John Gorman have helped us with the correspondence of Scheiner and his Jesuit colleagues, David Pankenier has secured images for us, and Franz Daxecker has generously shared his own research on Scheiner with us. Rob van Gent has been a source of astronomical and bibliographical information. We thank Thomas Hart for his help with Greek terminology. We are also especially grateful to Noel Swerdlow both for his support throughout this project, and for the technical expertise he provided in preparing the mathematical appendixes to the letters.

The extent of the debt we owe Mario Biagioli is difficult to convey to the reader. He was there at the beginning and never stopped challenging, advising, and encouraging us. In his recent book *Galileo's Instruments of Credit* he treats the sunspot argument from the perspective of Galileo, focusing on the Pisan's rhetorical and career strategies. Chapter 3 of his book is, as it were, a companion to our book.

We thank all these colleagues for their generous help; we remain responsible for any errors that may have crept into this work. We are grateful to Dmitri Sandbeck, Jennifer Howard, Abby Collier, Karen Merikangas Darling, Ryan Li, and Michael Koplow at the University of Chicago Press for their immense help in converting this manuscript to a book. We would like to acknowledge Martha Mayou for compiling an excellent index of the completed work. Our greatest debt by far, finally, is to Jim, Jimmy, and John English, and to Sara Booth, for living so long with sunspots.

1

Introduction

The Sun too, both in rising and in withdrawing
Beneath the waves, will give you signs; the Sun
Commands most certain signs, both those he brings
At break of day and when the stars are rising.
If hiding in a cloud he wears his morning guise
Flecked, and the centre of his disk concave,
Beware of showers: the south wind, from the deep
Driving, bodes ill for trees and crops and herds . . .
This too still more it will pay you to remember:
When, having spanned Olympus, he's departing
Often we see colours of varying hue
Wander across his face; purple means rain,
Flame-colour means east winds; but if the flecks
Begin to take a tinge of fiery red,
Then you will see a welter everywhere
Of winds and storm clouds both . . .
Who dares call the Sun a liar?

Virgil, *Georgics*[1]

1. "Sol quoque et exoriens et cum se condet in undas / signa dabit; solem certissima signa sequentur, / et quae mane refert et quae surgentibus astris. / ille ubi nascentem

The study of sunspots emerged in the second decade of the seventeenth century as an essential element of the debate about the nature of the heavens: if these evanescent phenomena were really located on the Sun, then the most splendid body in the heavens was imperfect, and with it the entire heavens. The instrument that made these and other celestial features visible, the telescope, was still a novelty, and the observations made with it were likewise subjects of controversy and negotiation. An entirely new dimension had been added to the study of the heavens; an instrument lately emerged from the mathematical or optical and mechanical traditions made available information that bore directly on philosophical and cosmological issues. Telescopic astronomy, as we might call it, was a subject that could not have been imagined in 1607—the year before the instrument came into use—and in 1612 and 1613, when the sunspot controversy raged, it had yet to find an institutional setting, an epistemological foundation, or even an adequate optical explanation. The initial arguments about the reality of the celestial phenomena originally observed through it, especially the uneven surface of the Moon and the satellites of Jupiter, were only just beginning to recede, giving way to debates about their interpretation, when the newly visible solar markings focused attention on the *precision* of telescopic observations, and added to the arguments concerning the nature of the heavens.

That Galileo's interpretation of sunspots prevailed over those of his opponents is indisputable; what one could perhaps more reasonably debate is the question of priority in their discovery.[2] This controversy, initiated by Galileo and his rival Christoph Scheiner themselves, was carried on with vigor by some

maculis uariauerit ortum / conditus in nubem medioque refugerit orbe, / suspecti tibi sint imbres: namque urget ab alto / arboribusque satisque Notus pecorique sinister. . . / hoc etiam, emenso cum iam decedit Olympo, / profuerit meminisse magis; nam saepe uidemus / ipsius in uultu uarios errare colores: / caeruleus pluuiam denuntiat, igneus Euros; / sin maculae incipiunt rutilo immiscerier igni, / omnia tum pariter uento nimbisque uidebis / feruere / Solem quis dicere falsum / Audeat?" Virgil, *Georgics*, translated with an introduction and notes by L. P. Wilkinson (London: Penguin, 1982), 1: vv. 437–444, 450–457, 463, p. 71.

2. The most exhaustive treatment of this controversy is Bellino Carrara S.J., "L''Unicuique Suum' nella scoperta delle macchie solari," *Memorie della Pontificia Accademia in Roma dei Nuovi Lincei* 23 (1905): 191–287; 24 (1906): 47–127. See also Antonio Favaro, "Sulla priorità della scoperta e della osservazione delle macchie solari," *Memorie del Reale Istituto Veneto di Scienze, Lettere ed Arti* 22 (1887): 729–790; and Walter M. Mitchell, "The History of the Discovery of the Solar Spots," *Popular Astronomy* 24 (1916): 22–31, 82–96, 149–162, 206–218, 290–303, 341–354, 428–440, 488–499, 562–570; Emil Wohlwill, "Zur Geschichte der Entdeckung der Sonnenflecken," *Archiv für die Geschichte der Naturwissenschaften und der Technik* 1 (1908–1909): 443–454. A judicious and appropriately short judgment on this matter can be found in Michele Camerota's recent *Galileo Galilei e la cultura scientifica nell'Età della Controriforma* (Rome: Salerno Editrice, 2004), 241–254.

nineteenth-century scholars, and as recently as 1974 John North weighed in with a defense of the priority of the English scientist Thomas Harriot.[3] The very notion of "discovery" needs some qualification, for sunspots have been observed since Antiquity, and were documented for almost two millennia before the advent of the telescope. What little that can be added to the tired question of priority is that the first record of a telescopic observation of sunspots was made by Thomas Harriot on 18 December 1610 (Gregorian), and that the earliest printed reference to its telescopic appearance came in a publication by the Frisian Johannes Fabricius in the summer of 1611. We prefer to let the voluble claims of Galileo and Scheiner emerge within the context of the controversy over the nature of the spots.

At the heart of the debate lay the question of the location and nature of the spots: those issues bore directly on the ongoing cosmological quarrel that had preoccupied European astronomers and philosophers for over a century. The question of the nature of the heavens had been raised well before Copernicus; it had been addressed both by the mapping of comets in the fifteenth century and by astronomers' novel claim that parallax measurements of those comets would show whether the original distinction between the perfect and unchanging heavens and the corrupt and ever-changing terrestrial region in Aristotelian cosmology could be maintained.[4] While the pace and intensity of the debate increased with the publication of Nicolaus Copernicus's *De Revolutionibus* in 1543, heliocentrism was by no means the only issue. As James Lattis has demonstrated, the prominent Jesuit Christoph Clavius exerted himself against Copernicus's heliocentrism, Tycho Brahe's geo-heliocentrism, Girolamo Fracastoro's revival of homocentric spheres, and neo-Stoic notions of a fluid heaven in which celestial bodies flew like birds in the air or swam like fish in the sea. In each of these alternative cosmologies, the nature of the heavens was a central issue.

3. John D. North, "Thomas Harriot and the First Telescopic Observations of Sunspots," in *Thomas Harriot: Renaissance Scientist*, ed. John W. Shirley (Oxford: Clarendon Press, 1974), 129–165. An even earlier claim was made by Adriaan Metius. In his *Institutiones Astronomicae* of 1614 Metius claims to have observed sunspots as early as 1608. See Tabitta van Nouhuys, *The Age of Two-Faced Janus: The Comets of 1577 and 1618 and the Decline of the Aristotelian World View in the Netherlands* (Leiden: Brill, 1998), 228.

4. Jane L Jervis, *Cometary Theory in Fifteenth-Century Europe* (Dordrecht: Reidel, 1985); C. Doris Hellman, *The Comet of 1577: Its Place in the History of Astronomy* (New York: AMS Press, 1944); William H. Donahue, *The Dissolution of the Celestial Spheres* (New York: Arno Press, 1981); William H. Donahue, "The Solid Planetary Spheres in Post-Copernican Natural Philosophy," in Robert S. Westman, ed., *The Copernican Achievement* (Berkeley: University of California Press, 1975), 244–275.

Several phenomena had already brought the solidity of the heavens into question.[5] By the first decade of the seventeenth century, an increasing number of astronomers believed that the new stars of 1572, 1600, and 1604,[6] as well as the comet of 1577, had been shown by the best measuring instruments to be located above the Moon. Philosophers, however, were slower to change their minds. In the wake of the Council of Trent (1546–1563) the connections between cosmology and biblical interpretation came increasingly under scrutiny, and in the religious orders the pressure for conformity became greater.

The resulting tension was especially apparent within the Society of Jesus, which had become prominent in education and had produced a number of talented investigators of nature. On the one hand, in the 1570s Father Robert Bellarmine, basing himself solely on biblical sources, had depicted an entire cosmology in his Louvain lectures; on the other, Father Clavius used Scriptural authority in arguments against alternative cosmologies and in support of the Aristotelian-Ptolemaic system. Jesuit teachers of the mathematical subjects in the provinces looked to their colleagues at the prestigious Collegio Romano for guidance, but it appears that they were often left to decide for themselves what they could and could not teach about the world system. A letter of 1607 from Father Johannes Lanz of the Jesuit University of Ingolstadt in South Germany to Father Clavius in Rome suggests that recent discussions of comets and of the New Star of 1604 were particularly perplexing. And the frequent queries of the erudite and well-connected Augsburg Humanist Marc Welser, closely associated with the Society, were doubtless flattering, but also served to increase the pressure on Jesuit astronomers both in Rome and beyond. A letter that Welser wrote in the fall of 1607 to his friend Joseph Scaliger suggests that he had rather high expectations of the sky watchers at Ingolstadt: "You may read in the attached passage what was written to me from Prague about the appearance of the heavens; our rather sleepy astronomers here have as yet seen nothing of this, but they will do so, once alerted by me. Let me know, please, what your astronomers think about it."[7]

5. Note that in the High Middle Ages and Renaissance there was no consensus among Aristotelians on whether the heavens were solid or fluid. See Edward Grant, *Planets, Stars, and Orbs: The Medieval Cosmos, 1200–1687* (Cambridge: Cambridge University Press, 1994), 324–370.

6. Those of 1572 and 1604 were supernovas; the new star of 1600, P Cygni, is classified as a variable star of the type "S Doradus."

7. "Quid de Praga de coeli facie ad me perscribatur, in scheda leges; nostri hic Mathematici somniculosiores, nihil hujus adhuc observarunt, sed observabunt, à me moniti: quid vestri sentient, fac, quaeso, sciam." Welser to Scaliger, in Marc Welser, *Opera Historica et Philologica, Sacra et Profana*, ed. Christoph Arnold (Nuremberg: Wolfgang Moritz, 1682), 812. The appearance of the heavens referred to here is in all likelihood the comet of 1607, Halley's Comet.

The telescope greatly exacerbated these issues. The older verbal explanation of the spots visible to the naked eye on the surface of the Moon—that they were caused by "rarer and denser" matter distributed within a smooth sphere—did not survive the advent of the telescope by more than two or three years. The satellites of Jupiter were an entirely unexpected revelation, and they proved, regardless of one's preference for a particular world system, that there was more than one center of motion in the universe. But the phenomena themselves, and the instrument that produced them, remained controversial for a year or more after the publication of *Sidereus Nuncius*, even in Italy, where the best telescopes were to be found.

The role played by the Jesuit mathematicians of the Collegio Romano in the reception of the new instrument, the verification of the new phenomena, and the exegetical problems these produced was therefore crucial, for mathematicians from that order from all over Europe, as well as those in the Far East and the New World, took their cue from their counterparts in Rome.[8] Within that institution, the cautious self-censorship of its senior mathematician, Christoph Clavius, was balanced by a certain readiness to refine their telescopes, to undertake celestial observations, and to consider new cosmological models among his younger colleagues, particularly Christoph Grienberger, Giovanni Paolo Lembo, Odo van Maelcote, and Paul Guldin.[9]

In South Germany, another group of Jesuit scholars soon became interested in the new celestial phenomena, and in a relatively short period began to produce research results. Christoph Scheiner emerged as the leading figure among these astronomers. Intelligent and ambitious, he did not want his publication of his observations of the novel phenomena of sunspots to be delayed by the

8. See in this connection Pasquale d'Elia, *Galileo in Cina: relazioni attraverso il Collegio Romano tra Galileo e i Gesuiti scienziati missionari in Cina, 1610–1640* (Rome: Gregorian University, 1947); Ugo Baldini, "The Portuguese Assistancy of the Society of Jesus and Scientific Activities in Its Asian Missions until 1640," in *Proceedings of the First Meeting History of Mathematical Sciences: Portugal and East Asia* (Camarante: Fundação Oriente, 2000), 49–104; and Henrique Leitão, "A Periphery Between Two Centres? Portugal on the Scientific Route from Europe to China (Sixteenth and Seventeenth Centuries)," in Ana Simões, Ana Carneiro, and Maria Paula Diogo, eds., *Travels of Learning: A Geography of Science in Europe* (Dordrecht: Kluwer Academic Publishers, 2003), 19–46.

9. On the mathematical and philosophical traditions at the Collegio Romano in the early modern period, see especially Corrado Dollo, "Le ragioni del geocentrismo nel Collegio Romano (1562–1612)," in Massimo Bucciantini and Maurizio Torrini, eds., *La diffusione del Copernicanesimo in Italia* (Florence: Leo S. Olschki, 1997): 99–166; Ugo Baldini, ed., *Christoph Clavius e l'attività scientifica dei Gesuiti nell'età di Galileo* (Rome: Bulzoni: 1995); James M. Lattis, *Between Copernicus and Galileo: Christoph Clavius and the Collapse of Ptolemaic Cosmology* (Chicago: University of Chicago Press, 1994); Michael John Gorman, *The Scientific Counter-Reformation: Mathematics, Natural Philosophy, and Experimentalism in Jesuit Culture, 1580–c. 1670* (Florence: European University Institute, 1998), 15–48.

bureaucratic machinery of the order; through the agency of Marc Welser in nearby Augsburg, in January and September 1612 he was able to publish his results and arguments, the *Tres Epistolae* and the *Accuratior Disquisitio*, under a pseudonym, and without submitting these works to censors. Scheiner was unable, however, either to maintain his secret identity or to impress his superiors with his publications, for he had proposed what was in essence a new cosmology, one in which innumerable planetary objects moved through fluid heavens and around an immaculate Sun. Thus while Galileo attacked Scheiner from a position of covert Copernicanism,[10] the rather unorthodox nature of the Jesuit astronomer's views proved troubling to conservative thinkers within his order and to the Catholic Church at large. It is therefore significant that the order's general, Claudio Aquaviva, circulated letters in 1611 and 1613 reminding all Jesuits to adhere to Aristotelian tradition in matters of cosmology, and that when Scheiner sought to publish later works under his own name, his texts were scrutinized closely by his superiors.

Scheiner was in some regards a tragic figure. Depicted as a dull and conservative thinker by Galileo, he appeared ungovernable to many within his own order. He was convinced by Galileo's arguments and went on to do an extended research project on sunspots, the *Rosa Ursina* of 1630, which remained the standard work on the subject for a century.[11] He was not, however, successful in Rome, where he went in the 1620s, and was condemned to spend the last two decades of his life in relative obscurity in the German provinces.[12] His *Prodromus pro Sole Mobili*, a refutation of Galileo's argument for the motion of the Earth on the basis of the motions of sunspots, completed shortly after the appearance of Galileo's *Dialogo* (1632), was published only posthumously in 1651 and read by few. Over the past two decades, interest in Scheiner and in his fellow investigators of nature in the Jesuit order has steadily increased.[13] Until recently, Scheiner was a figure of little moment to historians, while Galileo, whose subsequent career likewise took a tragic turn, has been a celebrated cultural icon for four centuries.

10. Note that only at the very end of his third letter on sunspots does Galileo cautiously express his approval of the Copernican theory. See p. 296, below.

11. For Scheiner's later scientific work, see below, pp. 311–316.

12. Marcus Hellyer, *Catholic Physics: Jesuit Natural Philosophy in Early Modern Germany* (Notre Dame: Notre Dame University Press, 2005), 122, 131.

13. See, for instance, Rivka Feldhay, *Galileo and the Church: Political Inquisition or Critical Dialogue?* (Cambridge: Cambridge University Press, 1995); Peter Dear, *Discipline and Experience: The Mathematical Way in the Scientific Revolution* (Chicago: University of Chicago Press, 1995); Hellyer, *Catholic Physics.*

2

Sunspots before the telescope

In the Western tradition, the scientific study of sunspots dates from the early seventeenth century, when the telescope made them apparent to observers. Though for the most part so small that they remain undetected in the bright field of the Sun, from time to time larger spots have been visible to the naked eye. Because of their interest in celestial omens, Far Eastern observers carefully noted such spots, and records dating back to at least 165 BC have come down to us.[1]

The observational record of the sunspots is much less complete in the West. Because in the Greek tradition the heavens were generally considered perfect and unchanging, technical astronomers and philosophers alike consigned phenomena such as meteors, comets, and sunspots to the imperfect and mutable realm of the sublunary world.

1. K. K. C. Yau and F. R. Stephenson, "A Revised Catalogue of Far Eastern Observations of Sunspots (165 BC to AD 1918)," *Quarterly Journal of the Royal Astronomical Society* 29 (1988): 175–197. Yau and Stephenson list no fewer than 235 naked-eye sunspot observations made in East Asia between 165 BC and AD 1918. The vast majority and the only ones before the twelfth century (except for the sole Japanese observations in AD 851), were made in China. The Korean record begins in AD 1151. See Yau and Stephenson, 178–197. See also Zhentau Xu, David W. Pankenier, and Yaotiao Jiang, *East Asian Archaeoastronomy: Historical Records of Astronomical Observations of China, Japan, and Korea*, Earth Space Institute Series, no. 5 (Amsterdam: Gordon and Breach Science Publishers, 2000), 147–181; and David H. Clark and F. Richard Stephenson, "An Interpretation of the Pre-telescopic Sunspot Records from the Orient," *Quarterly Journal of the Royal Astronomical Society* 19 (1978): 387–410.

Figure 2.1. Illustration of sunspots from the 1425 edition of the Ming dynasty work *Tianyuan baolixiang fu*. Cover illustration of Zhentau Xu *et al., East Asian Archeoastronomy*. Courtesy of the Nanjing University Library.

They were weather phenomena, and our word for the study and prediction of weather, *meteorology*, still reflects that classification.[2] The earliest known Western reference, attributed to Theophrastus of Eresus in the fourth century BC,

2. For a survey of various attempts to connect sunspots with short- or long-term weather patterns, see John A. Eddy, "Climate and the Role of the Sun," *Journal of Interdisciplinary History* 10, no. 4

merely states that "black spots in the Sun and the Moon foretell rain, and reddish ones wind."[3] Virgil's more ample presentation of sunspots appeared in the first book of the *Georgics*, composed around 36–29 BC. It is significant in that it describes such spots, portrayed as *maculae* or "flecks," as the best indices of approaching storms or showers, and treats phenomena that medieval and early modern observers would find rare and astonishing as quotidian, fleeting, and natural.

While Virgil's intention in the *Georgics* was to contrast these sorts of signs with the supernatural portents of Caesar's death, the two categories appear to have been conflated in some of the very few medieval observations of sunspots. Consider, in this connection, two poetic passages describing anomalous solar activity in late 813 and early 814, shortly before the death of Emperor Charlemagne on 28 January 814. Both occur in the work of the anonymous Saxon Poet, who presented around 890 familiar but somewhat conflicting impressions of the sunspots themselves.

> And as the shadow falls longer at day's end,
> And the sun hastening to bathe itself in the waves
> Rushes to its resting place, and hides,
> And gives strong warnings of the night to come
> With the iron-red of its light-bearing face,
> Its pallor having been mixed with dark spots...
> Not otherwise did prudent men then sense
> The coming storms of wars, rightly mourning
> The ruin of the kingdom of the Franks
> With the death of Charlemagne.[4]

> He died in the eight hundred and fourteenth year
> From Christ's birth, leaving this life
> On the fifth calend of the month in which Numa Rex

(1980): 725–747; on records and other long-term changes in the sun, see J. M. Vaquero, "Historical sunspot observations: A review," *Advances in Space Research* 40 (2007): 929–941.

3. Theophrastus, "De signis pluviarum, ventorum, tempestatis et serenitatis." In *Theophrasti Eresii opera, quæ supersunt,* ed. and tr. Friedrich Wimmer (Paris: A. Firmin Didot, 1866), II (27), p. 393. The Greek term which is here translated as "black spots" is σεμία μελανα.

4. *Poetæ Saxonis Annalium de gestis B. Caroli Magni libri quinque,* in *Patrologia Latina*, ed. Jacques-Paul Migne, 221 vols. (Paris: Garnier & J.-P. Migne, 1844–1865) vol. 99, cols. 0724A–0724B. The work was first published as *Annalium de gestis Caroli Magni Imp. Libri V. Opus auctoris quidem incerti, sed Saxonis & historici & poëtæ antiquissimi* (Helmstadt, Typis Iacobi Lucij, 1594). For a provisional list of early sunspot observations in Europe, see George Sarton, "Early Observations of the Sunspots?" *Isis* 37 (1947): 69–71.

Established the Festival of the Februa.[5]
There had been many portents of his approaching death,
And fitting signs of so great an event.
And thus it had happened that there had been
Three years before frequent eclipses
Of both the sun and the moon.
And for seven days a black color
Was seen on the sun, spotting its light-bearing face.[6]

The Saxon Poet's shift from a metaphorical mode in the first passage to an actual observation in the second, the ambiguity of his references to the many spots that signal storms and to the single spot that stayed for seven days, and the close coordination of portent and event, so unconvincing to the modern eye, may all serve to undermine his report of solar phenomena. The Saxon Poet had, in fact, based his account on the Frankish historian Einhard's *Life of Charlemagne*, composed around 830, where three years of solar and lunar eclipses and a sunspot of seven days' duration cluster about the great ruler's demise.[7] A fourth and quite important example from this era makes it clear that the inauspicious spots, as well as the eclipses, had been seen a few years earlier, and were only subsequently transferred to the more appropriate context of the emperor's death. This version of Charlemagne's life offered a host of astronomical information for the year 807. An early modern description of the work suggests that it was valued for a kind of naïve fidelity to events, for it was presented as *The Life of Charlemagne the King and Emperor of the Franks, described in large part, as it seems, by a monk of the monastery of Saint Eparchius of Angoulême, on the basis of annals composed in plebeian and rustic language*.

The previous year on the first of September [806][8] there was an eclipse of the Moon,[9] and the Sun was in the sixteenth part of Virgo, and the Moon in the

5. That is, on 28 January 814, the generally accepted date of Charlemagne's death.

6. *Poetæ Saxonis Annalium*, in *Patrologia Latina*, vol. 99, col. 0734D.

7. "Appropinquantis finis complura fuere præsagia, ut non solum alii, sed etiam ipse hoc minitari sentiret. Per tres continuos vitæque termino proximos annos et solis et lunæ creberrima defectio, et in sole macula quædam atri coloris septem dierum spatio visa." Einhard, *Vita Caroli Imperatoris*, in *Patrologia Latina*, vol. 97, cols. 0055C–0056A.

8. The dates are given in the Roman system of Kalends, Ides, and Nones. We have converted these to modern calendrical notation.

9. There was a total lunar eclipse from 1 September 20:48 U[niversal] T[ime] to 2 September 0:32 UT in AD 806. (This and the astronomical information in the next few notes were supplied by R. H. van Gent.)

> sixteenth part of Pisces. This same year, that is the 807th year from the Incarnation of Christ, on 31 January, the Moon was in the 17th day[10] when the star of Jupiter was seen as if passing in front of her.[11] And on 11 February there was an eclipse of the Sun in the middle of the day, both stars being in the 24th part of Aquarius.[12] Likewise, on 26 February there was an eclipse of the Moon,[13] and that night armies of marvelous size were seen [in the heavens], and the Sun was in the 11th part of Pisces and the Moon in the 11th part of Virgo. And on 17 March the star of Mercury was seen as a small dark spot slightly above the center of the Sun, and it was observed by us for eight days. But when it first entered and left the Sun, we could not observe it very well because of clouds. Likewise on 22 August an eclipse of the Moon took place at the third hour of the night,[14] the Sun being in the 5th part of Virgo, and the Moon in the 5th part of Pisces. And thus from the preceding year in September until the present year in September the Sun and the Moon were darkened three times each.[15]

What would become this text's central suggestion—that the apparent spot on the Sun's face in March 807 had been caused by a transit of Mercury—was to play an important role in the history of sunspot study, and we will return to it shortly.

Special conditions, as for example the haze caused by forest fires in Russia in 1365 and 1371, could make sunspots obvious to all, even when the Sun was high in the heavens. We read in the Niconovsky chronicle for 1371:

> During this year there was a sign in the sun. There were dark spots on the sun, as if nails were driven into it, and the murkiness was so great that it was impossible to see anything for more than seven feet. . . . Woods and forests were burning and the dry marshes began to burn and the earth itself burned, and great fright and terror spread among men.[16]

10. Actually the fifteenth day after the new moon of 17 January.

11. That is, behind her. From 1:46 to 2:51 UT, Jupiter was occulted by the Moon.

12. The solar eclipse was from 9:29 to 12:42 UT.

13. Total lunar eclipse from 1:23 to 4:27 UT.

14. Partial lunar eclipse from 21 August 21:12 UT to 22 August 0:35 UT.

15. *Vita Karoli Magni*, in Pierre Pithou, ed., *Annalium et Historiæ Francorum* (Paris: Claude Chappelet, 1588), 61–62.

16. St. Petersburg Series, Russian Chronicles, published in 1897, vol. II. The translation of this passage is in A. N. Vyssotsky, "Astronomical Records in the Russian Chronicles from 1000–1600," *Meddelande fran Lunds Astronomiska Observatorium* Scr.2. Nr. 126, Historical Papers Nr. 22, Lund. We have taken it from Schove, *Sunspot Cycles*, 36.

Another medieval instance, while even more apocalyptic in tone, suggests that similar conditions had prevailed in Bohemia in the summer of 1139, and had rendered the sunspots visible to many casual observers:

> On the eighteenth of July the air became dark; indeed smoke of unusual foulness blew about, almost like a cloud, except that it never ceased to produce smoke by day and by night, such that this obscurity having lasted a week, on the thirteenth [*sic*] of that same month, around noon, a blackness of exceptional density befouled the air with a rotten stench. It was as if it were billowing upwards from hell for men to smell. And there were even some who said they had seen something like a fissure in the sun.[17]

Examples such as these make plain that sunspots must have been seen by many through the Middle Ages and early modern period. These phenomena did not enter the astronomical record because Western astronomy, ruled by the Aristotelian dogma of the perfection of the heavens, provided no conceptual rubric for sunspots. The very few observations that did enter the record through Islamic sources were interpreted within the sole context astronomers could then imagine, that of the transits of Mercury and Venus.

In his discussion of the placement of Mercury and Venus in the *Planetary Hypotheses*, Ptolemy had contrasted the Platonic view, in which those bodies moved "above" the Sun, with the Aristotelian notion, in which they were "below" it. Those who favored Plato's order, Ptolemy noted, argued that if Venus and Mercury were below the Sun, they should have been observed from time to time in their passage across that body. The absence of such observations seemed to these thinkers strong evidence of those planets' position above the Sun. Ptolemy, however, inclined to Aristotle's arrangement (Earth—Moon—Mercury—Venus—Sun—Mars—Jupiter—Saturn), and he maintained that the infrequency of the transits and the minute size of the planetary bodies might well impede the expected observations.[18] This, then, is the context of both the Carolingian chronicle, and a report by the Islamic astronomer Al-Kindi in 840 AD.

17. J. Emler, ed., *Cosmae chronicon boemorum cum continuatoribus*, in *Fontes rerum bohemicarum*, 7 vols. (Prague: Nákl. N. F. Palackého, 1873–1932), 2: 230. See also L. Krivsky, "Naked Eye Observations of Sunspots in Bohemia in the Year 1139," *Bulletin of the Astronomical Institute of Czechoslovakia* 36 (1985): 60–61, at 60.

18. Bernard R. Goldstein, *The Arabic Version of Ptolemy's "Planetary Hypotheses,"* American Philosophical Society, *Transactions* 57 (part 4), 1967, p. 6. See also *Almagest* IX, 1; *Ptolemy's Almagest*, trans. G. J. Toomer (London: Duckworth; New York: Springer Verlag, 1984), 419.

> In the year 225 during the caliphate of al-Mu'tasim there appeared a black spot (*nukta*) close to the middle of the sun. This took place on Tuesday, 19 Rajab 225 [25 May 840], and when two days had gone from this date, i.e., after 21 Rajab, [calamitous] events occurred. Al-Kindi mentioned that this spot lingered on the sun for 91 days and soon thereafter Mu'tasim died. Before the death of al-Mu'tasim two comets appeared, as some had before the death of al-Rashid. Al-Kindi mentioned that this spot was due to the occulting of the sun by Venus, and their clinging together for this period.[19]

It was also within the context of planetary order that the eleventh-century philosopher Avicenna affirmed, with a greater measure of plausibility than Al-Kindi, that he had seen Venus "as a spot on the surface of the sun."[20] And in interpreting an observation of two black spots seen on the Sun in the time of Ibn Mu'adh (fl. ca. 1079), the twelfth-century Islamic philosopher Averroës argued that at that point both Venus and Mercury had indeed been transiting the Sun.[21] That large sunspots remain visible on the Sun for a number of days, while transits of Mercury and Venus take at most a few hours, seems to have been no impediment: the reports were never of accurate and systematic observations, and the length of time that these spots appeared on the Sun was thus subject to no discussion. Aristotelian cosmology could not be "falsified" this way.

Not all explanations of the simultaneous presence of two spots involved planetary transits. The *Chronicle of John of Worcester*, for instance, though careful in its location and description of two such spots observed in 1128 AD, has recourse neither to planets nor to portents:

> 8 December [1128], Saturday, from morning to evening two blackish spots appeared below the orb of the Sun. The one in the upper part was larger, the other, in the lower part, smaller. They were opposite each other, as in this figure [fig. 2.2].[22]

19. Bernard R. Goldstein, "Some Medieval Reports of Venus and Mercury Transits," *Centaurus* 14 (1969): 49–59; reprinted in *Theory and Observation in Ancient and Medieval Astronomy* (London: Variorum Reprints, 1985), no. 15, p. 52. The quoted words are from Nasir al-Din al-Tusi's report of Avicenna's observation.

20. Goldstein, "Some Medieval Reports," 52–53, 57–58.

21. Goldstein, "Some Medieval Reports," 53–55.

22. John of Worcester, *Chronicle*, ed. J. R. H. Weaver, 2: 8–1140 (Oxford: Clarendon Press, 1908; reprint, New York: AMS Press, 1989), 28: "vi° idus Decembris, Sabbato, a mane usque ad vesperam apparuerunt quasi duae nigrae pilae infra solis orbitam, una in superiori parte erat maior, altera in inferiori fuit minor; eratque utraque directa contra alteram. Ad huiusmodi figuram." See also R. W. Southern, *Medieval Humanism* (New York and Evanston: Harper & Row, 1970), 168–169,

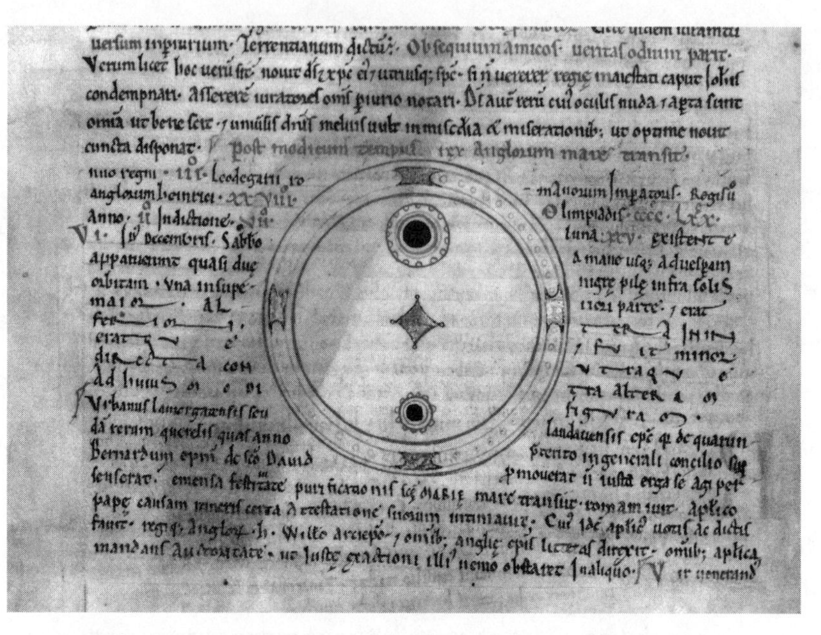

Figure 2.2. Sunspots seen in the reign of Henry I, from the *Chronicle of John of Worcester*, ca. 1130–1140. Corpus Christi College, Oxford, MSS CCC 157, p. 380. With permission from the Bridgman Art Library.

By contrast, the physician and scholar Giovanni Carrara of Bergamo relied on transits to explain the presence of two sunspots in the latter half of the fifteenth century: "At a certain time two drops of blood were seen on the sun and the populace was terrified . . . . However, my father, Guido Carrara . . . . considered and compared the planets, and found that Venus and Mercury were the cause."[23]

Among those who discussed the order of the planets, Averroës' statement remained authoritative. In his *Disputationes in astrologiam*, first published in 1494 or 1495, Giovanni Pico della Mirandola noted that "in his paraphrase of Ptolemy's *Almagest*, Averroës says that he once observed two almost black

and plate 7, as well as D. M. Willis and F. R. Stephenson, "Solar and Auroral Evidence for an Intense Recurrent Geomagnetic Storm during December in AD 1128," *Annales Geophysicae* 19 (2001): 289–302.

23. Cited in George Sarton, "Early Observations of Sunspots?" *Isis* 37 (1947): 69–71, at 70. On the Carrara family of Bergamo, see G. Ineichen, "Carrara (Alberti), Giovanni Michele Alberto," in *Dizionario biografico degli italiani*, ed. Alberto M. Ghisalberti, 65 vols. to date (Rome: Istituto della Enciclopedia Italiana, 1960–) vol. 20, 684–686. It is worth noting that neither father nor son, both renowned for their familiarity with Virgil's works, relied on the explanation offered in the *Georgics* I (see epigraph).

spots on the Sun, and when he considered the astronomy for that time Mercury was found opposed to the rays of the Sun."[24] It was in Pico's work that Nicolaus Copernicus read the reference to Averroës, incorporating it in the first book of his *De Revolutionibus* in 1543.

> [Mercury and Venus] do not eclipse the sun, because it rarely happens that they interfere with our view of the sun, since they generally deviate in latitude. Besides, they are tiny bodies in comparison with the sun.... Yet in his *Paraphrase* of Ptolemy, [Averroës] reports having seen something blackish when he found a conjunction of the sun and Mercury indicated in the tables. And thus these two planets are judged to be moving below the sun's sphere.[25]

Copernicus went on to demonstrate the feebleness of all arguments about the order of the planets in geocentric astronomy. In his own heliocentric system, this order was not a matter of dispute: it followed directly from the geometry of the solar system. The same was the case in the geo-heliocentric system of Tycho Brahe, and therefore in the new astronomy the context in which the discussion of dark spots on the Sun occurred disappeared. As we shall see, the argument would be revived in a new context by Johannes Kepler in 1607 (and discussed by Scheiner).

It is interesting to note that during the flowering of European astronomy following upon the work of Georg Peurbach and Johannes Regiomontanus in the second half of the fifteenth century and the geographical discoveries of Christopher Columbus and Vasco da Gama, the few astronomical discussions of spots seen on the Sun, inevitably interpreted as Mercury or Venus, mentioned those made centuries earlier, and usually referred back to Averroës' second-hand explanation of an observation made decades before his birth. Upon reading accounts of the darkening of the Sun in 1547 during the Holy Roman Emperor Charles V's defeat of Protestant forces at Mühlberg, Kepler suggested that the phenomenon had been caused by the interposition of cometary material, or perhaps that its source had been some newly generated and nebulous substance in the æther quite close to the solar body.[26] Though

24. Giovanni Pico della Mirandola, *Disputationes in astrologiam* X, 4, *Opera Omnia* (Basel, 1572–1573; reprint Turin: Bottega d'Erasmo, 1971), 1: 685.

25. Nicholas Copernicus, *De Revolutionibus Orbium Coelestium* (1543), I, 10; *Nicholas Copernicus on the Revolutions*, tr. Edward Rosen (Baltimore: Johns Hopkins University Press, 1978), 19.

26. *Ad Vitellionem paralipomena, quibus astronomiae pars optica traditur* VI, in *JKGW* 2: 226–227, and *De stella nova* XXIII, in *JKGW*, 1: 260–261. For an overview of the many accounts and later interpretations of this astronomical event, see Edouard-Albert Roche, *Recherches sur les offuscations du soleil et les météores cosmiques* (Paris: Leiber, 1868): 30–39. It should be noted that in

this latter hypothesis strongly resembles some of the eventual explanations of sunspots in the era of telescopic observations, it is probably crucial that Kepler did not, in this instance, begin with the need to account for something that he and many others were then scrutinizing. For most early modern Western observers, sunspots did not yet exist as a separate category of astronomical phenomena.

Were there then no prominent spots on the Sun in the late sixteenth and early seventeenth centuries? The Chinese record shows that there were, and we find occasional mention of spots on the Sun in the latter part of the sixteenth century in European sources as well, but not in European *astronomical* sources. In the Italian poet Torquato Tasso's casual mention of sunspots—*macchie del sole*—in a dialogue of 1582, they were associated with other predictable and passing indices of coming storms, precisely as they had been in Theophrastus' and Virgil's texts.[27] In 1590 English sailors off the coast of West Africa saw a spot "about the bignesse of a shilling" on the Sun; and in 1593 "two dark spots shaped like crows" were seen "inside the sun."[28] In a treatise of 1605 devoted to the celebrated apparition of a new star in the preceding year, the Florentine poet, playwright, and virtuoso Raffaello Gualterotti offered an intriguing pre-telescopic solar observation. Around sunset on 25 September 1604, he wrote,

> We saw a vast quantity of exhalations, and very tenuous and fine vapors rise and fill all the sky, and when the Sun struck them, the air towards the West was the color of fresh blood. Because in that period I spent every evening in the [Uffizi] Gallery of [Ferdinando I,] the Most Serene Grand Duke of Tus-

---

April 1547 Venus was nowhere near the Sun. On 13 April, at 3:48 UT, there was a superior conjunction of Mercury.

27. Torquato Tasso, "Il Gonzaga secondo overo del giuoco," in *Dialoghi*, ed. Ezio Raimondi, 2 vols. (Florence: G. C. Sansoni, 1958), 1: 491.

28. For the observation of 1593, see D. Justin Schove, *Sunspot Cycles*, Benchmark Papers in Geology, no. 68 (Stroudsburg, PA: Hutchinson Ross, 1983), 33, who lists as his source P.-Y. Ho, "Natural Phenomena Recorded in the Dai-viet Su'ky Toan-thu,'" *American Oriental Society Journal* 84 (1964):127–149. For the observation of 1590, see the original in the Hakluyt voyages. It is the voyage to Benin by John Newton and John Bird aboard the "Richard of Arundell." "The 7 [December] at the going downe of the Sunne we saw a great blacke spot in the Sunne, and the 8. day both at rising and setting we saw the like, which spot to our seeming was about the bignesse of a shilling, being in 5 degrees of latitude, and still there came a great billow out of the southerboord." Richard Hakluyt, *The Principal Navigations Voyages Traffiques & Discoveries of the English Nation*, 12 vols. (Glasgow: James MacLehose and Sons, 1903–1905), 6: 462. On 16 December they saw "another spot in the Sunne at his going downe" (463).

> cany, I saw in the setting Sun a black spot *somewhat bigger in size than Venus.* I saw this spot in the same place for several evenings: it was between the center and edge of the Sun.[29]

While a rather undisciplined and sometimes unintelligible writer with scant regard for convention, Gualterotti nonetheless presented the sunspot in terms of the transit theory, even as he implicitly rejected that explanation. Sunspots remained, even for followers of Copernicus, part of the authoritative textual tradition going back to Averroës. The only exception was Johannes Kepler, who, typically, turned the problem on its head. He set out to observe a transit of Mercury and instead found what later turned out to be a large sunspot.

In Ptolemaic as well as Copernican astronomy, the model of Mercury was by far the most complex and least satisfactory. Since it never strays from the Sun by more than about $23\frac{1}{2}°$, the planet can be observed rather infrequently, and only just before sunrise or shortly after sunset. Its positions can therefore be determined with any degree of accuracy only at two points of its orbit, its greatest elongations from the Sun. Catching Mercury as it crossed the face of the Sun would very accurately fix the planet's position at another point in its trajectory. Such an observation would also pinpoint the location of Mercury's nodes—the intersection of its orbital plane with the ecliptic—and offer a crucial parameter in its orbital theory. This is the problem Kepler undertook in his *Astronomiae pars optica* of 1604. Silently paring away the dubious bits of astronomical information offered in the "plebian and rustic" account of Charlemagne's life, Kepler analyzed the most plausible section of the text, the statement about the transit of Mercury. First he attacked what he saw as more of a philological problem than an astronomical issue, the spot's supposed visibility over the course of eight days. Recognizing that this could not have been maintained of Mercury, he changed the chronicler's *octo dies* to the obscure *octoties*, or "eight times." Next, he turned his attention to the theory of Mercury, but found that, unfortunately, the planet had not been near the Sun in March 807. According to his calculations, however, a transit of Mercury had occurred on 17 March 808, and Kepler thus corrected the year as well. The entry now read that on 17 March 808 a spot was seen on the Sun

29. Raffaello Gualterotti, *Discorso sopra l'apparizione de la nuova stella* (Florence, Cosimo Giunti, 1605), 28. Gualterotti's Stoicizing interpretation of the new star and sunspot are discussed in Eileen Reeves, *Painting the Heavens* (Princeton: Princeton University Press, 1997), 78–81; Thomas B. Settle, "Danti, Gualterotti, Galileo: Their Telescopes?" *Atti della Fondazione Giorgio Ronchi* 61 (2006): 625–638.

eight times. Kepler concluded, "We now do not have to believe only Averroës about this phenomenon, after a man of the Christian religion also adds his calculation."[30]

In a tract that is now lost, Kepler's teacher Michael Mästlin disagreed with his student, but Kepler defended his textual emendations, certain that Mercury had been seen on the Sun and should be visible to his generation as well.[31] The fact that one of the two works in which he had encountered the extraordinary report of astronomical events of 807 also contained other and conflicting descriptions of the much disputed spots may have further encouraged him to modify what had long been rather malleable material.[32] The opportunity to test his conclusion, in any case, soon presented itself. According to Kepler's calculations, Mercury would transit the Sun on 28 May 1607. On that date, he observed the Sun in the attic of his house, projecting its image through a small hole in the roof onto a piece of paper. On the solar image, a little less than an inch in diameter, he saw a dark spot, "about the size of a small fly, . . . dilute like a scattered cloud."[33] Kepler moved the paper several times to make sure that the spot was not the shadow of something between the aperture and the paper, suspended from the thread of a spider's web, and he then held the paper up to a little hole in another part of the roof. The dark spot still appeared on the image of the Sun. The rector of the University of Prague, in whose house Kepler lived at that time, witnessed the observations and wrote next to the record of the observation, "I, Master Martinus Bachazek, who was present at this observation, acknowledge that the observation was thus."[34]

Kepler then hurried to the court in Prague's Old City in order to acquaint

30. Johannes Kepler, *Ad Vitellionem paralipomena, quibus astronomiae pars optica traditur* (1604), in *JKGW*, 2: 264–265.

31. The lost book is *Disputatio de multivariis Motuum Planetarum in Coelo apparentibus irregularitatibus, seu regularibus inaequalitatibus, earumque Causis Astronomicis. Quam praeside Michaele Mästlino . . . defendere conabitur die 21. et 22. Februarii . . . M. Samuel Hafenreffer* (Tübingen, 1606). See also *JKGW*, 4: 489, and Edward Rosen, tr., *Kepler's Somnium* (Madison: University of Wisconsin Press, 1967), 141–142, note 385. Kepler defended his argument in a letter to Samuel Hafenreffer, 16 November 1606, *JKGW*,16: 359–362, at 360.

32. Kepler had available to him the 1594 Frankfurt edition of Pierre Pithou, ed., *Annalium et Historiæ Francorum* (Paris: Claude Chappelet, 1588), and Justus Reuber, ed., *Veterum Scriptorum, qui Caesarum et imperatorum Germanicorum res per aliquot secula gestas, literis mandarunt, Tomus unus* (Frankfurt, 1584). See *JKGW*, 4: 489. The latter also contained Einhard's chronicle and the very popular *Historia Karoli Magni et Rotholandi*, a twelfth-century account purportedly by Charlemagne's contemporary Turpin, which states that among the signs that occurred three years before the emperor's death was the fact that the Sun and the Moon both turned black for the space of seven days.

33. Johannes Kepler, *Phaenomenon Singulare Seu Mercurius in Sole* (1609), in *JKGW*, 4: 92.

34. *JKGW*, 4: 93.

the Emperor Rudolph II with his observations, and on his way ran into a Jesuit with whom he had discussed the observation earlier in the day before the clouds had parted. He told him what he had observed, but the Jesuit had used an aperture that was too large and had seen nothing. Next Kepler stopped in the workshop of Jost Bürgi, the imperial clock-maker, but found him gone. With two of Bürgi's assistants, Kepler now covered a window in a stairwell leading down into the armory, and projected the Sun's image through a hole one-ninth or one-tenth of an inch in diameter, holding the paper about fourteen feet away. It produced a solar image about two inches in diameter. Again, the spot was seen in the same place on the Sun. As clouds moved across the Sun, they saw them encroach on the immobile spot. This time Heinrich Stolle, a journeyman clock-maker, signed the observation record. When Bürgi returned, the interference of the clouds was too great for him to distinguish with any clarity the immobile spot amid the moving clouds on the Sun. After sunset Kepler questioned Matthias Seiffard, one of the late Tycho Brahe's students, whom he had advised to observe the Sun for four days around the calculated date of the transit. When asked if he had noted anything through direct observation of the Sun, Seiffard said he had seemed to see something near the edge of the solar body, but added that he had been almost blinded and could not verify it. The location of this half-glimpsed mark, did, however, agree with Kepler's observation.[35]

Kepler had satisfied himself that he had seen something that did not originate in the observing setup—it is difficult to call such a rough *camera obscura* an instrument, and Kepler himself only referred to a hole in the roof and a sheet of paper—and that it was above the clouds and appeared as a spot on the Sun. Moreover, it could not be argued that it was a product of his imagination or of his admittedly poor eyesight: others had seen it with him. He was convinced that he had seen Mercury on the Sun. Though then current estimates of planetary angular diameters predicted that Mercury would occupy about one-tenth of the Sun's diameter, however, Kepler's figure showed the spot covered only about one-twentieth of that length. He tried to explain this difference, as well as the dilute appearance of the spot, by discussing the size and shape of the apertures in the several setups he had used.[36] Being convinced that he had seen Mercury on the Sun, and knowing that such a transit lasted only a few hours, Kepler did not repeat the observation the next clear day. Had he done so, he would have been in for a surprise.

In going back through the record of sunspot sightings during the pretelescopic era, then, we see that although such sightings were more frequent and

35. *JKGW*, 4: 93–94.
36. *JKGW*, 4: 95.

better documented in the Chinese tradition, Western records are by no means lacking, and that such spots must have been sighted on many other occasions. The latitudes of China and Western Europe are roughly the same, so that we cannot argue that the discrepancy was due to a geographic difference. A recent test has shown that even small sunspots can be seen with the naked eye near sunset under normal conditions.[37] The single most significant impediment to sustained sunspot observation in the West appears to have been the lack of a clear conceptual rubric. Just how strong these cosmological strictures were is illustrated by the observation made in 1607 by Kepler, surely the most liberated and creative astronomer of his time, if not of all history.

Sixteen months after Kepler's observation, Hans Lipperhey, a spectacle-maker in the city of Middelburg in the United Provinces, set off on a journey to The Hague to petition the States General for a patent on a new device for seeing faraway things as though they were nearby.[38] By the spring of 1609 spyglasses were available in France, Germany, Italy, and England, and that summer Thomas Harriot was observing the Moon through a six-powered instrument while Galileo was preparing an eight-powered one that he presented to the Venetian senate. That autumn Galileo turned such a spyglass —and then even more powerful ones—to the heavens. The age of telescopic astronomy had begun, and with it the scientific study of sunspots in the West.

37. J. E. Mossman, "A Comprehensive Search for Sunspots without the Aid of a Telescope, 1981–1982," *Quarterly Journal of the Royal Astronomical Society* 30 (1989): 59–64; Peter Wade, "Naked Eye Sunspots, 1980–1992," *Journal of the British Astronomical Association* 104 (1994): 86–87.

38. For a discussion of the various names given to the new instrument, see Edward Rosen, *The Naming of the Telescope* (New York: Henry Schuman, 1947).

3

Turning the telescope to the Sun: Thomas Harriot and Johannes and David Fabricius

Much attention has been devoted to the question of priority in telescopic sunspot observations, the principal candidates being Thomas Harriot, Johannes and David Fabricius, Galileo Galilei, and Christoph Scheiner. The origins of these controversies must be found in the writings of Galileo and Scheiner themselves, and for the most part learned men of the early modern period believed Galileo's account. The flowering of historical scholarship, often in the context of nineteenth-century nationalist agendas, led to a reexamination of the claims of both astronomers, as well as the advocacy of those of Harriot and of Johannes and David Fabricius. Usually the verdict correlated well with the author's nationality and/or religious affiliation. In the case of Galileo, nationalism merged with hagiography, his claims being supported by a broader and more ecumenical range of scholars.

We do not believe that we can add anything to what has already been written on this subject, and we do not believe that a definitive answer—if that were possible—to the question of priority would add anything important to our historical knowledge. The phenomena that we call sunspots had, after all, been observed for millennia in various parts of the world, including the West, and it is thus difficult to call the first telescopic observation of them a "discovery." For that reason we will content ourselves with stating that the scientific study of spots on the Sun by European astronomers began in 1611. Over the next several years they formulated a theory of sunspots that is the linear ancestor of our current concept. In doing so, they separated

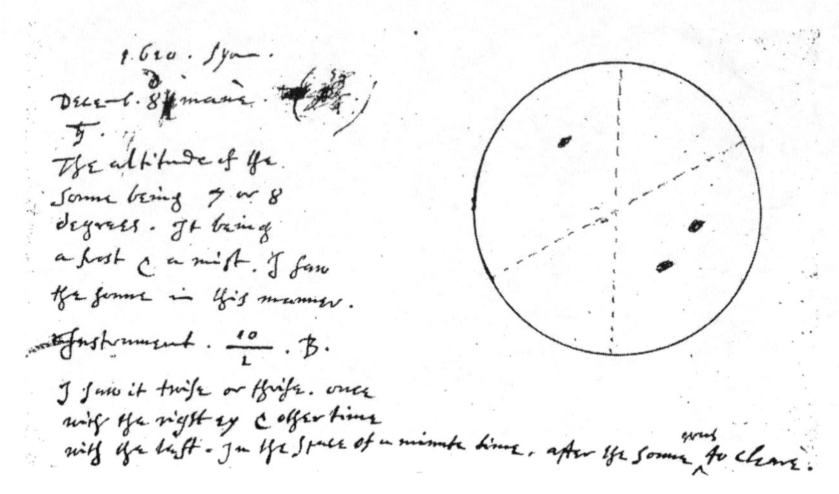

Figure 3.1. Thomas Harriot's observation of the Sun, 18 December 1610. Harriot MSS, HMC 241/8, p. 1. Courtesy of Lord Egremont.

the study of these phenomena from transits of Mercury and Venus, which were pursued with great profit in separate research programs made possible by the telescope. It is at this point of separation that we must begin.

On 18 December 1610 (Gregorian), the English explorer, mathematician, and astronomer Thomas Harriot observed the Sun through a ten-powered telescope and made the following entry in his observational records:

> 1610. Syon [house]
> Decemb. 8. *mane* [Saturday] The altitude of the Sonne being 7 or 8 degrees. It being a frost & a mist. I saw the sonne in this manner. Instrument. 10/1. B.[1] I saw it twise or thrise. once with the right ey & other time with the left. In the space of a minute time, after the Sonne was to cleare.[2]

This is the oldest surviving record of a telescopic observation of spots on the Sun; it is also one of the most interesting ones. Harriot had been observing the nighttime heavens through a telescope for eighteen months—since at least early August 1609—and we may well ask why he and others had waited so long before turning their instruments to the Sun. Part of the answer surely lies in the

1. Harriot nearly always indicated the instrument he used; in this case a ten-powered telescope designated "B." England was still on the Julian calendar: 8 December Julian = 18 December Gregorian.

2. Harriot MSS, HMC 241/8, p. 1

time and conditions of observation. It was early on a December morning, the Sun was still close to the horizon, and the air was misty. At such times more of the solar rays are scattered during their long trip through the thick atmosphere, and the Sun appears as a pale reddish-yellow disk. Conditions such as these were optimal for the direct observation of sunspots. With the telescopes used in 1610—instruments with apertures of perhaps 1 to $1\frac{1}{2}$ cm—it was just possible to manage a very brief look at the Sun. In the space of a minute, Harriot was able to peek at the Sun two or three times, first with his right eye, then with his left. Beyond this point, presumably because the Sun was rising and the mist was dissipating, the Sun was too bright and the observation became too painful. Observing the Sun directly through a telescope, then, required special conditions not very different from those that prevailed when sunspots were seen fortuitously during the pretelescopic era; it also brought with it the risk of permanent damage to the observer's eyes.

Given the painful nature of solar observations, it is logical to inquire why the attempt would have been made in the first place. Were these three spots observed on 18 December so large that Harriot saw them with the naked eye and subsequently verified them? We cannot be sure that their size in the drawing is accurate, but this was a period of exceptionally high sunspot activity, and in 1612 Galileo accurately represented sunspots so large that under the right observing conditions they must have been visible with the naked eye.[3] It is more likely, however, that Harriot was initially interested in Venus, which was in the morning sky. The planet had come into the best observing position, being at its greatest elongation from the Sun, on 18 December. On 5 December 1610 Benedetto Castelli had written Galileo saying that the Copernican theory predicted that Venus should appear horned when far removed from the Sun,[4] and on 11 December Galileo had sent an anagram to Giuliano de' Medici containing his verification of the phases of Venus.[5] Harriot, too, was a Copernican,[6] and he may very well have been pursuing the same problem when

3. Kunitomo Sakurai, "The Solar Activity in the Time of Galileo," *Journal for the History of Astronomy* 11 (1980): 164–173.

4. *OGG*, 10: 481: "If the position of Copernicus is correct (as I believe), that Venus revolves about the Sun, it is clear that it would be necessary that it would sometimes be seen by us horned and at other times not."

5. *OGG*, 10: 483. Castelli's suggestion was in reference to Copernicus's discussion of the order of the inferior planets in *De Revolutionibus*, I, 10 (see *Nicholas Copernicus on the Revolutions*, 18). But see also Neil Thomason, "1543: The Year that Copernicus Didn't Predict the Phases of Venus," in Guy Freeland, ed., *1543 and All That: Image and Word, Change and Continuity in the Proto-Scientific Revolution* (Dordrecht: Kluwer, 2000), 291–332.

6. Edward Rosen, "Harriot's Science: The Intellectual Background," in John W. Shirley, ed., *Thomas Harriot: Renaissance Scientist* (Oxford: Clarendon Press, 1974), 1–15, at 4.

on 18 December conditions presented themselves that allowed him to peer briefly at the Sun through his telescope.

The comments accompanying his sketch speak only of the circumstances of the observation, the weather, the instrument used, and the number of times Harriot was able to take a look at the Sun, alternately with his right and left eye. They say nothing about the Sun's appearance itself; Harriot does not even mention the spots. The visual message is uncoupled from the verbal one, and as tempting and natural as it is for us to interpret the mute sketch in terms of subsequent events, we must admit that we do not know what he thought about his observation. He returned to the Sun on 29 January 1611, when there was "a notable mist," and recorded that "I observed diligently at sundry times when it was fit. I saw nothing but the cleare sonne both with right and left ey."[7] No sketch was needed. After this, Harriot recorded no sunspot observations until December of that year. His surviving observational record is a fair copy, and it may be that he did monitor the Sun during the intervening period but made no drawings or notes, or that he lost or ignored those observations, but neither scenario is likely. It appears more plausible that after a second painful observation revealed nothing unusual about the Sun, Harriot engaged in no routine study of the solar body.

Why, then, did he return to the Sun in December 1611? His record of that observation is as follows:

> Syon [house].
> 1611. Decemb. 1. [Sunday] *mane. ho. 10.0. per horologium solare[m].* I saw three blacke spots in such order as is here expressed, as nere I could iudge. observed by 10/1. Sr. Wm. Lower with Chr[istopher Tooke] also saw the same. at sundry times all three seen & observed at once. for halfe an houre space. at which time and all the morning before it was misty. The greatest was that which was most oriental appearing somewhat ragged & was of apparent angle about 2′. The other two, were nere of one bignes: & of 1′ magnitude. or there aboutes.[8]

These comments differ in tenor from those accompanying the sketch made almost a year earlier. Harriot specifies the time precisely—judging from subsequent observations, to the nearest quarter hour—and tells us how he measured it. He managed to observe for half an hour, which must have taken some determination, and he also mentioned his friend Sir William Lower and his

7. Harriott MSS, HMC 241/8, p. 1
8. Harriot MSS, HMC 241/8, p. 2

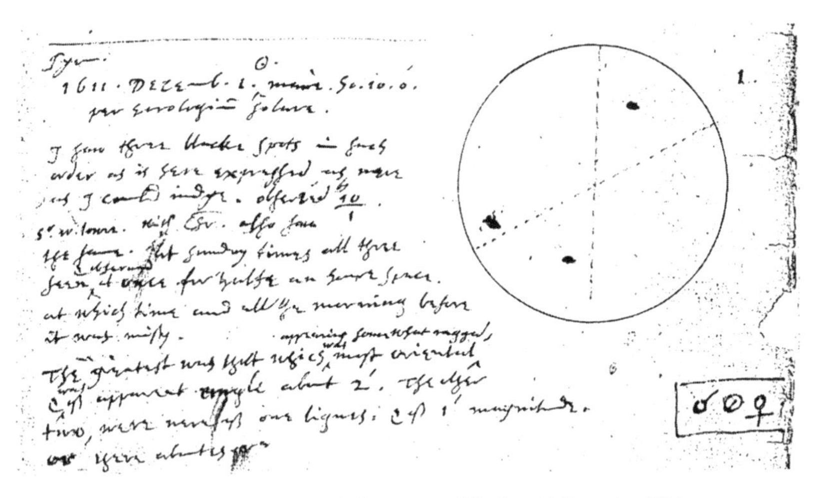

Figure 3.2. Thomas Harriot's observation of the Sun, 11 December 1611. Harriot MSS, HMC 241/8, p. 2. Courtesy of Lord Egremont.

instrument-maker and factotum, Christopher Tooke, as participants and witnesses to the observation. The spots are clearly identified as "blacke spots," and their sizes and shapes are described.

This was no casual observation, and perhaps the astrological symbols in the box at the lower right (see fig. 3.2) provide the reason for this care, for they indicate that according to the ephemerides on this day there was a conjunction of Venus with the Sun. In fact, the Italian astronomer Giovanni Antonio Magini had predicted that at this superior conjunction Venus's ecliptic latitude would be less than the Sun's radius (see appendix 4). This meant that if Venus's orbit lay entirely below the Sun, as in the Ptolemaic scheme, the planet would appear as a dark spot on the Sun. Christoph Scheiner would make this transit into an acid test of the Ptolemaic system, and it appears that Harriot perhaps had something similar in mind. We are left once again with a sunspot observation made in the context of the ancient question of planetary order. Beginning with this observation, however, Harriot undertook a prolonged and sophisticated research project on sunspots in which, as John North has shown, he came to conclusions about the rotation periods of spots that were well ahead of those of his contemporaries.[9] Regrettably,

9. John D. North, "Thomas Harriot and the First Telescopic Observations of Sunspots," in John W. Shirley, ed., *Thomas Harriot: Renaissance Scientist* (Oxford: Clarendon Press, 1974), pp.129–165. See also Richard B. Herr, "Solar Rotation Determined from Thomas Harriot's Sunspot Observations of 1611 to 1613," *Science* 202 (1978): 1079–1081.

these remained undiscovered until the late eighteenth century and unpublished until 1833.[10]

In the meantime, others had begun observing sunspots. In the little town of Osteel, East Frisia, David Fabricius shepherded his Lutheran congregation by day and observed the heavens at night. In March 1611, his son Johannes had just returned from studies in the Netherlands; while there, he had obtained some spyglasses powerful enough for celestial observations. During his stay at Osteel, Johannes explored the heavens with these instruments, observing the uneven surface of the Moon, hoping to see the triple appearance of Saturn, and trying but failing to observe the moons of Jupiter discovered by Galileo.[11] Shortly after sunrise on 9 March, Johannes turned one of his instruments to the Sun. He noticed unevenness in the Sun's limb, something, as he reported, his father had already mentioned to him in a letter. As he was observing this feature, he suddenly saw a dilute blackish spot on the Sun, of significant size with respect to the solar body. As Johannes would later relate, he was at the outset skeptical of this observation, and attributed it to passing clouds. He repeated the observation as many as ten times, and with different spyglasses, until he was certain that the spot was not produced by clouds. He then called in his father, who joined him in the observations:

> Each applying a spyglass to our eye, we received the rays of the Sun: first those emitted from the edge, and then gradually progressing to the middle,

10. F. X. von Zach, "Beobachtungen des Uranus; . . . und Anzeige von den in England aufgefundenen Harriotschen Manuscripten," in J. E. Bode, ed., *Astronomisches Jahrbuch für das Jahr 1788* (Berlin, 1785), 139–155; Stephen P. Rigaud, "Account of Harriot's Astronomical Papers," in *Supplement to Dr. Bradley's Miscellaneous Works: with an Account of Harriot's Astronomical Papers* (Oxford: Oxford University Press, 1833), 1–70.

11. *Joh. Fabricii Phrysii De Maculis in Sole Observatis, et Apparente earum cum Sole Conversione Narratio* (Wittenberg: Impensis Johan Borneri Senioris & Eliae Rehefeldij 1611), C2r–C2v. What little information that is available about Johannes Fabricius has been gathered by Wilhelm Olbers, "Materiellen zur einer Lebensbeschreibung der beiden Astronomen David und Johannes Fabricius," *Astronomische Nachrichten* 31 (1850):129–142, reprinted in *Olbers, Sein Leben und seine Werke* (Berlin: Springer Verlag, 1894), vol. 1, 200–211; Gerhard Berthold, *Der Magister Johann Fabricius und die Sonnenflecken, nebst einem Excurse über David Fabricius* (Leipzig, 1894); Diedrich Wattenberg, *David Fabricius: Der Astronom Ostfrieslands (1564–1617)* (Berlin-Treptow: Archenhold Sternwarte, 1964). See also Ernst Zinner, *Astronomie: Geschichte ihrer Probleme* (Freiburg/Munich: Verlag Karl Alber, 1951), 241–245. Both Wattenberg and Zinner offer German translations of the few pages of observational detail contained in the *Narratio*, and both present an illustration of a camera obscura that is not found in *Narratio*, but comes rather from Kepler's *Phaenomenon Singulare*. See *JKGW*, 4: 94. On the career of David Fabricius, see Menso Folkerts, "Der Astronom David Fabricius (1564–1617): Leben und Wirken," *Berichte zur Wissenschaftsgeschichte* 23 (2000): 127–142.

until, when our sight had become accustomed to the profusion of rays, it allowed the entire globe to be seen. We then saw the spot more distinctly and certainly.[12]

At this point, cloudiness and the increasing brilliance of the Sun's rays as it climbed higher in the sky curtailed their observation. Even the weaker rays from the early morning Sun had caused redness in their eyes that lasted two days "not without distortion in the images of objects."[13] That evening, father and son speculated about the explanation of this phenomenon. Johannes was uncertain, and he argued that a spot actually on the solar globe would reappear in the morning, while a body suspended between the Earth and Sun, or an effect of passing clouds, would not. At sunrise the two men directed their instruments to the Sun again, and they saw that the spot was still there.[14] Presumably, then, it was on the Sun, the option Johannes had favored, but they also noticed that the spot seemed to have moved a little from its previous position.

In order to continue their observations without further pain and damage to their eyes, Johannes and David Fabricius prudently moved to a camera obscura. Projecting the solar image onto a paper in a darkened room, they observed sunspots over the course of what appears to be several months. They saw as many as three spots at a time, followed their trajectory across the face of the Sun until they disappeared, and witnessed their eventual return on the opposite limb. In doing so, they noticed that the path of the spots was inclined to the ecliptic, that their motion was slower near the edges than in the middle, and that their shapes were foreshortened at the edges. Johannes estimated that a large spot took ten days to reemerge in the east after disappearing from the western edge.[15]

In the late spring of 1611, Johannes wrote up his observations in a little tract entitled *De Maculis in Sole Observatis, et Apparente earum cum Sole Conversione Narratio* (Narrative about spots seen on the Sun and their apparent revolution with the Sun). In this work, he rehearsed his observations—accompanied neither by time nor by date—and then discussed possible explanations of the phenomena. These observations made it "likely that the spots adhere to the body of the Sun, which is spherical, round, and solid."[16] As their varying speeds and foreshortened appearance also strongly supported this hypothesis, Johannes judged it "manifest that the matter could not be otherwise if the spots were

12. Fabricius, *De Maculis in Sole Observatis*, C2v–C3r.
13. Fabricius, *De Maculis in Sole Observatis*, C3r.
14. Fabricius, *De Maculis in Sole Observatis*, C3r–C4r.
15. Fabricius, *De Maculis in Sole Observatis*, D1r–D1v.
16. Fabricius, *De Maculis in Sole Observatis*, D1v.

moved around the Sun."[17] He adduced the testimony of Giordano Bruno and Kepler that the Sun rotates on its axis, and concluded, "for what we would make of the spots if we did not place them on the Sun itself, I don't know."[18] He did not, however, firmly conclude that the Sun rotated on its axis, saying that he preferred to keep a fitting silence about matters still obscure to him.

> But if someone . . . is convinced of its rotation he will not be opposed by me, neither will he be opposed by others, because many will not only not deny it but will also assert . . . that they do not fear it. For this rotation is the cause of the other heavenly bodies' motion because of the solar rays going around in a circle[19] and the light spreading everywhere. For indeed, as this is the one principle and instrument of Aristotle for moving these inferior parts, so also it is without doubt the cause of motion in the superior regions. For it is very true, as Aristotle says in his *Problems*, that "the Sun is the father and author of all motion."[20]

This is as far as Johannes Fabricius allowed himself to go in his conclusion; the remaining discussion concerned other matters. He did not declare himself a follower of Copernicus, but indicated a preference for the diurnal rotation of the Earth, and the system of Tycho Brahe was not mentioned in this connection.[21] A recent discovery suggests that Johannes's cautious stance about the Sun's rotation derived from a disagreement with his father. In a letter addressed to Michael Mästlin on 11 December 1611—that is, shortly before Scheiner's first publication on sunspots—David Fabricius had written,

> Finally, I add that together with my son Johannes Fabricius I observed this year some spots on the disk of the Sun through a Dutch perspective [telescope]. Indeed, this summer I often observed ten or eleven spots scattered on the Sun's disk at one time. A true and marvelous thing. I do not believe that they are on the Sun's body, but that at all events they pass over the Sun's disk, which happens in ten or twelve days at most. The center of motion of these spots is in the Sun, and it is carried around by it in its annual motion. Therefore, when these spots appear to us, they are situated on the lower part of [the

17. Fabricius, *De Maculis in Sole Observatis*, D1v–D2r.

18. Fabricius, *De Maculis in Sole Observatis*, D2v.

19. This may be a reference either to Kepler's *Mysterium Cosmographicum* or *Astronomia Nova* of 1609.

20. Fabricius, *De Maculis in Sole Observatis*, D3r. In the *Problemata* (Book 11, 899a22), Aristotle describes the Sun in passing as "the source of all movement."

21. Fabricius, *De Maculis in Sole Observatis*, D3r–D4r.

> Sun's] circle and are moved in the direction of precession, close to the line of the ecliptic. My son has published a little treatise on these things in Wittenberg, where he studies medicine, for the last [Frankfurt] fair.[22]

David Fabricius thus did not share his son's opinion that the spots were on the Sun itself; all he could conclude was that their center of motion lay in the Sun. Subsequent publications on sunspots by others did nothing to change the elder man's mind. His only published allusion to the subject appeared in 1615, in his prognostication for that year:

> I should also mention some other phenomena [*Meteoris und Apparentiis*], especially spots or black blotches that are at present found and observed on the Sun, such as were observed for the very first time, in my presence, by my son Johannes Fabricius, a medical student, in the year 1611, on 27 February, old style [9 March, Gregorian], through the Dutch spectacles, and about which he published a Latin tract in quarto in Wittenberg before the end of the year. But after this, the Anonymous Apelles [i.e., Christoph Scheiner] has wonderfully extended this speculation and has come to the conclusion that the air is filled with such shadowy bodies, because almost daily such spots in continuously changing shapes run under the Sun and are seen there. What they actually are, and to what end such things were created by God, is difficult to fathom.[23]

Clearly, then, if Johannes Fabricius inclined toward an interpretation that put the spots on the Sun itself, his father did not agree with him. In his treatise on the sunspots, Johannes neither offered a strong argument about the matter nor provided detailed observational data and visual material. From a remark in his tract we gather that, unlike Scheiner and Galileo, he feared others would soon tire of these tedious observations, and he "considered it enough to have presented a narrative demonstration and to have challenged others to further investigation."[24] As is clear from his own annotation, Mästlin did not obtain a copy of the book until May 1612; Kepler saw the tract announced in the

22. David Fabricius to Michael Mästlin, 1 December 1611 (old style). An extract of the letter, in Mästlin's hand, is glued into Mästlin's copy of Johannes Fabricius's *De Maculis in Sole Observatis* (between pp. C3v and C4r), and preserved in the Schafhausen Municipal Library. In a note at the bottom of the letter, Mästlin wrote that he received the letter on 2 May 1612. His copy of the book has the note that he bought it on 21 May, that is, eight days before Mästlin observed sunspots for the first time. We thank Robert S. Westman for providing us with the extract and information.

23. Cited in Gerhard Berthold, *Der Magister Johannes Fabricius* (Leipzig, 1894), 6. Part of this passage is cited in *JKGW*, 17: 450.

24. D1v–D2v.

Frankfurt catalog for 1611 but did not obtain a copy until after he had read Scheiner's *Tres Epistolae*.[25] There is no evidence that Johannes Fabricius sent copies to other astronomers: Mästlin, for instance, had had to buy his own, and though he had corresponded earlier with Kepler, Johannes failed to send the treatise directly to this potential ally. The fact is that Johannes Fabricius neither availed himself of the correspondence network of his father, nor enjoyed the aggressive patronage system supporting Galileo in Italy and Scheiner in both Germany and Italy. An isolated figure, he disappeared from the astronomical scene soon after publishing his tract, despite the fact that many others in the United Provinces and the Spanish Netherlands—Willebrord Snellius, Nicolaus Mulerius, Adriaan Metius, Charles Malapert S.J., Leonard Lessius S.J., and François Aguilon S.J.—were discussing sunspot observations by 1612 or 1613.[26] Nothing further is known about Johannes except his father's reference to his death on 19 March 1616.[27]

The very nature of Johannes Fabricius' tract also contributes to the indifference with which it met. Of its forty-odd pages, a scant ten treat the sunspots. The language is at times digressive rather than soberly scientific, and Johannes provided little in the way of observational detail. Kepler, though likewise much given to digression, could manage only lukewarm praise for the work, telling his friend Matthäus Wacker von Wackenfels, "If you can overcome the tedium of reading these winding swamps of verbiage, you will see an agreement [with my ideas] that is not to be scorned."[28] It is hardly surprising that Fabricius' tract was overtaken by events.

25. In a letter to Matthäus Wacker von Wackenfels in 1612, Kepler wrote that he had seen the book announced but that no copy of it was yet available in Prague. In a postscript to that same letter, however, he mentioned that he had at last received a copy. See *JKGW*, 17: 7, 15.

26. On Snellius, who authored the anonymous *De Maculis in Sole Animadversis . . . Batavi Dissertatiuncula* in 1612, and on Mulerius, who appears to have observed sunspots by early 1613, having learned of them from Adriaan Metius, a professor of mathematics and brother of the instrument maker Jacobus Metius, see Rienk Vermij, *The Calvinist Copernicans: The Reception of the New Astronomy in the Dutch Republic, 1575–1750* (Amsterdam: Edita, Royal Dutch Academy of Arts and Sciences, 2002), 43–47, 104. On the Flemish Malapert, who began observing with Christoph Scheiner in 1613, see Luigi Guerrini, "Tradizione astronomica e cultura matematica in un'orazione gesuitica del 1619," *Giornale critico della filosofia italiana* 79 (2000): 209–235, and Luigi Guerrini, "Nel Dedalo del Cielo: Gassendi e le macchie solari in un inedito del 1633," *Nuncius* 16:2 (2001): 643–672. The Flemish Jesuits Leonard Lessius and François Aguilon both discussed the sunspots in their publications of 1613; see Lessius, *De Providentia Numinis et Animi Immortalitate* (Antwerp: Plantin, 1613), 18–19, and Aguilon, *Opticorum Libri Sex* (Antwerp: Plantin, 1613), 421.

27. Prognostication for 1618, now lost. See W. Olbers, "Materiellen zu einer Lebensbeschreibung der beiden Astronomen David und Johannes Fabricius," *Astronomische Nachrichten* 31 (1850): 129–142. Reprinted in *Olbers, Sein Leben und seine Werke* (Berlin 1894), 1: 200–211.

28. *JKGW*, 17: 15.

4

Christoph Scheiner

Little is known about Christoph Scheiner's early life.[1] He was born in Wald, near Mindelheim in Swabia, on 25 July 1573, and attended the Jesuit Latin school in Augsburg, continued his studies at the Jesuit college in Landsberg, and entered the order in 1595. Having completed his preparatory study, Scheiner enrolled in the Society's University in Ingolstadt in 1600, where he focused on metaphysics, and studied mathematics with Johannes Lanz.

In September 1600, Scheiner wrote to Matthäus Rader, prefect of the Jesuit gymnasium in Augsburg, and a mentor to many members of the order in Bavaria; he had promised Rader a quadrant, and he feared it was inferior to other such instruments, and perhaps not without flaws.[2] Here he took up what would become a crucial pseudonym, that drawn from Pliny's portrait of the greatest of Greek artists:[3]

> In the meantime, however, this consoles me: the most noble of painters, Apelles, displayed his errors (if any such existed) in a

1. Anton von Braunmühl's biography, *Christoph Scheiner als Mathematiker, Physiker und Astronom*, Bayrische Bibliothek, vol. 24 (Bamberg: Buchnersche Verlagsbuchhandlung, 1891), has recently been supplemented by Franz Daxecker, *The Physicist and Astronomer Christoph Scheiner: Biography, Letters, Works* (Innsbruck: Leopold Franzens University of Innsbruck, 2004), and by Luigi Ingaliso, *Filosofia e cosmologia in Christoph Scheiner* (Soveria Mannelli: Rubbettino, 2005).

2. On Rader, see Rita Haub, "'Ich bin eines armen Bäckers Sohn!' Matthäus Rader S.J. (1561–1634)," *Archivum Historicum Societatis Iesu* 139 (2001): 173–180.

3. Pliny, *Historia Naturalis* 35: 84.

Figure 4.1. Portrait of Christoph Scheiner, ca. 1735, by Christoph Thomas Scheffler, SJ. With permission from the Ingolstadt Stadtmuseum.

painting so that they might be corrected by the people. So that he might emerge in this way always more perfect, he would never go out without displaying his works. My works, in fact, would not please me if ever I saw that they were displeasing to others. Let this reason suffice for my willingness to expose my errors time and again to the scrutiny of others, though there is another motive as well: I have known from the outset the kindness and modesty of Your Reverence.[4]

Upon receiving his M.A., Scheiner was sent to Dillingen in 1603 to teach grammar at the Jesuit gymnasium there, and by 1605 he was teaching *Human-*

4. Christoph Scheiner to Matthäus Rader, in Rader, *P. Matthäus Rader S.J., Briefwechsel, Band I: 1595–1612*, ed. Helmut Zäh, Silvia Strodel, and Alois Schmid (Munich: C. H. Beck'sche Verlagsbuchhandlung, 1995), 140–141. It is worth noting that references to Apelles are frequent in the correspondence directed to Rader; in this volume see 441, 450, 477, 527.

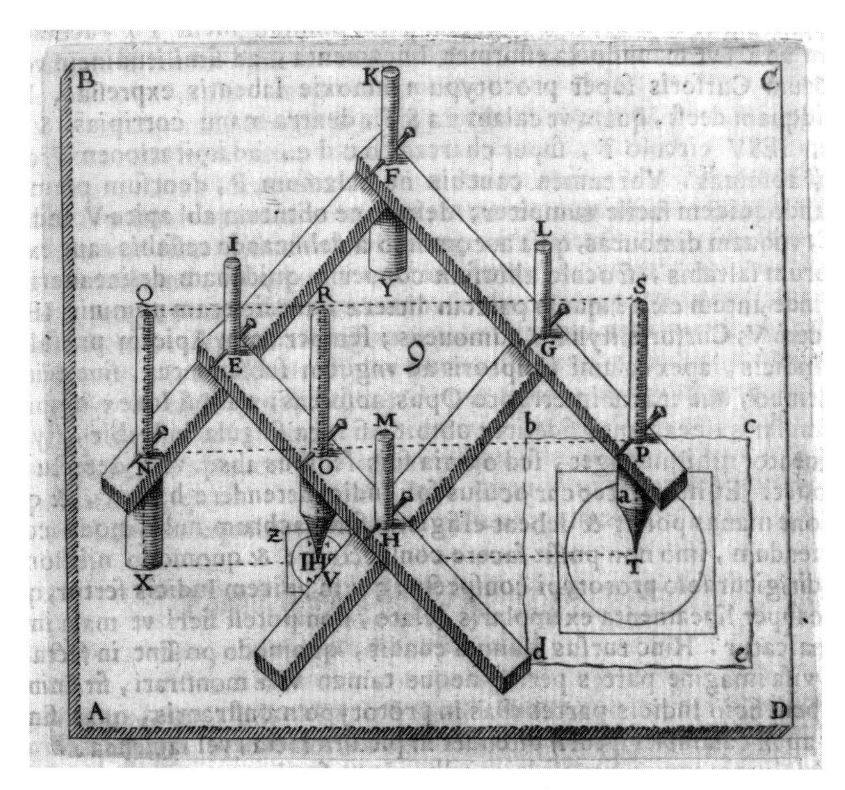

Figure 4.2. Christoph Scheiner's design of a pantograph. From *Pantographice* (Rome, 1631), 29. Courtesy Posner Memorial Collection, Carnegie Mellon University Libraries, Pittsburgh, PA.

iora as well. In this same period, he also taught mathematics at the academy connected with the gymnasium. In these years, moreover, Scheiner invented a pantograph, a device for copying, enlarging, or reducing diagrams. Scheiner's story of its conception is instructive. In Dillingen, he recounted, he often talked with a painter, who told him about a mechanism he used to draw the outline of things on paper. Because this was a trade secret,[5] the painter naturally refused to show it to him; wholly obsessed with the matter, Scheiner dreamt of the copying device, and on the night of 19/20 January 1603, the mechanism was revealed to him.[6]

5. In later years, in fact, the Jesuit Mario Bettini claimed to have seen a pantograph brought from Belgium to the Duke of Parma, whose tutor he was, before he knew of Scheiner's device. See Mario Bettini S.J., *Apiaria* (Bologna: Typis Io. Baptistae Ferronij, 1642), 5: 56.

6. *Pantographice seu Ars Delineandi Res quaslibet per Parallelogrammum lineare seu Cavum Mechanicum mobile* (Rome, 1631), 29. On this and other pantographs see Martin Kemp, *The Science*

In 1605 Scheiner returned to the Jesuit University of Ingolstadt. Here he continued to study the principles of the pantograph and to refine the device, and was also engaged in gnomonics. In the summer of 1605 his teacher Lanz wrote,

> Master Scheiner in Dillingen recently wrote Father Ferdinand Crendel that he has discovered some new and easy method of describing sundials. He does not write what it is, but I hope that it is something excellent, especially since he says that he worked several years in inventing it and would excel in inventing new things. Without doubt with this method of his he will far surpass the trifles I have accomplished.[7]

Father Lanz depicted an ambitious young man who saw in his ability in mathematics and mechanical arts the road to advancement. The sundial and the pantograph did indeed cause Scheiner to be noticed among a wider circle within the Jesuit community,[8] as well as by highly placed persons close to the order. In 1606 he was summoned to Munich by Duke William V of Bavaria, who had an avid interest in painting and was curious about Scheiner's pantograph, and Scheiner spent some considerable time with the duke.[9]

Scheiner's years as a student at Ingolstadt came at a time of great uncertainty among mathematicians in the Jesuit order, for the increasingly critical debate about the nature of the heavens placed the Society's teachers in a difficult position. We have some index of the interest and confusion occasioned by the new star that appeared in October 1604, just after a conjunction of Mars and Jupiter in Sagittarius, in a letter written by Scheiner's friend Georg Stengel S.J. to Father Rader on 14 November of that year.

> And here around us, nothing new is heard, but a great deal is seen. What is it? A new god. Which one? A Stoic one. Where? In that part of the sky close to Jupiter. A star, I say, bearing a torch, of which there has been frequent and long discussion, and it has even been fully described to us through the offices of My Lord [Marc] Welser. Some say it is a meteor, others believed it an unformed comet, for the failing and scattered fire of the material never appeared, but will

of Art: Optical Themes in Western Art from Brunelleschi to Seurat (New Haven: Yale University Press, 1990), 180–183.

7. Lanz to Guldin, 27 July 1605, Dillingen Studienbibliothek, MSS 2° vol. 247, ff. 244–247, at 246.

8. Crendel notified Johannes Lanz, who passed the news to Paul Guldin in Rome. Later in 1605 Lanz described Scheiner's design to Guldin (Dillingen Studienbibliothek, MSS 2° vol. 247, 172–175, 180).

9. Braunmühl, *Christoph Scheiner*, 6.

> return in a future time as the flare of a comet. May God make it a bright and favorable omen.[10]

We have no record of Scheiner's position on the new star, but the fact that it did not fade away entirely until October 1605, a full year after its emergence, meant that he may well have been involved in discussions of the material, location, and nature of the nova after his return to Ingolstadt. In 1606 and 1607 Johannes Lanz wrote at length of these and related matters first to Christoph Grienberger, a prominent Jesuit mathematician at the Collegio Romano, and upon receiving no reply, to Grienberger's celebrated senior colleague, Christoph Clavius. He asked Clavius about the location of comets in view of all the accounts that had appeared claiming that comets were above the Moon, and also about the nova of 1604, about which he had conflicting information. "I request that Your Reverence think it worthy to respond to these trifles of mine," Lanz wrote. "I also ask especially that if you have something useful for some instruction in these studies of ours, you communicate it to me."[11]

If Clavius answered Lanz's barrage of questions and speculations, that letter has not survived. It is clear, however, that Lanz and his colleagues must have been in some difficulty in teaching mathematical astronomy and cosmology to their more advanced students. As Scheiner matured professionally during these years, he became an active participant in the debate about the nature of the heavens, emerging finally as one of the most serious advocates of the fluidity and imperfections of the heavens, a cosmological idea against which Clavius had battled for much of his career.[12]

Upon finishing his studies at the university, Scheiner spent a probationary period at Ebersberg and then, in 1610, was appointed professor of mathematics and Hebrew at Ingolstadt. During his first year as a professor, Galileo's *Sidereus Nuncius* appeared, and this little book changed Scheiner's life almost as much as it did Galileo's. No correspondence of German Jesuit mathematicians between February 1610—the month before the publication of the *Sidereus Nuncius*—and February 1611 survives, but we may assume that the news spread to the fathers in Ingolstadt through the well-established networks of the Society

10. Georg Stengel to Matthäus Rader, in Rader, *P. Matthäus Rader*, 317. On Stoicism and the new star of 1604, see Reeves, *Painting the Heavens*, 57–90. On Stengel, see Carlos Sommervogel, *Bibliothèque de la compagnie de Jésus*, 12 vols. (Brussels: Gregg, 1960), 7: cols. 1546–1559.

11. Johannes Lanz S.J. to Christoph Clavius S.J., 10 July 1607, Graz, Universitätsbibliothek, MS 159, 17, 1; Clavius, *Christoph Clavius: Corrispondenza*, ed. U. Baldini and P. D. Napolitani, 6 vols. (Pisa: Università di Pisa, Dipartimento di matematica, 1992) 6: 59–60.

12. Lattis, *Between Copernicus and Galileo*, 94–102.

of Jesus. As early as 12 March 1610—the very day on which Galileo signed the dedication of *Sidereus Nuncius*—Marc Welser wrote with a certain measure of caution to Clavius about the astronomical news that had reached him from Italy.

> On this occasion I cannot neglect to tell you that I have been written from Padua as a certainty that Signor Galileo Galilei, mathematician of that university, has found four planets—new as far as we know, and to our knowledge never seen before by mortal man—and also many fixed stars, until now wholly unknown and unseen, and around the Milky Way marvelous things, all with that new instrument, which many people call a *visorio*, and of which he makes himself the author. I realize that *the sinew of wisdom is believing slowly*,[13] and therefore I have not made up my mind. But I ask Your Reverence to tell me freely and confidentially your opinion about this fact.[14]

Since Welser was in very close touch with the Jesuit scholars of the area,[15] we may assume that the news of the discoveries reached Ingolstadt in March, but it may have been some time before copies of *Sidereus Nuncius* followed. It appears that the text was not available in Germany until it had been reprinted in Frankfurt, without Galileo's permission, in time for the autumn fair.[16] Though the Archduke Maximilian III, ruler of Tyrol, had had a telescope since midsummer 1609, high-quality instruments with which to verify Galileo's discoveries were not available;[17] it is, however, quite possible that the South German mathematicians, armed with cruder spyglasses, were able to observe some of the lunar phenomena that Galileo had described in his book.

The focus of these initial discussions was the question of what such observations meant for the alleged perfection of the heavens. In October 1610

13. Cicero, *Commentariolum petitionis* X [39]. The expression was a common one, and occurs with some frequency, for instance, in the works of Francis Bacon.

14. Welser to Clavius, 12 March 1610, *OGG*, 10: 288.

15. In *Germaniae aliorumque Quorundam Locorum Europae Poliometria* (Augsburg, *ca*. 1770) by Tobias Conrad Lotter (1717–1777), the distance is given as nine (geographical) miles. Ingolstadt today is about 75 km by road from Augsburg. In the seventeenth century, it would have been a day's journey.

16. In his argument about the height of lunar mountains, Brengger used the Frankfurt reprint as his source. See *OGG*, 10: 461. The nominal dates for the Frankfurt fair were 15 August (St. Mary's assumption) and 8 September (St. Mary's birthday). See H. Grotefend, *Zeitrechnung des deutschen Mittelalters und der Neuzeit*, 2 vols. (Hanover: Hahn, 1891–1898), 1: 68, 122.

17. See Daxecker, *The Physicist and Astronomer Christoph Scheiner*, 11, and Girolamo Sirtori, *Telescopium: sive ars perficiendi* (Frankfurt, 1618), 29–30.

Welser, writing directly to Galileo, included a rather technical argument by Johann Georg Brengger, a physician of Augsburg, about the height of lunar mountains.[18] Welser wrote that he was convinced that Galileo would not be displeased to learn that his writings were read with increasing attention beyond the Alps,[19] and in addition to the mathematical issue of the lunar peaks, he also pursued the more philosophical question about the nature of the Moon. Some time in the fall of 1610 Welser turned likewise to the Jesuit fathers at Ingolstadt for their views. As he had done numerous times before, he wrote Jakob Gretser[20] and asked him for Johannes Lanz's opinion. Because Lanz had moved to Munich, Gretser asked his replacement, Father Scheiner, to respond to Welser's request.[21] Scheiner complied, suggesting that there was not yet definitive proof that the mountains projected beyond the lunar sphere, and Welser sent the argument to Galileo, ascribing it to "a certain other friend" of his.[22]

Scheiner's views about the lunar sphere were entirely conventional,[23] if somewhat confusingly expressed, and in February 1611 Galileo duly responded to them in a letter to Welser. Asking his correspondent's indulgence for his inability to follow all of Scheiner's argument, he insisted on his original comparison of the rough surfaces of the Earth and the Moon. His rebuttal, written in Latin for the benefit of his adversary, was accompanied by drawings, and the letter ended with a general complaint in Italian about the pretensions and strategies of his detractors.[24] The two letters can be found in appendix 1 herein. In his

18. For the argument about the height of lunar mountains, see Stillman Drake, *Galileo at Work*, 167–168, 183–184; Drake, *Telescopes, Tides, and Tactics*, 37–40.

19. Welser to Galileo, 29 October 1610, *OGG*, 10: 460.

20. Jakob Gretser S.J. (1562–1625) was a theologian, liturgist, polemicist, patristic scholar, philologist, and playwright. From 1589 to his death he was on the faculty of the University of Ingolstadt, where he taught metaphysics (1589–1592), scholastic theology (1592–1605), and moral theology (1609–1616). See Bernard Duhr, *Geschichte der Jesuiten in den Ländern deutscher Zunge*, 1: 668–671, 679–680; Sommervogel, *Bibliothèque de la compagnie de Jésus*, 3: cols. 1743–1809.

21. Lanz to Guldin, 14 May 1611, Graz, Universitätsbibliothek MS 159, no 17,6.

22. Welser to Galileo, 7 January 1611, *OGG*, 11: 13–14. Antonio Favaro doubted that the "other friend" was Christoph Scheiner because neither in his *Tres Epistolae* nor in his *Accuratior Disquisitio* did Scheiner deal with the nature of the lunar surface (*OGG*, 11: 13 n 3). But it was indeed Scheiner; see Lanz to Guldin, 14 May 1611, Graz, Universitätsbibliothek MS 159, no 17,6.

23. In 1611 Jesuits in Mantua, perhaps Mario Bettini or his teacher Jean Verviers, proposed a similar problem concerning the lunar mountains; see *OGG*, 3 (1): 303–307, and Denise Aricò, "'In Doctrinis Glorificate Dominum': Alcuni aspetti della ricezione di Clavio nella produzione scientifica di Mario Bettini," in *Christoph Clavius e l'attività scientifica dei Gesuiti nell'età di Galileo*, Ugo Baldini, ed. (Rome: Bulzoni: 1995), 189–207, especially 191–196.

24. Galileo to Welser, February 1611, *OGG*, 11: 38–41.

reply, dated 25 March 1611, Welser identified the author of the opinion as a "Jesuit Father."[25]

Scheiner did not reply to Galileo's answer, but he had by no means abandoned the new discoveries. At the end of 1610 the problem that he and his fellow Jesuits in Germany faced was not only that instruments powerful enough to verify all of Galileo's observations were not yet available, but also that some news but less guidance from the mathematicians at the Collegio Romano had reached them.[26] By mid-February 1611 they learned that Galileo's observations had at last been verified by their colleagues in Rome, where Fathers Grienberger and Lembo had been working for months to refine their telescopes. A more elaborate account came a few weeks later, when a letter from Paul Guldin to Lanz, dated 13 February 1611, arrived in Munich. As Guldin was studying mathematical subjects at the Collegio Romano at this time, it seems only reasonable to assume that he was writing with Clavius's permission. Guldin explicitly raised the issue of the general cosmological implications of the new discoveries at the end of his letter:

> But in what place will we put these new planets? Which and how many orbs and epicycles will we assign to them? Is Tycho's opinion to be embraced? Will we allow the stars to move freely like fish in the sea? And surely Aristotle will hardly be alone in arguing why this cannot or ought not to be done. Who will prohibit each individual star from being assigned an intelligence as a guide and director of its journey? But these things are beyond what a wretchedly poor novice in physics such as I, utterly confused by shortcomings, could appropriately discuss about the material, and the form.[27] I would like to hear Your Reverence, your Rector, and other experts converse about these issues.[28]

As Guldin's letter to Lanz shows, the new discoveries again brought to the forefront the problem of the constitution of the heavens. Should the planets be allowed to make their course through the heavens freely, "like birds in

25. Welser to Galileo, 25 March 1611, *OGG*, 11: 74.

26. The progress made by the Jesuits at the Collegio Romano was passed on in reports throughout this period; see Galileo to Gualdo, 17 December 1610, *OGG*, 10: 484; Welser to Gualdo, 7 January 1611, *OGG*, 11: 15; Welser to Clavius, 7 January 1611, *OGG*, 11: 14; Clavius to Welser, 29 January 1611, *OGG*, 20: 600–601; Welser to Clavius, 11 February 1611, *OGG*, 11: 45; Gualdo to Galileo, 10 February 1611, *OGG*, 11: 43. On relations between the Jesuits of Ingolstadt and the Collegio Romano, see also Ingaliso, *Filosofia e cosmologia*, 60–67.

27. I.e., the material and formal causes.

28. Guldin to Lanz, 13 February 1611, Dillingen, Studienbibliothek, MSS 2°, 247, p. 220–222.

the air, or fish in the sea," as the Stoics had maintained?[29] On 1 March Lanz wrote to Adam Tanner at the University of Ingolstadt to relate what he had learned from his colleague in Rome, and he asked that the contents of Guldin's letter be communicated "to Father Scheiner and others interested in these things."[30]

Questions regarding the nature of the Moon and the overall constitution of the heavens would continue to occupy the order. It is clear from Lanz's letter of 1 March to Tanner that Tanner and Scheiner were working hard trying to make or to obtain improved telescopes, doubtless with the intention of seeing the new phenomena for themselves and, in the case of Scheiner, of joining other observers at the forefront of astronomical research. If the satellites of Jupiter offer an index of the quality of one's instruments, we may infer that Scheiner and his student Johann Baptist Cysat had caught up with their Roman colleagues in the spring of 1611, for on 14 May Lanz wrote to Guldin that "Father Scheiner and Master Baptist Cysat continue steadily to observe the motion of those four planets of Jupiter, having sent me some such observations recently."[31]

In the meantime, Scheiner and Cysat had already turned a telescope briefly to the Sun. "In March [1611]," Scheiner recalled much later, "led not by an advance rumor but by a spontaneous desire, having ascended the tower of our church, and having directed a telescope to a Sun rather weakened by moderate clouds, I saw sunspots for the first time with my associate Johann Baptist

29. The earliest known use of this simile was Andalo di Negro's (ca. 1270–1340) commentary on Sacrobosco's *Sphere*. At about the same time it was used by John Buridan and Pierre d'Ailly. See Edward Grant, "Cosmology," in *Science in the Middle Ages*, ed. David C. Lindberg (Chicago: University of Chicago Press, 1978), 280, 298 n. 55. For a list of medieval thinkers who used the simile, see Grant, *Planets, Stars, and Orbs: The Medieval Cosmos, 1200–1687* (Cambridge: Cambridge University Press, 1994), 274, n 16. The citation here is taken from Robert Bellarmine's "Lectiones Lovanienses"; see Ugo Baldini and George V. Coyne, *The Louvain Lectures (Lectiones Lovanienses) of Bellarmine and the Autograph Copy of His 1616 Declaration to Galileo* (Vatican City: Specola Vaticana, 1984), 18–19. Clavius cites the expression in his refutation of the fluidity of the heavens in *Commentary*, 24. See also James M. Lattis, *Between Copernicus and Galileo: Christoph Clavius and the Collapse of Ptolemaic Cosmology* (Chicago: University of Chicago Press, 1994), 94–102. Note also that Erasmus Reinhold, an early follower of Copernicus, used the expression in 1542. See Westman, "The Astronomer's Role in the Sixteenth Century: A Preliminary Study," *History of Science* 18 (1980): 105–147, on 113, 139 n. 45.

30. Lanz to Tanner, 1 March 1611, Graz, Universitätsbibliothek, MS 159, no. 17,2. The theologian Adam Tanner S.J. (1572–1632), a student of Jakob Gretser, taught at the University of Ingolstadt from 1603 to 1618. On 21 October 1611 he was one of the first to verify Scheiner's observations of sunspots (*De Creatione Mundi*, vol. I, Disp. 6, quaest. 3, § 5). This is mentioned in *RU*, "Ad Lectorem," second unnumbered page. Tanner also mentioned his part in these observations in an earlier work, the *Dissertatio peripatetico-theologica de coelis* (Ingolstadt: Gregor Haenlin, 1621), 152. On Tanner see Sommervogel, *Bibliothèque de la compagnie de Jésus*, 7: cols. 1843–1855.

31. See note 21.

Cysat, then a student of theology." Scheiner went on to relate how Cysat urged him to buy colored lenses or pieces of plane glass [*vitris coloratis*], which he finally did in October, when the two began their research project on sunspots.[32] The delay suggests that Scheiner was at that time more interested in exploring the night sky; he may have trying to determine the periods of Jupiter's moons.

To summarize Scheiner's situation just before he turned his attention to the Sun, then, he had been a professor of the mathematical sciences and Hebrew at Ingolstadt for a year, and was even then regarded as a promising scholar by his colleagues and superiors. His mechanical interests and abilities had turned him to instruments, beginning with the pantograph and sundials and increasingly after 1610, the telescope, for he and his student Johann Baptist Cysat were scanning the night sky and observing what Galileo had discovered a year earlier. He was also advancing in the patronage networks of southern Germany, having already been summoned by the Duke of Bavaria for his pantograph and subsequently having been given his provincial's permission to communicate anonymously with learned men outside the Society of Jesus through contact with Marc Welser. Scheiner would use this acquaintance to enter the most closely scrutinized arena of scientific research.

His expressed views on the constitution of the heavens were conventional, as was befitting to a young member of his order. In his comments on Galileo's lunar observations, Scheiner based himself on orthodoxy; the apparent inequalities in elevation, whether seen with the naked eye or with the telescope, were to be explained by assuming rarer and denser parts inside an otherwise perfectly spherical Moon. As he was to do with great frequency in his publications, Scheiner appealed to the "general opinion of the philosophers."

Scheiner's sunspot study

The Jesuits of Bavaria and their colleagues in Rome began gradually to mention sunspots in their correspondence. In a letter of 30 April 1611, Paul Guldin related to Johannes Lanz that Galileo was often at the Collegio Romano, and that he had showed the appearance of Venus. There was no mention of sunspots. On 21 May, Guldin wrote once again to Lanz, reporting briefly on the official reception for Galileo at the Collegio Romano, and alluding this time

32. *RU*, Ad Lectorem, 2nd (unnumbered) page. On this claim, and on Adam Tanner's failure to mention Scheiner's activities in his *Astrologia Sacra* of 1615, see Ingaliso, *Filosofia e cosmologia*, 148–150.

to the appearances of Saturn and Venus.[33] Though the solar phenomena do not figure in these letters, it is nonetheless certain that during his stay in Rome Galileo did show sunspots to at least one of the Jesuit mathematicians, Odo van Maelcote.

Maelcote's account is illuminating; writing from Brussels in December 1612, he told Johannes Kepler that, "even though Galileo showed [these sunspots] to me in Rome, and others did so in Germany, I neglected to observe them, being afraid, like you, that the [solar] image might burn my eye."[34] Such fears had surely been a crucial factor in Harriot's failure to pursue his initial sunspot observation for almost a year, in Scheiner's delay of six months, and in Galileo's reluctance to devote serious attention to solar phenomena until April 1612, after showing them to observers in Rome the previous spring. It was difficult just to get a glimpse of sunspots through a telescope; it would have been excruciatingly painful to make a sustained study of the phenomena by looking at them directly through a telescope for an extended period of time. Thus, after the efforts of a single day, Johannes and David Fabricius, almost certainly influenced by Kepler's *Phaenomenon Singulare*, conducted their research by projecting the solar image through a pinhole camera. In his letter to Kepler—which opened with a flattering reference to his acquaintance with the German astronomer's works—Maelcote, too, showed that he eventually adopted such an observational procedure, though not until after his visit to Galileo in February 1612 and to his hosts in Germany shortly thereafter.

> Having read of your device, and having modified it slightly, I learned to contemplate [the sunspots] easily on a canvas or paper, while turned away from the Sun, truly as if transmitted by a ray of the Sun itself, through the tube of a telescope, when its length and both concave and convex lenses have been properly adjusted.[35]

It also seems that the Jesuits in Germany, likewise inspired by the procedure outlined in the *Phænomenon Singulare*, used a camera obscura method as well, though perhaps somewhat inconsistently. Scheiner's friend Georg Stengel S.J., writing to his brother Karl, a Benedictine in Augsburg, on 9 January 1612, asked,

33. Guldin to Lanz, 21 May 1611, Dillingen 2°, vol. 247, 234–235.

34. Odo van Maelcote to Kepler, *OGG*, 11: 445; *JKGW*, 17: 37.

35. *OGG*, 11: 445. Maelcote visited Galileo in Florence in early 1612 on his way home to Belgium, presumably just before proceeding to Germany, where unknown observers looked directly through the telescope at the Sun.

> Have you heard about the spots we saw in the Sun? The means of seeing them are several: for they can be seen with Galileo's instrument, when the Sun is hiding beneath a not especially thick cloud, and also through the instrument itself when the Sun is bright. But in order that it not burn your eyes, a thick blue-green glass is added. And thus when the chamber has been entirely closed [to light], and solar rays are admitted through a single small aperture the size of a marble, if these fall on blank paper in the more distant parts of the room, they will depict these very spots. Other marvels can be seen no less precisely with this art, when the Sun illuminates an object but not the aperture, if the paper is moved away from the aperture. And when the face of the observer comes between the paper and the aperture [i.e., turned away from the aperture and toward the paper], then all the things beyond the window come together on the paper clearly positioned and with their true colors, and thus they can be made out exactly and in their proper relation, and they can even be painted.[36]

In the late nineteenth century Antonio Favaro, basing his argument on the *post eventum* account of Paul Guldin reported to Giovanni Pieroni, suggested that Scheiner and his associates in Germany had first heard about sunspots from Guldin.[37] Guldin's recollections, however, date from two decades after the fact, and he did not say exactly when he conveyed this information to Scheiner. But as we have seen, there is no mention of sunspots during the summer and autumn of 1611 in the surviving correspondence between the Jesuit mathematicians in Rome and their colleagues in Germany, and it is entirely likely that Guldin notified Lanz and Scheiner of the new phenomena only when others in Rome had begun observing them. Such observations did not begin until September 1611,[38] and it is conceivable that Guldin immediately wrote

36. Georg Stengel S.J. to Karl Stengel, in Fidel Rädle, "Die Briefe des Jesuiten Georg Stengel (1584–1651) an seinen Bruder Karl (1581–1663)," in *Res Publica Litteraria: Die Institutionen der Gelehrsamkeit in der frühen Neuzeit*, ed. Sebastian Neumeister and Conrad Wiedemann (Wiesbaden: Otto Harrassowitz, 1987) 2: 525–534, on 530–531. On Georg Stengel, see Sommervogel, *Bibliothèque de la compagnie de Jésus*, 7: cols. 1546–1559.

37. Antonio Favaro, "Sulla priorità della scoperta e della osservazione delle macchie solari," *Memorie del Reale Istituto Veneto di Scienze, Lettere ed Arti*, 22 (1887): 729–790, at 748–749. Pieroni wrote to Galileo on 4 January 1635 that Guldin, whom he does not identify, "has told me several times that he remembers, as ever one can say that one remembers by human certainty, that he was the first who advised the said Father Scheiner that spots were seen on the Sun, first discovered by you" (*OGG*, 16: 189). In his letter to Galileo of 17 October 1637, Pieroni identified Guldin as the person who told him this (*OGG*, 17: 193).

38. The first mention of sunspots in Galileo's correspondence is in a letter from Lodovico Cardi da Cigoli, dated 11 September 1611, in which Cigoli acquainted Galileo with the fact that his fellow artist Domenico Cresti da Passignano stated that he had already observed the morning, noon, and evening Sun many times through a recently acquired telescope (*OGG*, 11: 208–209).

to Scheiner, who received the news in October and began his observations on 21 October.

Lodovico Cigoli's failure to mention in 1611 any connection between sunspots and members of the Collegio Romano in his gossipy letters to Galileo is perhaps a significant silence. The sole contemporaneous reference we have to Scheiner's activities in the fall of 1611 is in the correspondence directed to Matthäus Rader in Augsburg: his nephew Zacharias Rader, writing from Dillingen on 20 November 1611, described Scheiner as waiting eagerly for Rader's *Chronicon Paschale*, which Scheiner, Crendel, Gretser, and Lanz had seen in manuscript form.[39] Curiously, Zachiaras included almost immediately after his reference to Scheiner a short poem alluding to the unblemished solar body's intolerance for dark shadows,

> It is enough for me, Matthäus, to be your shadow,
> For it is said that a body is followed by a dark shadow.
> Though even Phœbus itself casts no shadow,
> It drives away shadows wherever it turns.[40]

We have no context for Zachiaras' plaintive verses, but they suggest, at the very least, some of the conceptual difficulties many of the Society would have in accepting any form of darkness in proximity to the luminous solar body.

The scenario for the Society of Jesus seems, therefore, to be as follows. Scheiner, his student Cysat, and Adam Tanner were working away in early 1611, trying to obtain or to improve telescopes with which to observe the satellites of Jupiter. As they began to explore the heavens in March, one morning at sunrise they sought to look at the solar body through a telescope. Undecided if the dark spots they noticed were caused by the glasses or the atmosphere, they went on with their other observations, for the first order of business was to check the known celestial phenomena. They made a study of Jupiter's satellites and of Venus, exploring also Mars, Mercury, and Saturn, but Cysat, in particular, remained preoccupied months later by the dark markings on the Sun. When he persuaded Scheiner to reexamine the phenomena, they used colored lenses or plane pieces of glass put in the telescope, a technique which Willebrord Snellius (also known as Snel) suggested was already used by sailors taking the altitude of the Sun, but one which might have distorted the images

39. Rader, *P. Matthäus Rader S.J.*, 575.

40. "Esse tuam Matthaee tibi, mihi sufficiat, umbram: /Atra suum corpus dicitur umbra sequi. / Quanquam etiam Phæbus de se non proijcit umbram, / Umbras, sed quaquà se rotat ille, fugat." *P. Matthäus Rader S.J.*, 575.

of the sunspots.[41] This was on 21 October 1611, as Scheiner would tell the reader in his *Rosa Ursina*,[42] a date verified by the plate of sunspot observations in *Tres Epistolae* (see p. 64, below). Somewhat more experienced as observers at this point, they soon decided that the phenomena could be attributed neither to clouds nor to the flaws in their instruments, and began serious and sustained study of the sunspots.

As Georg Stengel's letter of January 1612 shows, and as Scheiner related two decades later in the *Rosa Ursina*, others in Ingolstadt soon became aware of his observations, and shortly thereafter, Marc Welser approached him for information. After receiving several letters from Scheiner, it occurred to Welser to publish them, before the charm of novelty vanished or the glory of their discovery was appropriated by someone else. In his account in the *Rosa Ursina*, Scheiner recalled,

> But since this matter, not only new but also difficult, was perceived to be in many ways contrary also to the opinion of the philosophers, lest something would be hastily and rashly published by someone in that university whose retraction would then be difficult and unseemly, my superiors ordered me to proceed cautiously and slowly, until the phenomenon itself had been corroborated by the experience of others, and not to stray casually from the trodden path of the philosophers without evidence to the contrary, and not to publish my observations in the letters to Welser under my name. For in this way they would leave anyone with greater freedom to judge, and no envy would be created. Because of these precautions, it came about that many fewer letters were published under the pseudonym of Apelles than I had in fact written to Welser.[43]

The letters dated 12 November, and 19 and 26 December in *Tres Epistolae de Maculis Solaribus Scriptae ad Marcum Welserum* are the only ones that survive.[44] They were published *ad insigne pinus*, Welser's private press, and were accom-

41. [Willebrord Snellius], *De Maculis in Sole Animadversis . . . Batavi Dissertatiuncula* (Leiden: Plantin, 1612), 8–9. For the identification of Snellius as the author of this tract, see Rienk Vermij, *The Calvinist Copernicans*, 44–45, and pp. 316–318 below.

42. *RU*, Ad Lectorum, 2nd (unnumbered) page. On the possibility of distortion from colored lenses, see Mario Biagioli, *Galileo's Instruments of Credit: Telescopes, Images, Secrecy* (Chicago: University of Chicago Press, 2006), 182 n. 120.

43. *RU*, Book I, Ch. 2, 6–7.

44. It appears that Scheiner sent the earlier letters, and perhaps some others, to Guldin. See Scheiner to Guldin, 17 January 1612, Graz, Universitätsbibliothek, MS 159, fasc. 1, no. 1. See also Giuseppe Biancani to Giovanni Antonio Magini, 17 May 1613, *OGG*, 11: 509.

panied by an engraving by Alexander Mair of Augsburg.[45] Because Scheiner was urged to proceed with caution by his superior, Father Provincial Theodor Busaeus S.J., no author was indicated on the title page, and the last letter was signed *Apelles latens post tabulam*, "Apelles hiding behind the painting."[46] Scheiner was henceforth to refer to this publication as "Apelles' first painting." In a second letter to his brother, dated 25 January 1612, Georg Stengel likewise alluded to Apelles' ambition and desire for anonymity, and to the scope of the project:

> I saw those sunspots publicized by Welser, but I also saw them before they were made public, and I assisted that man who calls himself "Apelles beneath the canvas" in seeking out those solar blemishes. I saw others, which have not yet been exposed to public view. If they emerge again, you will see new observations and conjectures. We cannot yet manage demonstrations, because everything is a crude beginning. I won't name the author, and in fact this Apelles does not as yet want to be dragged forth from his canvas.[47]

The first letter of Scheiner's *Tres Epistolae* related his initial discovery of the sunspots. This occurred "seven or eight months ago," counting backward from 12 November, when he tried to measure the angular diameter of the Sun in comparison to that of the Moon with a telescope that magnified twenty-five to twenty-eight times. "Yet, while attending to this matter, we noticed on the Sun some rather blackish spots like dark specks."[48] Because Scheiner was allowed to mention neither the Society of Jesus nor the names of any Jesuit, it was not until *Rosa Ursina* that he would identify his fellow observer as Johann Baptist Cysat.[49]

The two men soon decided that neither atmospheric conditions, nor flaws

45. On this printing consortium, see "Appendix librorum typis elegantissimis typographicis Velserianis ad insigne Pinus ab anno MDXCIV usque ad annum MDCXIV impressorum," in Georg Wilhelm Zapf, *Annales typographiæ Augustanæ ab ejus origine MCCCCLXVI usque ad annum MDXXX* (Augsburg: A. F. Bartholomæi, 1778), 103–114, and Leonhard Lenk, *Augsburger Bürgertum im Späthumanismus und Frühbarock (1580–1700)* (Augsburg: H. Mühlberger, 1968), 221–223. On the presentation of information in Mair's engraving, see Edward R. Tufte, *Envisioning Information* (Cheshire, CT: Graphics Press, 1990), 19, 21.

46. Daxecker, *The Physicist and Astronomer*, 36. On Busaeus, see Sommervogel, *Bibliothèque de la compagnie de Jésus*, 2: cols. 442–443.

47. Georg Stengel to Karl Stengel, in Rädle, "Die Briefe," 532.

48. P. 61, below.

49. Zinner (*Entstehung*, 494–496) reproduces an affidavit by Franz von Paula Schrank (1747–1835), who claimed to have read a two-volume astronomical diary by Cysat, in which Cysat states that the discovery happened on 6 March 1611 (presumably Gregorian) and that he and Scheiner climbed the tower at dawn to make observations: Scheiner to measure the angular diameter of the Sun and Cysat to find the satellites of Jupiter. Now the ecliptic longitude of the Sun on 6 March 1611

in their lenses, nor defects in their own eyesight explained the phenomena. As Scheiner related in the first letter, they assumed therefore that the dark marks were either on the solar surface or "in some heaven outside the Sun." Both because the generation of dark marks seemed, from an Aristotelian viewpoint, unsuitable for the brightest body in the heavens, and because he had never seen any spots return unaltered after rotation about the solar axis, Scheiner concluded that the phenomena were bodies partially eclipsing the Sun. In his view, the question at hand was whether these star-like bodies moved below or around the Sun, and he suggested he would resolve the issue shortly.[50]

Since the spots moved across the Sun and disappeared at the edge, Scheiner expected them to reappear at the other edge after an equal interval of time; Johannes Fabricius had entertained similar beliefs, and that was what he had observed.[51] But in this first letter Scheiner did not speak of changes in the shapes of the spots, and presumably he anticipated that sunspots reappearing on the eastern limb would have the same figure and formation they had had upon disappearing on the western. Spots of unchanging shapes on the surface of the Sun would mean that the Sun rotates, as Kepler had already postulated in his *Astronomia Nova* of 1609, but since spots of the same shape did not reappear after the appropriate time, Scheiner concluded that they could not be on the Sun. Thus, while Johannes Fabricius started with rotation time and found the spots reappearing, Scheiner began with constant shapes and saw no return.

The fact that the sunspots seemed to him not to reappear presented an obvious difficulty. For the twenty-three days between their first observation on 21 October and 12 November, Scheiner and Cysat, perhaps with Stengel's assistance, had been able to observe the Sun on all days except for 3, 4, and 11 November. As Scheiner's figure shows, the motion of the spots is apparent enough, especially if one mentally rotates the figures to account for the different times of observation, and yet in this period he had not seen any spot return, and it is apparent from his figures that no obvious candidate for a reemergent spot appeared. The shapes of the spots simply changed too much. Presumably bodies orbiting the Sun would take longer to return.

At this point in his first letter, Scheiner interrupted his narration to call the reader's attention to a possibility raised by his observations. The order of the

was 346° and that of Jupiter 105°. Cysat could thus not have observed Jupiter at dawn because it had set in the west at about 2 a.m. This fact casts great doubt on Schrank's affidavit.

50. P. 62, below.

51. Fabricius, *De maculis in sole observatis*, D1r.

planets had always been a convention in the Ptolemaic system. As mentioned earlier (p. 14), Ptolemy had suggested that if Venus and Mercury were below the Sun, the absence of observations of their passage across the solar face might be due to the infrequency of transits in the same plane and to their small size.[52] To this argument, Copernicus had added the point that Venus was too small to be seen on the Sun.[53] Upon observing the sunspots, however, Scheiner realized that even if Venus only obscured one hundredth part of the Sun during a transit, as Copernicus had stated, it would still be visible through the telescope. The issue might settle the question as to whether Venus, and likewise Mercury, were always "above," "below," or alternatively above and below the Sun. Scheiner was to return to this important matter in his second letter.

Scheiner next briefly discussed his observations as they appeared on the engraved plate, pointing out that while he had made his best effort, the drawings were not very exact. Moreover, their size with respect to the Sun was not accurate: in order to make them more conspicuous on paper, he had drawn the spots larger than they were, especially the smallest ones. Although he mentioned the spots' emergence and disappearance, Scheiner did not make clear whether he intended their appearance on and departure from the limb, or in the middle of the Sun.

This first letter ended with some advice on how to observe sunspots. Scheiner noted that one might observe the Sun through a telescope without danger for about a quarter of an hour at sunrise and sunset, or when it was covered by transparent mist or clouds. A piece of colored glass, when suitably dark, could be placed in front of the ocular to permit observations when the Sun was higher in the sky and not shielded by transparent clouds. Scheiner's recommendation that the reader begin at the Sun's limb and move gradually towards its center suggests that such observations would still have been painful. There is no explicit mention of the dark room method described by Stengel.

While the first letter was based on observations made over less than a month, and contained rather modest information about the shapes and motions of the spots, the next two letters, written in December 1611, drew more extensively on his subsequent findings. In that of 19 December, he returned, with unfortunate results, to his conjecture about a transit of Venus. In his *Ephemerides*, Giovanni Antonio Magini had predicted a conjunction of Venus and

52. *Ptolemy's Almagest*, trans. G. J. Toomer (London: Duckworth; New York: Springer Verlag, 1984), 419.

53. Copernicus, *Nicholas Copernicus on the Revolutions*, tr. Edward Rosen (Baltimore: Johns Hopkins University Press, 1978), 19.

the Sun for 11 December, at which time Venus's (ecliptic) latitude would be less than the Sun's diameter.[54] If Venus were always below the Sun, it would then appear for several days as a black spot slowly traversing the solar disk. Mistakenly thinking that Magini had meant that the conjunction began on the 11th, Scheiner had observed for several days, but he had seen no trace of the planet on the solar surface. "Venus evidently blushed and ran away to keep her nuptials secret," he explained by way of accounting for the discrepancy between expectation and outcome.[55] Scheiner ended the discussion on an apologetic note, insisting only that during the conjunction of 11 December, Venus had been above the Sun.[56]

Scheiner's argument was marred by several serious errors.[57] To begin with, he had mistaken Magini's time for the middle of the conjunction for the time of the start, a difference of several days,[58] but not one that would have prevented him from seeing the transit, had it occurred. But what did this absence of evidence prove? It could equally well be argued that Venus was always "above" the Sun, as Plato had supposed.[59] The pure Ptolemaic scheme predicted that at superior conjunction Venus would be "below" the Sun.[60] In January 1612, Scheiner would correct his error about the timing of the conjunction; he and

54. See appendix 4.

55. Scheiner's peculiar discussion of Venus' behavior was later echoed by Charles Malapert S.J., born in Mons, in what is now Belgium, an astronomer who had begun observing sunspots with Scheiner in Ingolstadt in 1613. "For a long time we were deceived by Venus, whose womanly nature it has long been to be watched exulting as she invites and tempts us with her blandishments. Amongst all our telescopic phenomena, in my view, there is nothing more charming than seeing Venus undergo her increases and decreases of light, taking on all the shapes we see in the Moon, being now crescent, now halved, now resplendent with a full orb." See *Oratio habita Duaci* (Douai: Typis Baltazaris Belleri sub Circino Aureo, [1620]), 30, cited in Luigi Guerrini, "Tradizione astronomica e cultura matematica in un'orazione gesuitica del 1619," *Giornale critico della filosofia italiana* 79 (2000): 209–235, on 229.

56. Pp. 66–67, below.

57. For a full discussion of these calculations, see appendix 4.

58. Depending on Venus's ecliptic latitude, a superior conjunction can last up to fifty-five hours. (We thank H. R. van Gent for calculating this for us.)

59. *Timaeus* 38d.

60. In the so-called "Capellan" variation (after Martianus Capella) of the Ptolemaic scheme, Mercury and Venus went around the Sun, while the Sun and every other heavenly body revolved around the central Earth. See Bruce S. Eastwood, "'The Chaster Path of Venus,' Orbis Veneris castior, in the Astronomy of Martianus Capella," *Archives Internationales d'Histoire des Sciences*, 32 (1982): 145–158; Eastwood, "Plato and Circumsolar Planetary Motion in the Middle Ages," *Archives d'Histoire Doctrinale et Littéraire du Moyen Age*, 68 (1993): 7–26. Note that the Capellan variation was the predominant version of the geocentric arrangement of the heavens at Dutch universities in the late sixteenth century. See Vermij, *The Calvinist Copernicans*, 39–42.

Welser sent a separate sheet addressing this problem to those who had received copies of the tract.[61]

Scheiner returned to the subject of sunspots in his third letter, dated 26 December 1611. By this time he had become much more confident in his conclusion: "I am now unafraid of affirming [the things I timidly touched on some days ago], trusting sure and confirmed principles . . . [for] it pleases me to free the body of the Sun entirely from the insult of spots."[62] He had found that spots take at most fifteen days to complete their journey across the face of the Sun, and had thus concluded that if they were on the solar body, they would reappear after another fortnight. After more than two months of observations, he had yet to see this happen.

Having discarded the hypothesis of spots on the solar body, Scheiner also eliminated the possibility that they were in the heavens of the Moon, Mercury, and Venus. Concluding, finally, that they could be nowhere but in the heaven of the Sun, he argued that they moved around it "either in a fixed or wandering fashion," that is, either as if fixed in a sphere, or in various individual orbits. The reasons supporting this conclusion were several. First, Scheiner had observed that the spots were foreshortened at the limb; a spot that was circular near the center of the Sun's disk appeared as a thin line near the edge. Second, loose formations of spots in the center seemed to him compressed as they neared the limb. Third, the spots appeared to move more slowly near the edges than in the middle. Scheiner neglected to inform his audience that these arguments would apply equally if the spots were on or near the surface of the Sun; indeed, in order for his conjectures to be valid, the spots or satellites would have to be in very tight orbits around the Sun!

At length Scheiner addressed the issue of the spots' substance. They could not be clouds, he stated, because such masses would be huge, and the motions of the spots were very regular, while clouds above the Earth moved in a haphazard fashion. Nor could they be comets, for reasons Scheiner did not elaborate.[63] They were, he reasoned, therefore "either the denser parts of some heaven—and

61. Welser to Galileo and Johannes Faber, 13 January 1612, *OGG*, 11: 263.

62. P. 67, below.

63. As the 1607 letter from Lanz to Clavius shows, the question of comets, brought to the fore by measurements of the comet of 1577, exercised the southern German Jesuit mathematicians. Clavius, who died on 6 February 1612, never gave clear guidance on this problem, and the Society's mathematicians and philosophers conformed to the Aristotelian notion of sublunary comets. See Lattis, *Between Copernicus and Galileo*, 26, 156–60. But note that five years later, in his anonymously published *De tribus Cometis anni MDCXVII Disputatio Astronomica* (Rome, 1619), Orazio Grassi S.J., placed these three comets in the heavens. See *OGG*, 6: 19–35.

according to the philosophers that is what stars are—or solid and opaque bodies existing in themselves."[64] But if they were heavenly bodies, Scheiner noted, they would be like Venus and the Moon, which were dark where not illuminated by the Sun. In a not very successful geometrical demonstration supported by a diagram, Scheiner tried to explain the foreshortening at the limb in terms of phases like those of Venus and the Moon. (Note in the figure on p. 71, below, how closely the sunspot orbit encircles the Sun.)

He ended his third letter with a number of conclusions. First, the spots were very near to the Sun, immense, quite thick, and opaque. Second, were it not for the great brightness of the Sun, one would be able to see their phases, and in theory, Scheiner concluded, the same would hold true for the other stars. Third, the satellites of Jupiter were similar in nature to sunspots, and there were more than four of them. In this connection, Scheiner explained that the different orbital planes in which these satellites moved accounted for their sometime departure from a rectilinear path through Jupiter. Finally, he conjectured that perhaps something like sunspots could serve to elucidate the varied guises of Saturn, which appeared "sometimes as an oblong shape, and at other times accompanied by two stars touching its sides."[65]

Scheiner repeated the now familiar conclusion that sunspots were either fixed or wandering stars that moved about the Sun, and he inclined toward the latter hypothesis. Since he might by this time have already adopted the Tychonic view of the arrangement of the heavens, or at least the Capellan variation of the Ptolemaic scheme, it appears that he envisioned the Sun surrounded not only by the five planets, but also by innumerable planetary bodies moving in heliocentric orbits at least as far out as Venus. Similar swarms of planetary bodies also appeared to surround Jupiter and perhaps Saturn as well.

While his colleagues in Rome had been skeptical in 1610 about the four satellites that Galileo claimed to have discovered about Jupiter and took considerable time to be convinced about them, a little more than a year later Scheiner was boldly filling the universe with such bodies. He had discovered the seventeenth-century equivalent of our dark matter, and in his adaptation of two lines from the *Georgics*—"The Sun will also give signs: who would dare call the Sun false?" (p. 73 below)—Scheiner suggested that he saw the solar markings not as the meteorological indices Virgil had described, but as the beginnings of a profound cosmological revision. Rather than solid spherical shells revolving around the Earth, housing the seven "wandering stars," there would now be a fluid medium through which perhaps thousands of bodies orbited

64. Pp. 69–70, below.

65. P. 72, below.

around at least three centers, the Sun, the Earth, and Jupiter.[66] It is no surprise that his superiors in the Society of Jesus wanted the names neither of the order nor of any of its members mentioned in this tract. Through the offices of Marc Welser, *Tres Epistolae* was published without permission from censors; had the work been submitted to a rigorous vetting, it is entirely unlikely that it would have been printed in this form.

66. And perhaps around Saturn as well, as observations of the strange appearances of this planet were interpreted by Galileo to be bodies that moved around the planet. See Galileo's first and third sunspot letters, pp. 102 and 295–296, below.

5

Tres Epistolae

First letter

To Marc Welser, Prefect and Magistrate of Augsburg

Since you have asked me, dear Welser, about the solar phenomena that I have observed, I am presenting new and almost incredible findings. At first these produced astonishment not just in me but also among my friends, and subsequently an intellectual delight, for by means of these phenomena we may consider it clearly established that many things thus far doubted, ignored, or perhaps even altogether denied by astronomers can now be brought into the resplendent light of the truth, thanks to the Sun, the source of illumination and the commander of heavenly bodies.

About seven or eight months ago,[1] with a certain friend of mine,[2] I directed to the Sun the optical tube[3] that I still use now and which

1. Since this letter is dated 12 November, and was presumably some days in writing, this observation must have taken place sometime between the beginning of March and the middle of April. In his later *Rosa Ursina*, Scheiner states that it happened in March (preface, second unnumbered page). Ernst Zinner reproduces a document written by Franz von Paula von Schrank (1747–1835), who had access to now lost observation records by Johann Baptist Cysat, wherein Schrank claims that the date in question was 6 March, and that it was Cysat and not Scheiner who first observed spots. See Zinner, *Entstehung und Ausbreitung der Coppernicanischen Lehre* (Erlangen: Mencke, 1943), 494–496.

2. Johann Baptist Cysat.

3. The expression *tubus opticus*, or simply *tubus*, was a common expression for the telescope in the seventeenth century.

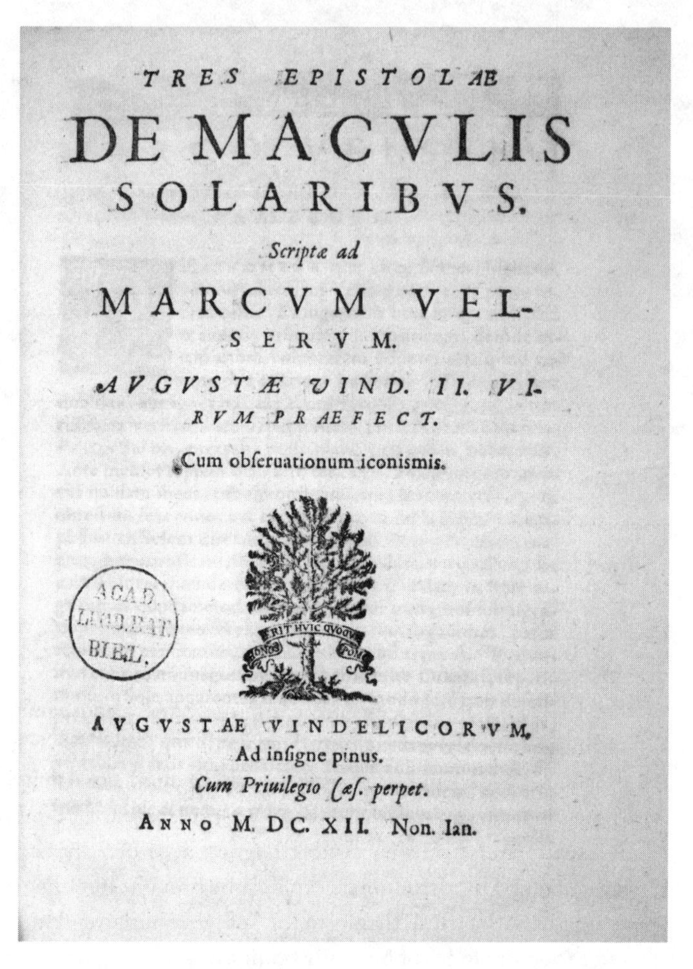

TRES EPISTOLÆ
DE MACVLIS
SOLARIBVS.
Scriptæ ad
MARCVM VEL-
SERVM.
AVGVSTÆ VIND. II. VI-
RVM PRAEFECT.
Cum obſeruationum iconismis.

AVGVSTAE VINDELICORVM,
Ad inſigne pinus.
Cum Priuilegio Cæſ. perpet.
ANNO M. DC. XII. Non. Ian.

Figure 5.1. Frontispiece to Scheiner's *Tres Epistolae*. Translation: Three letters about solar spots, written to Marc Welser, magistrate and prefect of Augsburg. With illustrations of the observations. / By Apelles hiding behind the painting. / Augsburg, at the famous pine tree. With perpetual privilege of the Emperor. The nones of January 1612.

magnifies an object six hundred or even eight hundred times in surface.[4] We were intending to measure the visual magnitude[5] of the Sun in comparison to that of the Moon, and we found them about equal. Yet, while attending to this

4. That is a linear magnification of 25–28. Here Scheiner is following Galileo in expressing magnification in terms of area. See *Sidereus Nuncius*, in *OGG*, 3: 61; *SN*, 37–38.

5. That is, its apparent, or angular, diameter. Note that with these "Galilean" telescopes with their concave ocular, angles could only be estimated. True micrometers, in which the measuring instrument is superimposed on the observed image within the telescope itself, became possible only with the shift to the "astronomical" telescope with its convex ocular later in the seventeenth century.

matter, we noticed on the Sun some rather blackish spots like dark specks. As we had not intended to investigate this, we put it off for another time, considering it a matter of little importance. We returned to the business this past October[6] and found the spots appearing on the Sun as you see them shown in the drawing. But because this was almost beyond belief, we were at first uncertain whether our results might have occurred because of a hidden defect in our eyes, in the tube, or in the air, so we called on the eyes of many different men,[7] all of whom without exception saw the same spots in the same place, arrangement, and number. We concluded, therefore, that the defect was not in our eyes, for how could the eyes of such different men all be afflicted together, and then on certain days unanimously take on another condition? And further: if these things resulted from a defect of the eye, the spots should traverse the Sun as the eye does, but this never happened. We therefore correctly concluded, to a man, that it is absolutely not because of an ocular error that these spots can be proposed [to be] on the Sun.

Next, we worried about a flaw in the glass, fearing the tube might have deceived us. To investigate this, we used eight tubes of very different powers, and each according to its measure showed the same spots on the Sun. If, as time passed, one revealed to us something new or changed, the others likewise revealed the same thing. Moreover, neither rotating nor shaking any of the tubes ever moved the spots along with it, which ought to happen if the tube produced this phenomenon. From this we all correctly concluded that the tube was free of any blame in the matter.

There remained the air, to which, however, these appearances could not be attributed. First of all, [they could not be] because these phenomena rise and set together with the Sun in the diurnal motion that the Sun receives from the *primum mobile*. But it is unheard of that the air, or something in the air, should revolve so constantly, especially below so small a body as the Sun, that is less than one degree, or about 30 minutes [of arc in apparent diameter].[8] Second, because these phenomena suffered no parallax, which would have happened between morning and evening had they been revolving in the air with the Sun.[9]

6. 21 October, at 9 a.m. See Scheiner's illustration (p. 64), and *RU*, preface, second unnumbered page.

7. In *RU*, loc. cit., Scheiner states that from 9 to 10 a.m. he showed the spots to many other fathers and students or interested men (*studiosis*), and that at about that time Father Adam Tanner also observed the spots.

8. What Scheiner means here is that with its apparent diameter of $30'$, the Sun occupies only a very small portion of the heaven. Since bodies in the air do not move with uniform motion, as the Sun does, such bodies can never match the angular speed of the Sun so exactly that they will appear to remain on the Sun.

9. If these bodies were in the air, that is, closer to us than the Moon, their daily parallax with respect to the Sun would be more than $1°$, while the apparent diameter of the Sun is only $\frac{1}{2}°$.

Third, because these phenomena, turning with their own constant motion either below or along with the Sun, were observed in different places on the Sun, before vanishing entirely from it after many days, and moved in a westward or slightly southward direction (as it seemed to me).[10] In regard to this motion, however, more extensive and accurate observations will yield more certain results. Fourth, because these phenomena seemed to be unchanged when observed through thin clouds sailing turbulently below the Sun. These spots, then, are not in the air, so I may pass over several other valid objections in silence.

They must therefore be either on the Sun or in some heaven outside the Sun. It has always seemed to me unfitting and, in fact, unlikely, that on the most lucid body of the Sun there would be spots, and that these would be far darker than any ever observed on the Moon, except for one very small one.[11] Moreover, if they were on the Sun, it would necessarily rotate on its axis and cause them to move, and those spots first seen would at length return in the same arrangement and in the same place with respect to each other on the solar body. So far they have never returned, and yet successive new ones have run their course across the solar hemisphere visible to us.[12] This proves that they are not on the Sun; indeed, I would judge that they are not true spots but rather bodies partially eclipsing the Sun from us and are therefore stars[13] either below[14] or around the Sun. Which of the two is the case I will in due time and with God's help bring to light.

We now have a means of acquiring clear knowledge of whether Venus and Mercury are carried[15] sometimes above or always below[16] the Sun. They will demonstrate this during their diametral conjunctions with the Sun,[17] since with

10. Since the Sun's equatorial plane is inclined $7\frac{1}{4}°$ to the ecliptic (see pp. 315–323, below), the motion of the spots has a north-south component with respect to the ecliptic.

11. Presumably Scheiner is referring to a sunspot that was diffuse, not a facula, a phenomenon first observed by Galileo. See p. 281, below.

12. Scheiner here assumes that the Sun is a solid body and that if the spots were on its surface they would be permanent markings. Their motion would mean that the Sun rotates on its axis, and this rotation would periodically bring the same spots, unchanged in shape, back into view. Since this does not happen, the spots cannot be on the Sun. He ignored the possibility that the Sun is not a solid body and that spots on or near its surface could therefore be of changing shapes.

13. In Aristotelian cosmology and terminology, all heavenly bodies were stars; most were fixed, some wandered.

14. In the Aristotelian cosmos, all directions were referred to the center of the cosmos. Thus, if spots were in the heavens between us and the Sun, they would be said to be below the Sun.

15. This terminology derives from the belief that planets and fixed stars were embedded in transparent spheres that rotated about the central Earth.

16. See note 14.

17. In the Ptolemaic theory, the epicycles of Venus and Mercury are entirely below the Sun. True superior conjunction occurs when the planet is in the line from the Earth to the Sun near the apogee

their bodies they will produce spots on the Sun[18] and at the same time they will show their motions to us. In fact, the portal through which we can begin freely to consider their sizes with respect to the Sun is now wide open.[19] Several other things that I here pass over will also become known, but it seemed best to put forward these few findings to be sampled now, and if they prove tempting, we will very soon endeavor to elicit something more from this kernel, provided the clouds do not begrudge us the splendor of the Sun. For the more brightly it shines, the more agreeable its appearance to our eyes, even at noon. Indeed, we observe it no differently than we do the Moon.[20]

I have the following admonitions about the observations shown here. 1. They are not terribly exact, but rather are hand-drawn on paper as they appeared to the eye without definite and precise measurements,[21] which, due to the inclement and inconstant weather, or sometimes the lack of time, or elsewhere other impediments, could not be made. 2. Those spots which are the most notable and constant in their appearance are marked by the same letters. 3. Wherever I have skipped over some days, the Sun was then covered by clouds and could not be observed. 4. Where I added some spots without letters, these were either not continually visible because of turbulent air or, if they did appear constantly, they did not need to be observed in comparison to the others because of their smallness. But this should be noted as well: the proportion of the spots to the Sun should not be taken from the drawing, for I made them larger than they ought to be so that they would be more conspicuous, especially in the case of some very small ones that otherwise could scarcely be the object of our vision. Frequently many small spots were conflated into one so as to appear as one long or even triangular spot, as is done with spots A and C, which can, however, be distinguished through tubes of great power.[22] Spot A I have conflated from three spots, C from five, and D from four, and I have designated with single letters just as I have done with the rest of the connected

---

of its epicycle, and true inferior conjunction occurs when the planet is in the line from the Earth to the Sun near the perigee of its epicycle.

18. See pp. 14–22.

19. A canonical system of planetary sizes and distances had been handed down from Ptolemy, and although in the Copernican and Tychonic system planetary distances were different, planetary sizes had remained more or less the same. No doubt Scheiner took his planetary sizes from Clavius's *Commentarius in Sphaeram Ioannis de Sacro Bosco* (*Opera Mathematica*, 3: 100–102, 117), which were virtually identical to Ptolemy's sizes. See Albert Van Helden, *Measuring the Universe*, 27, 30, 32, 50, 53.

20. Scheiner means that he and his colleagues at Ingolstadt were observing sunspots by pointing their telescope directly at the Sun, even when it was high in the sky, just as one observed the Moon. Presumably they used colored glass to protect their eyes.

21. See the discussion on p. 53, above.

22. The references are to the figure on p. 64.

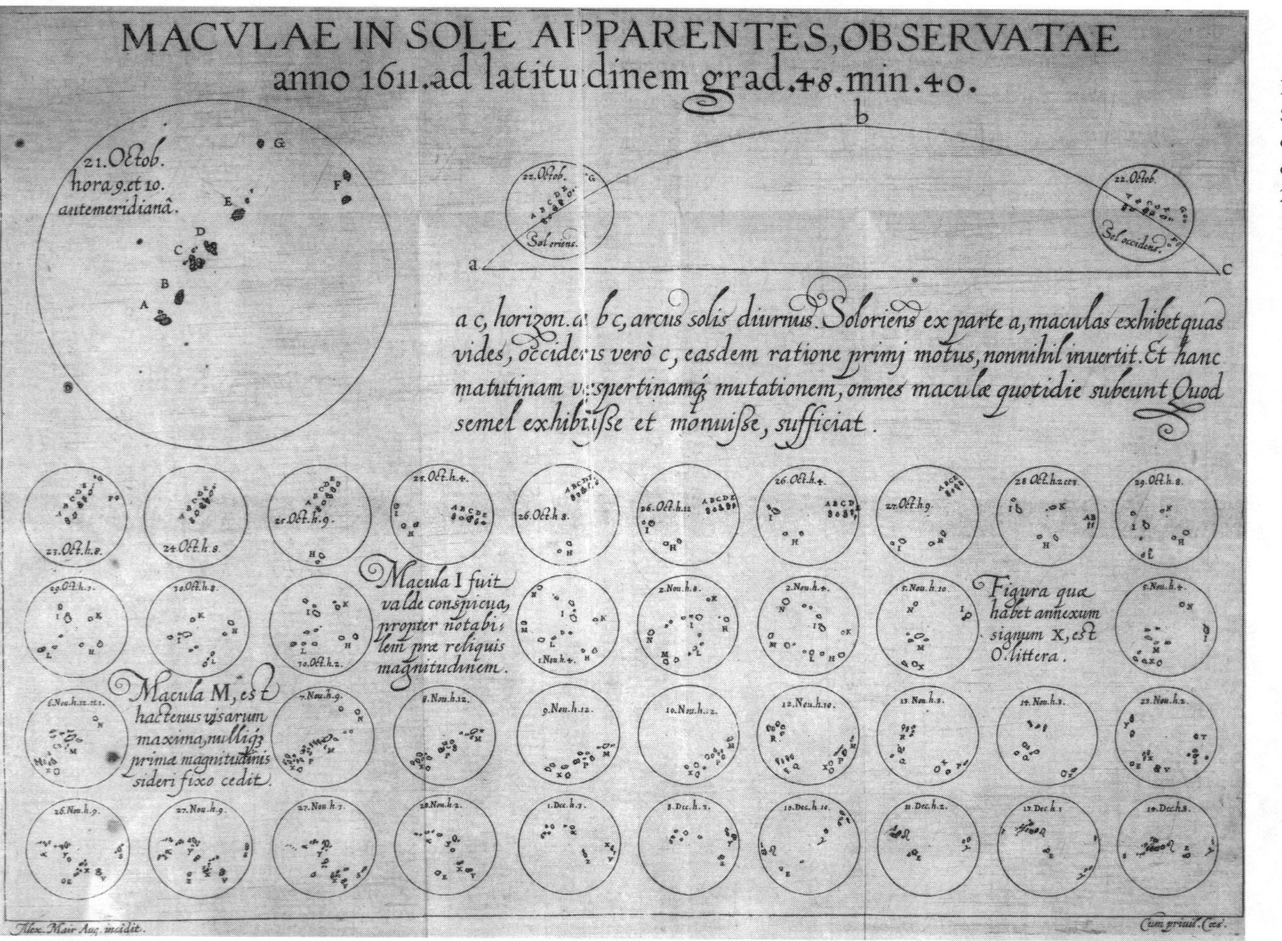

Figure 5.2. Scheiner's observations of sunspots, 21 October–14 December 1611.

spots. The spots that always retain the same letters next to them are always the same, although they are depicted as they appeared at the time they were drawn. When some spots and their letters are no longer drawn, these had ceased at that time to appear on the Sun. But when different spots are designated by different letters, these are different and newly emergent spots. When, however, spots not designated by any letters are at times represented and at times not, these either have entirely set[23] and so are not drawn, or, as often happens, they have not appeared due to thick air, since these kind of spots only offer themselves to view when the Sun is very bright and the air very pure.

Since I remember that you at one time asked me who were those young eagles who dared to look directly at the Sun, I will also briefly show the method that may be safely used by mathematicians who in such a great matter would trust their own eyes rather than another's. 1. A bright and unclouded Sun may, without any danger whatsoever, be observed on the morning or evening horizon for a quarter of an hour with a simple tube,[24] but it must be a good one. 2. Whenever the Sun is covered by mist or a suitably transparent cloud, it can be observed with a simple tube without damage to the eyes. 3. Whenever the Sun is uncovered it can be observed through a tube equipped with a convex and concave lens, with an added lens that is plane on both sides, of suitable thickness, and dark blue or green, mounted at the end near the eye. Such a tube protects the eyes from injury even when the Sun is on the meridian and even more so when a thin vapor or mist in the air, like a veil over the Sun, is added to any insufficiently tempered dark glass. 4. The observation should begin at the perimeter and gradually extend to the middle, and remain there for a little while, since the surrounding light does not immediately permit the shadows [to be seen].[25] Now use and enjoy these findings. More of them will follow, God willing. Farewell. 12 November 1611.

Second letter

According to the *Ephemerides* of Magini,[26] on 11 December, which was a Sunday, at the eleventh hour of the night, a conjunction of Venus with the Sun be-

23. That is, disappeared from the edge of the Sun.

24. In view of the fact that in no. 3 Scheiner specifies lenses and well as colored glass, it may be the case that in nos. 1 and 2 he is referring to a tube without any lenses at all. From the context of the argument, however, it would appear that he is referring to a telescope without a colored glass.

25. This procedure was also recommended by Johannes Fabricius. See pp. 30–31.

26. Giovanni Antonio Magini, professor of mathematics at the University of Bologna, published several ephemerides. Scheiner is most likely referring to *Tabulae secundorum mobilium coelestium*.

gan, which will be considered in due course. According to Magini's calculation, it lasted at least forty hours. From this it follows that it was not over before the third hour of the following Tuesday.[27] So I reasoned as follows: if, as the general school of astronomers has hitherto taught, the heaven of Venus is below the Sun, it follows that at every conjunction with the Sun, Venus is between us and the Sun. And since this conjunction happens at nine minutes of latitude,[28] Venus must conceal from us some portion of the Sun, must appear to us as a spot much larger than any ever seen (since its diameter is at least 3′), and, moreover, must pass under the Sun from west to east against the motion of the spots. All that was lacking were the clear skies that would permit observation. A cloudy Monday made me very anxious, and I lamented that such a ripe occasion for searching out the truth would be snatched away from me, an occasion that, if I am not mistaken, would not return for many years. But Tuesday, wholly clear from early dawn to late evening, gladdened me again, as I had not seen a more beautiful day for two months and at that time of the year could not have wished for a better one. So I greeted a most limpid rising Sun with joy, and observed it assiduously. Nor was I alone, for many others celebrated with me the conjunction of the Sun and Lucifer[29] the entire day. And what are you expecting? Nowhere did we see Venus under the Sun, though according to the calculation it was below the Sun. Venus evidently blushed and ran away to keep her nuptials secret. What follows from this I do not say; you must find your own way. And even if we were lacking other arguments, one argument would prevail, namely, that the Sun is encircled by Venus. And there is no doubt that the same is done by Mercury,[30] which I will not neglect to investigate in a similar manner as soon as an opportune conjunction offers itself.[31] Nothing contrary can be said unless it was that we had observed it carelessly, which is scarcely the case, or that Magini's calculation deviated from the truth by 7 minutes and very many hours,[32] but it is absurd to

27. Scheiner is using the civil calendar here, in which the twenty-four-hour day begins at midnight.

28. "... et cum haec conjunctio fiat in 9 latitudinis gradu...." The ecliptic latitude of the Sun is, of course, zero by definition, and therefore Scheiner is referring to the ecliptic latitude of Venus. But since the Sun's angular diameter is about 30′, if the latitude of Venus were 9° at the conjunction it would not be seen on the Sun. Therefore Scheiner means a latitude of 9 arcminutes. This is confirmed by the information cited by Scheiner in his *Accuratior Disquisitio*; see p. 190, below.

29. Lucifer was one of the names of Venus appearing as the morning star, that is, rising in the morning in the east, ahead of the Sun. Since Venus was in conjunction at this time, it was neither a morning nor an evening star.

30. Since Venus was not seen "below" the Sun, it could equally well be concluded that it, and Mercury as well, were always "above" the Sun, as in the planetary order ascribed to Plato.

31. See note 17.

32. If Magini's ephemeris was in error by 7′ in Venus's latitude, the planet would have been removed 16′, or more than the Sun's angular radius, from the center of the Sun, and therefore even if Venus was "below" the Sun, there would have been no transit.

think this of so famous a mathematician,[33] and we will fully explore the question in due course, or that the star of Venus does not appear to us as a shadow or spot because it is does not receive light from the Sun, but is endowed with its own, as is the case with the Moon. But experience, reason, and the general opinion of all mathematicians, ancient and recent, will cry out against this.[34] What remains, therefore, is this: if Venus was in conjunction with the Sun it either had to be seen by us or, since it was not seen, the conjunction must have occurred in the upper hemisphere of the Sun.[35] Farewell. December 19, 1611.

Third letter

It is amazing how success panders to audacity. You remember those things that I timidly touched on some days ago: I am now unafraid of affirming them outright, trusting sure and confirmed principles that I submit to your judgment. It pleases me to liberate the Sun's body entirely from the insult of spots, and I am persuaded that this can be done by the following argument:

By observing the spots accurately, it is established that, at most, they spend no more than fifteen days on the Sun. Drawing on common knowledge, assuming the Sun's visual diameter to be 34′, we will see at most 179°26′ of the circle of the Sun.[36] Now if some spot travels 179°26′ below the Sun in the

33. Scheiner used Magini's ephemerides for the meridian of Venice (longitude 12° 34′ E) without any adjustment to the meridian of Ingolstadt (longitude 11° 27′ E). The difference of about 1° would introduce an error of about 4 minutes of time. His calculation of 11:40 p.m. was almost ten hours late according to modern computations. See appendix 4.

34. In his *Almagest*, Ptolemy tacitly assumed that Venus and Mercury shine with their own light (IX, 1). The assumption was shared by Copernicus (*De Revolutionibus*, I, 10). In his commentary on the *Sphere* of Sacrobosco, Clavius stated that since all heavenly bodies (except the Sun) are of the same nature as the Moon, and since the Moon receives its light from the Sun, as is illustrated by its eclipses and phases, therefore "according to the astronomers and philosophers, all stars and planets received their light from the Sun." (*Opera Mathematica*, 3: 45). This judgment was not, however, universal. Until he was notified by Galileo of Venus's phases, Johannes Kepler thought that because of its brightness Venus shone with light of its own; see his letter to Galileo of 28 March 1611, *OGG*, 11: 78.

35. In other words, Venus was on the other side of the Sun from us.

36. The observer at O does not quite see an entire solar hemisphere. Since the Sun's angular radius is 17′, angle SOA is 17′. Since visual ray OA grazes the Sun, SA is perpendicular to OA. Therefore

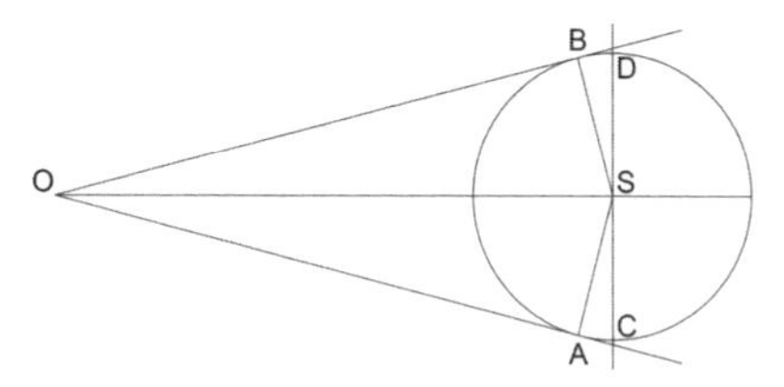

space of fifteen days, it will revolve on the opposite part of the Sun 180°34′ in fifteen days, two hours, and twenty-two minutes. Therefore, if we assume that this spot is on the Sun it must, after it begins to revolve on the other side, return after fifteen days, two hours, twenty-two minutes. But so far, as a course of observations of about two months makes apparent, no spot has returned to the same place and arrangement. It is therefore impossible for any spot to be on the Sun. What then?

1. That the spots are not in the air, I demonstrate thus: if they revolve in the air, they will have a larger parallax than the Moon either at apogee or perigee,[37] but they do *not* have a greater parallax. It follows that they are not in the air. The major premise is evident; the minor is consistent with experience. Now a spot almost on the periphery of the Sun, such as γ or δ, occupies the same position for the entire day without sensible change, and this would be impossible if it suffered as much parallax as the Moon, since the Moon's parallax, even at apogee, is nearly a whole degree. It would then be necessary for any given spot to desert the Sun at one time or another during the day and yet appear under it again the following day.[38] This is contradicted by experience. They are therefore not in the air.

2. That spots are not in the lunar heaven, I demonstrate in the following fashion. First, from parallax: for the above-mentioned parallaxes would occur, contrary to experience. Second, from the motion of the Moon and the spots: for the latter are uniformly carried to the west, while each and every orb of the Moon[39] is daily carried *per se* or *per accidens* to the east much faster than the Sun.[40] Third, on the basis of experience itself, for otherwise these spots would appear illuminated and would be shining at night in the opposite part of the lunar heaven, but this does not happen.[41]

---

angle ASC is 17′. The observer thus sees 34′ less than 180°, or 179°26′ of the circumference CABD of the Sun, and the invisible circumference is 180°34′.

37. Its greatest and least distances from the Earth.

38. Scheiner's point is that since a body below the Moon would have a parallax greater than 1°, in the Sun's journey from the horizon to its zenith, a spot apparently situated on the Sun but in reality situated in the air below the Moon would show a parallactic displacement comparable to the angular diameter of the Sun ($\frac{1}{2}$°).

39. *Lunae orbes omnes et singuli.*

40. The heaven or sphere containing the Moon partakes of the twenty-four-hour diurnal rotation from east to west, but in addition has a west-to-east motion of about 13° per day. Sunspots move across the disk of the Sun from east to west in about thirteen or fourteen days, but any body in the lunar heaven would have a west-to-east motion, and this movement would be much faster than the slow progression of sunspots in the opposite direction.

41. Bodies in the lunar heaven would, presumably, be of the same nature as the Moon. Because of the different motions of the spheres of the Sun and Moon, a body could not remain exactly below

3. The spots are not in the heaven of Mercury for the same reasons adduced concerning the heaven of the Moon, but in the proportion of the former.[42]

4. They are not in the heaven of Venus because of the last two reasons adduced about the Moon, and as Venus's parallax is about the same as the Sun's, it is perhaps not a very pressing point.[43]

What remains is that these shadows[44] must revolve in the heaven of the Sun. Since they cannot be in the Sun's eccentric circle, as its motion and that of the Sun are one and the same, and since they cannot be in two eccentric circles *secundum quid* or in any other (if there were another circle of the Sun),[45] the only possibility that remains is that they are moved by their own proper motions either in a fixed or wandering fashion. Which of the two it is, I cannot yet say. This much is certain: they move around the Sun, and for this I offer three convincing arguments. First, every spot observed by itself near the edge of the Sun becomes more slender at ingress or egress.[46] This phenomenon can be explained only by the motion of a spot around the Sun. Second, near the edge of the Sun, two, three, or more spots seem to come together into one large one, while in the middle of the Sun they split into many. This can be explained only by their motion around the Sun. Third, they move more quickly in the middle than they do near the perimeter of the Sun. This too can only be explained by a motion around the Sun: *ergo*. For want of time I pass over many other arguments.

But, finally, what are they? Not clouds, for who would imagine clouds there? And if they were clouds, how large would they be? Why would they always be moved in the same manner and with the same motion? How could they produce such large shadows? They are thus not clouds. But for the same reason and for others that I omit at the moment, they are not comets.[47] It remains that they

---

the Sun for very long. As it diverged from the Sun, it would reflect sunlight to the Earth and thus, like Venus, be visible to us.

42. I.e., in proportion to the greater size of Mercury's orbit. The traditional Ptolemaic order of heavenly bodies was Moon, Mercury, Venus, Sun, Mars, Jupiter, Saturn, and the fixed stars.

43. Since there could be no empty spaces in the cosmos, Venus's greatest distance had to equal the Sun's least distance from the Earth.

44. Up to this point, Scheiner has referred to them as spots. Here he begins referring to them as shadows, i.e., the shadows of satellites of the Sun.

45. If the spots were exactly in the same orbit as the Sun, their motion would equal that of the Sun, and they would never be seen to move with respect to the Sun, as sunspots do. Therefore, if the spots were on the Sun there would be a logical inconsistency: on the one hand they would exactly share the orbital motion of the Sun, while on the other they would have to have an additional motion in a second orbit. But then the Sun would have two orbits.

46. The spots are foreshortened near the limb of the Sun.

47. Comets and the novae of 1572 and 1604 were the only precedents for changing phenomena in the heavens.

are either the denser parts of some heaven—and according to the philosophers that is what stars are—or they are solid and opaque bodies existing in themselves. In that case they will also be stars, no less than the Moon and Venus, which appear black where they are turned away from the Sun. (And just the other day N.[48] affirmed that about twelve or more years ago he and his father observed Venus under the Sun in the form of a spot on the Sun).[49] That these spots are solar stars[50] is demonstrated by the foregoing and by the following: as they produce exceedingly dense and black shadows, it is credible that they strongly withstand the light of the Sun, and it is thus probable that they are very much illuminated by it.[51] Since, as we have argued, the spots grow thinner at the edge of the Sun, the phenomenon cannot be explained by a single circular motion:[52] *ergo*. Another reason ought to be added: this illumination diminishes for us the opaque part of the spots and thus makes the shadow thin. This is demonstrated in the following way.

Let the Sun be ABCDE, whose center is A and perimeter BCDE. About the center let the circle FGHIK be described, and let spot L be carried from G to H and from H to K.[53] Let the Sun illuminate this spot by rays BG and OM when it is at G; by rays CN and DH when at H; and by rays PQ and EI when at I. Let the eye, situated on the Earth, R, see spot L at G by the rays RG and RM; at H by rays RN and RH; and at I by rays RQ and RI. Constant observation shows that spot L is seen under a smaller angle at G and I than at H, and that it is also thin and elongated at G and I, but round at H. This happens for the following reason: spot L facing the Sun is vigorously illuminated, and at G and I offers to the eye a large part of its illumination, but its unilluminated part is presented obliquely because it is being carried on the circle FGHIK. At H, however, it directly offers its dark part to us, and so it follows that less of the dark part appears, and under a smaller angle, when the spot is at G and I than at H, and likewise, other things being equal, it will appear thin and elongated at G and I but round at H, as can be seen in the figure.

48. We have not been able to identify this person.

49. As there was no transit of Venus in 1599, this was doubtless a large sunspot. See above, pp. 14–17. Transits of Venus occur very infrequently; the previous one had taken place in 1526, and the next one would be in 1631.

50. See note 13.

51. Since they are close to the Sun, they are highly illuminated, yet they throw a dark shadow. Hence they must be solid bodies.

52. That is, a single circular motion about the Earth.

53. From the drawing and the subsequent text, it is clear that Scheiner means point I, not K. Note also that R is at the eye of the observer, and that M is missing from the figure. It is positioned near G, as Q is near I.

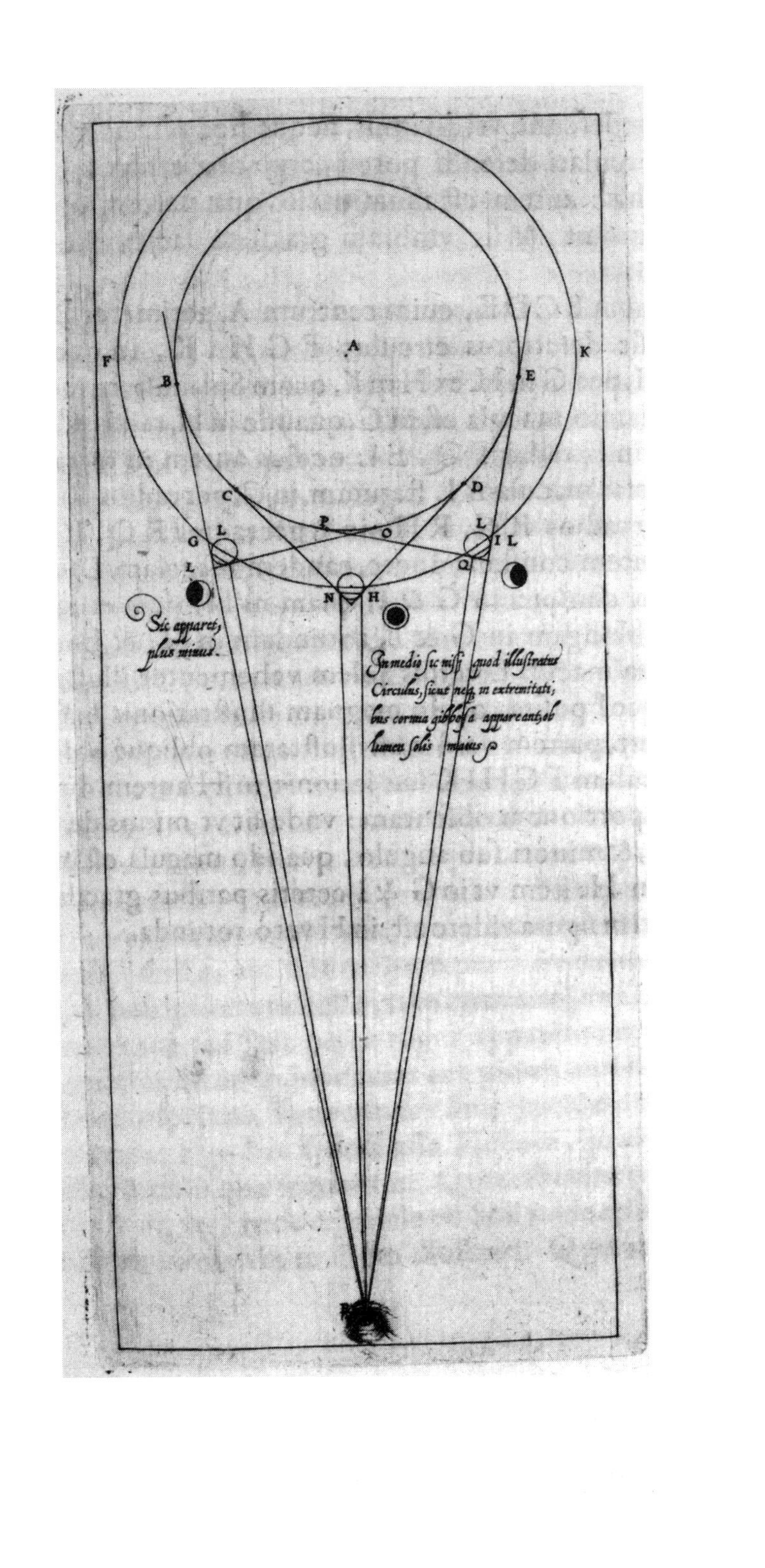
A
F
K
B
E
C
D
P
O
G
L
L
I L
L
Q
N
H
Sic apparet, plus minus.
In medio sic nisi quod illustratus Circulus, sicut neque in extremitati; bus cornua gibbosa appareant, ob lumen solis maius.

From all this, these corollaries are derived:

1. These spots do not recede much from the Sun.
2. They are very large; otherwise the Sun, illuminating them, would entirely absorb them by its great size.
3. They are exceedingly opaque and thick, to the point that, so close to the Sun and so vigorously illuminated on the part facing the Sun, they nevertheless cast such black shadows to such a distance, that is, all the way to us.
4. If [unimpeded by] the brightness of the Sun it were possible to distinguish their illuminated parts from their unilluminated parts, they would appear to us as many miniature moons—horned, gibbous, new, and perhaps even full.
5. The same deduction concerning their illumination might be made about other stars.[54]
6. Hence it is also reasonable to assume that the companions[55] of Jupiter have a similar nature as far as motion and place are concerned; wherefore we consider it almost certain that there are not just four of them but many, and that they are carried not in just one circle[56] about Jupiter, but in many. Given this, any objections can be easily answered, and many conflicts about their motions can be resolved, for they [i.e., their orbits] appear inclined to Jupiter, sometimes a bit to the south and sometimes a bit to the north.
7. And I am not altogether afraid of believing something similar about Saturn. For why does Saturn appear sometimes as an oblong shape, and other times accompanied by two stars touching its sides?[57] But I will restrain myself for now about this issue.

In the meantime, I maintain that these are either wandering or fixed stars; I incline, however, toward wandering stars, since a number of arguments, although rather obscure ones, militate for this. But there will be time for all these matters: their motion, shape, magnitude, distance from the Sun, and other conditions. It occurs to me to conjecture that from the Sun to Mercury and Venus,

54. This is perhaps an allusion to Venus and Mercury.

55. I.e., satellites.

56. Scheiner apparently thought that some were of the opinion that Jupiter's four moons all went around the planet in the same orbit. See Galileo's first letter, p. 102 below.

57. In July 1610 Galileo had discovered that Saturn does not appear as a simple round disk, but as three disks—two smaller ones touching a larger central disk on opposite sides. Scheiner and others sometimes saw that appearance, and at other times an oblong one. See p. 102, below, and Albert Van Helden, "Saturn and His Anses."

at proper distances and proportions, there are many wandering stars revolving, though we have learned only of those whose motion follows after the Sun. If it could be managed—and I have not yet utterly despaired of it—that we might contemplate stars even closer to the Sun, this entire dispute would be resolved. Farewell. 26 December 1611.

Yours
Apelles hiding behind the canvas

In all sciences a great journey remains and what has already been discovered must be counted as the smallest part of what will be discovered. In this connection,

The Sun will also give signs: who would dare call the Sun false?[58]

The second letter, a sketch of the conjunction of Venus with the Sun, was incomplete, and made the inference about the 13th[59] on the basis of the hypothesis that the beginning of the conjunction happened on the 11th. But if the most learned Magini were to propose it as more probable that the *mean* conjunction had happened on the 11th, the letter, provided that it is referred to that other day,[60] is actually complete and thus ends in complete agreement with his calculation.

Apelles[61]

58. Sol quoque signa dabit: Solem quis dicere falsum / Audeat? Virgil, *Georgics*, book I, lines 463–464; see also lines 438–439. The passage is based on Aratus's *Phaenomena*, lines 819–891, which concern the prediction of weather by the Sun's appearance.

59. I.e., the end of the conjunction.

60. . . . [E]pistolae in illum ipsum diem versa.

61. This page was added after the printing of the tract had been completed and copies had already been sent out. On 13 January 1612 Marc Welser sent copies of the page to Galileo and others, including Christoph Clavius. See *OGG*, 11: 263. Scheiner sent the page to Paul Guldin four days later; see Scheiner to Guldin, 17 January 1612, Graz Universitätsbibliothek, MS 159, no 1. For Scheiner's calculations of the conjunction, see appendix 4.

6

Galileo answers Apelles

The observers in Rome, both within and outside of the Collegio Romano, proved as crucial to Galileo's early work on sunspots as to that of Scheiner. On several occasions during his momentous visit to the city in the spring of 1611, Galileo had shown sunspots to various people, among them his friend Lodovico Cigoli, then completing a vast fresco in the dome of the Pauline chapel of Rome's Santa Maria Maggiore. By the fall of that year, Cigoli and another Florentine painter, Domenico Cresti—better known as "il Passignano"[1]—were observing sunspots through their telescopes, and sending their findings and drawings to Galileo, now back in Florence. On 16 September, Cigoli reported that Passignano had observed the Sun "in the morning, midday, and evening."[2] Cigoli himself, although interested, had little time to observe the Sun, and he emphasized the difficulty of Passignano's experience: "he observes, then turns away, and for a little while he can't see, but when he comes back [to the task], he sees very clearly and with ease."[3] Passignano was certain that the spots rotated about the body of the Sun, and Cigoli advised him to observe and draw the Sun for a week, and then to send his observations to Galileo.[4] Unaware that the Sun's daily motion caused the top of the face in the morning to become the bottommost portion

1. In the draft of his third letter on sunspots, Galileo mentioned Passignano, but his name does not appear in the printed version (*OGG*, 5: 191).

2. Cigoli to Galileo, 16 September 1611, *OGG*, 11: 208.

3. Ibid., 209.

4. Ibid.

in the evening, Passignano postulated instead that the spots moved in a spiral path.[5] Galileo replied,

> I am pleased that Signor Passignano is observing the Sun and its revolutions; but you must tell him that he should know that the part of the Sun that is the lowest at sunrise is the highest at sunset. Because of this it could appear to him that the Sun has some turning motion on itself other than that which I truly believe it has, and which I think I discern through the changes of its spots. I would very much like to have the observations made about this by Cavalier [Passignano], in order to compare them with mine.[6]

It does not appear that Galileo was yet making sustained observations of the sunspots. In the autumn of 1611, he was increasingly embroiled in the controversy about floating bodies,[7] and he was also busy continuing his research on Jupiter's moons. Now that these moons had been accepted as real, the next problem was to determine their periods, and here Galileo had some competition from Giovanni Battista Agucchi, who had sent him estimates of those periods in October 1611.[8] He was further handicapped by a serious illness, which severely restricted his activity well into 1612.[9]

News about the activities of German observers reached Galileo in November 1611. On the 11th of that month, the Vicentine scholar Paolo Gualdo wrote from Padua,

> These days reports have come to Signor [Lorenzo] Pignoria from Signor Welser that in Germany there are those who have also begun observing the Sun. Having heard this, Pignano [Gasparo Pignani], who has great enthusiasm for these *occhiali* [spyglasses],[10] and a gentleman named [Vincenzo] di Dotti[11] have observed, and they find that at the center of the Sun there are no

5. Cigoli to Galileo, 23 September 1611, *OGG*, 11: 212.

6. Galileo to Cigoli, 1 October 1611, *OGG*, 11: 214. Cigoli had written a treatise on perspective: *Prospettiva Pratica*, ed. Filippo Camerota (Florence: Leo S. Olschki, in press).

7. Stillman Drake, *Galileo at Work*, 169–174, and *Discourse on Bodies in Water*, pp. xii–xx. See also Mario Biagioli, *Galileo Courtier* (Chicago: University of Chicago Press, 1993), pp.159–209.

8. Agucchi to Galileo, 14 and 29 October 1611, *OGG*, 11: 219–220, 225.

9. See, e.g., Maffeo Barberini to Galileo, 11 October 1611, *OGG*, 11: 216.

10. In the *Difesa*, Gasparo Pignani was described as an excellent manufacturer of all sorts of mathematical instruments, and as a man with an uncommon grasp of the discipline itself. See *OGG*, 2: 562.

11. Vincenzo Dotti, also known as Vincenzo de' Dotti, an architect with interests in cosmology and geography, was then engaged in the design of the Scala dei Giganti in Padua. See A. Bevilacqua, "Dotti, Vincenzo," in *Dizionario biografico degli italiani*, 41: 548–549.

> rays—such that one can look at it—but that the greatest amount are near the circumference. And [I hear that] they have observed in that center two spots similar to two eyes, and one lengthwise that appears in the exact shape of a nose. [The Sun] had this appearance shortly after noon. They want to observe it also at sunrise and sunset to see if the same spots are visible there.[12]

By the end of that month, Galileo had received the report about activities in Germany more directly through the German botanist and Lincean Johannes Faber, who quoted verbatim the passage from Welser's letter, presumably written in December 1611. "Certain friends of mine," Welser had related,

> have observed with the telescope some spots appearing so consistently in the Sun that they consider them a matter beyond doubt. But Your Lordship should notice that I say that they *appear* to be in the Sun, not that they *are* there, because they are persuaded by a number of strong arguments that these are stars that, encountering our line of sight because they are under or beside the Sun, take on such appearances. I would like to know if there's news of this in your area, and if any one has made any observations.[13]

A jocose letter written by the Tuscan poet Alessandro Allegri further suggests that sunspots were observed with the telescope or "improved *occhial di Fiandra*" in very late 1611 or early 1612 in Galileo's own circle. Allegri's work, dedicated to a kinsman of Galileo's close friend Filippo Salviati, appeared in 1613, and the letter portraying the sunspots as "filth on the cheeks of the Sun" was dated 4 January 1612.[14] Observations recorded by Thomas Harriot likewise show a number of large spots on the Sun in this period.[15] Galileo's earliest public pronouncements on the subject of sunspots came in the form of a brief paragraph included in the first edition of his *Discourse on Bodies in Water*, which appeared in the spring of 1612. His allusion to the solar phenomena follows a report on his recent determination of the periods of Jupiter's satellites,

12. Gualdo to Galileo, 11 November 1611, *OGG*, 11: 230–231. See also Fortunio Liceti to Galileo, 16 December 1611, *OGG*, 11: 244. On Gualdo, see Giorgio Ronconi, "Paolo Gualdo e Galileo," in *Galileo e la cultura padovana*, ed. Giovanni Santinello (Padua: CEDAM, 1992), 359–371.

13. Johannes Faber to Galileo, 15 December 1611. *OGG*, 11: 239. On Faber, see David Freedberg, *The Eye of the Lynx* (Chicago: University of Chicago Press, 2002); and Antonio Favaro, "Dal carteggio di Marco Velser con Giovanni Faber," *Memorie dell' Istituto Veneto* 24 (1891): 79–100.

14. Alessandro Allegri, *Lettere di Ser Poi Pedante* (Bologna: Vittorio Benacci, 1613), 14. On Allegri, see Alberto Asor Rosa, *Dizionario biografico degli italiani*, 2: 477–478, and Brendan Dooley, *Morandi's Last Prophecy and the End of Renaissance Politics* (Princeton: Princeton University Press, 2002), 24–27, 32.

15. Harriot MSS, HMC 241 / 5–7.

presented here in the mid-seventeenth century translation made by Thomas Salusbury:

> I adde to these things the observation of some obscure Spots, which are discovered in the Solar Body, which changing . . . position in that, propounds to our consideration a great argument either that the Sun revolves in itselfe, or that perhaps other Starrs, in like manner as *Venus* and *Mercury*, revolve about it, invisible in other times, by reason of their small digressions, lesse than that of *Mercury*, and only visible when they interpose between the Sun and our eye, or else hint the truth of both this and that; the certainty of which things ought not to be contemned, nor omitted.[16]

This cautious statement was written some time between September 1611, when Galileo began this discourse, and the spring of 1612, when it emerged in print. All that Galileo could or would say at this time was that there were some small dark spots moving about the Sun, possibly on the Sun itself or revolving around it like Venus and Mercury. There were no details about the direction and varying speed of their motions, the foreshortening at the limb, or their changing shapes, all of which were mentioned in Christoph Scheiner's *Tres Epistolae*, published early in January 1612. From this silence we may surmise that Galileo wrote the passage in the fall of 1611, when he had yet to make a serious study of the solar phenomena.

In late December 1611, Galileo moved for reasons of health to Villa delle Selve, the home of the Florentine nobleman Filippo Salviati, and this was to be the setting for much of his work on sunspots.[17] He received Scheiner's *Tres*

16. *A Discourse presented to the Most Serene Don Cosimo II. Great Duke of Tuscany, concerning the Natation of Bodies Upon And Submersion In, the Water. By Galileus Galilei: Philosopher and Mathematician unto His most Serene Highnesse*. Englished from the Second Edition of the Italian compared with the Manuscript Copies, and reduced into Propositions: By Thomas Salusbury, Esquire (London, 1663), 2. Reproduced in Stillman Drake, ed., *Discourse on Bodies in Water* (Urbana: University of Illinois Press, 1960). In the second edition of this book, Galileo firmly dismissed the possibility that the spots were satellites. In Salusbury's English (2): "Continuall observation hath at last assured me that these Spots are matters contiguous to the Body of the Sun, there continually produced in great number, and afterwards dissolved, some in shorter, some in a longer time, and to be by the Conversion or Revolution of the Sun in it selfe, which in a Lunar [Month], or thereabouts, finisheth its Period, carried about in a Circle, an accident great of it selfe, and greater for its Consequences."

17. Galileo seems to have gone to Villa delle Selve in late December to escape the damp air of the city, to have left it briefly around 9 January, and to have returned to it for a longer stay a few days later. See *OGG*, 11: 248, 258, 265. His host Filippo Salviati's long and somewhat violent quarrel with a member of a minor branch of the Medici family, Bernadetto de' Medici, reportedly resulted in a brief imprisonment for both antagonists in December 1611, but appears to have been resolved in early

Epistolae some time in late January. Among his surviving sunspot observations, the first is dated 12 February; here the face of the Sun is represented by a compass-drawn circle, and the several small spots are marked 1, 2, and 3. Within the circle is the Latin inscription "12 February 1612, at sunset," and besides the figure is a note: "1 and 2 [are] perfectly circular, and black. 3 [is] not as black, nor so [clearly] delineated."[18] Galileo went on to observe the Sun on 17 and 23 February, and in March he made nine observations of about the same quality. The notes that accompany those of 2 and 4 March, while no sunspots appeared, offer interesting insight into his observing procedure:

> 2 March, on the East of the Sun no spot appeared; and it took 270 pulse beats to rise. 4 March, no spot appeared, and it took 250 pulse beats to rise.[19]

On these occasions, Galileo was evidently observing the Sun exactly at sunrise, when its rays had the least power.[20] While he does not appear to have repeated the timing of the sunset after 9 March, he observed invariably at sunrise or sunset, and the spots were usually sketched as little dots. Not until the beginning of April did he describe them as cloudlike and irregular in shape, and to use shading in order to portray them. On 6 April he singled out a large irregular spot, first shown within the circle, and then enlarged and off to the side (fig. 6.1). He was to use this sketch in his first letter on sunspots. Galileo followed this spot until 10 April, after which it disappeared at the limb. His observations for the rest of April were perfunctory, until the emergence of a large spot at sunset on 26 April. He followed this spot more closely, making observations at sunset from 28 April to 1 May, and finding it again on 3 May.[21]

Through this three-month series of observations, there is evidence of substantial improvement in Galileo's technique. The size of the figures became more standard, and the drawings of individual spots became more differentiated

---

January 1612; on the much-discussed conflict see Vatican Library, Fondo Urb. Lat. 1077, fol. 260r, Fondo Urb. Lat. 1079, fol. 827v–828r, 834v, 849v, 858r, 872r, Fondo Urb. Lat. 1080, fol. 20r, 26r, 51r. On Salviati himself, see Mario Biagioli, "Filippo Salviati: A Baroque Virtuoso," *Nuncius* 7, no. 2 (1992): 81–96.

18. *OGG*, 5: 253.

19. Ibid. Thomas Harriot likewise noted with some surprise that there were no spots visible to any of his three instruments between 21 and 25 February 1611 (O.S.). By 26 February he was able to make out two spots. See Harriot MSS, HMC 241/8 pp. 15, 16.

20. The Villa delle Selve is situated on a bluff overlooking the Arno valley, near present-day Lastra a Signa, about 15 km west of Florence. There are no mountains nearby in the westerly direction that would interfere unduly with (near) horizon observations.

21. *OGG*, 5: 253.

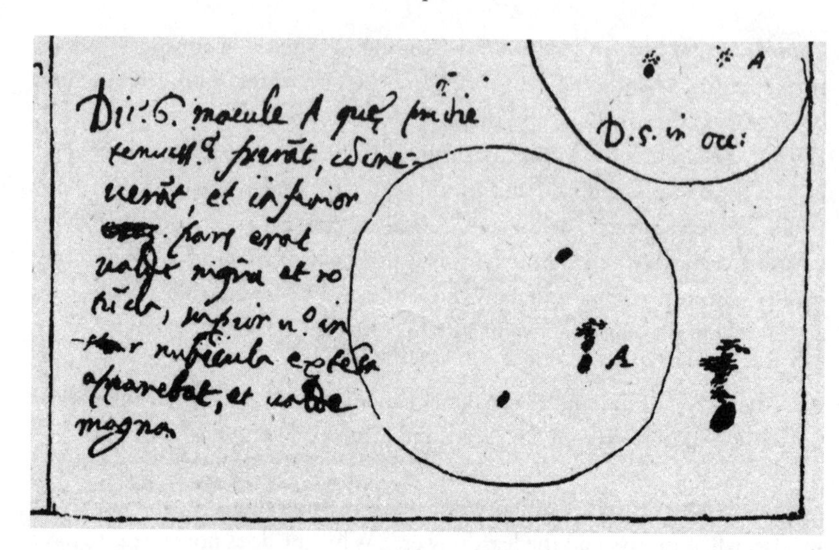

Figure 6.1. Galileo's sunspot observation of 6 April 1612. *OGG*, 5: 253.

and subtle. As Horst Bredekamp has pointed out, "It can be observed in detail how through modulations of the pen strokes Galileo found certainty about the character of the observed. The movements of his hand that determined the style of the representation had direct consequences for the understanding of the represented phenomena. Mode of thought and style of representation mutually reinforced each other. In the style lay the alternative to Scheiner's theory."[22]

Galileo was also composing his first response to the *Tres Epistolae* in this period. He finished the letter on 4 May, and it included illustrations of individual spots observed a few days earlier. Unlike Scheiner's *Tres Epistolae*, however, his letter contained no series of systematic observations; there were hardly two dozen made between 12 February and 3 May. Fortunately, in early May Galileo learned of a technique that made more systematic research on sunspots possible. His student and friend, the Benedictine scholar Benedetto Castelli (fig. 6.2), then engaged in overseeing the printing of the *Discourse on Bodies in Water* in Florence, was also observing sunspots in this period. He discovered that one could project the solar image through a telescope, which made it possible to observe the Sun and its spots anytime the Sun was above the horizon

22. Horst Bredekamp, *Galileo Galilei der Künstler: Der Mond, die Sonne, die Hand* (Berlin: Akademie-Verlag, 2007), 238.

Figure 6.2. Benedetto Castelli, 1578–1643. Oil on canvas, Italian School, seventeenth century. Courtesy of the Bridgeman Art Library.

and not hidden behind clouds. Because observations could then be made at the same time of day every day, the orientation of the ecliptic, and thus the alignment of the spots, remained constant, and direct comparison of the spots on successive days became much more accurate. Second, the new technique allowed measurements, and all this took place without inconvenience or damage to the eye.

Castelli also standardized this method of observation. The image of the Sun was projected onto a paper on which a circle of predetermined size had been drawn in advance, and the distance between the telescope and the paper was adjusted so that the solar image coincided precisely with this circle. The spots were then simply traced on the paper. After informing Galileo during the first few days of May of the telescopic projection method, on the eighth of the month Castelli wrote,

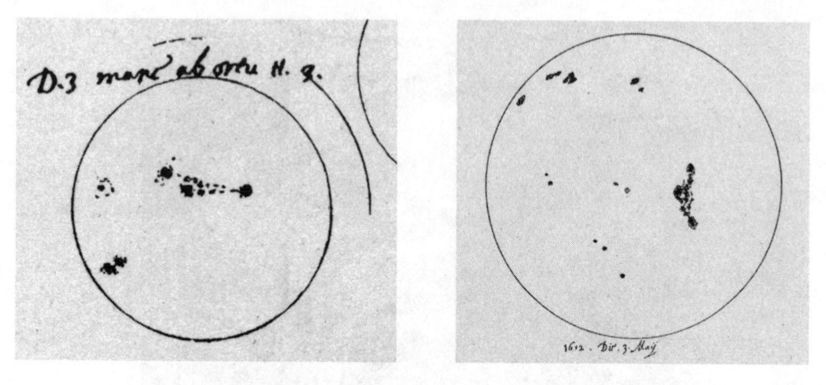

Figure 6.3. (a) 3 May 1612, three hours after sunrise. *OGG*, 5: 254.
(b) 3 May 1612, early afternoon. *OGG*, 11: 307. Reduced.

> I am sending you the enclosed observations,[23] depicted both as well as I could draw them and as the telescope showed them to me. But I think that the centers of the spots are in their proper places, especially because the diameters of the circles they describe if they move on the body of the Sun have been depicted separately, and the semicircles have been divided into fifteen parts, so that their changes in twenty four hours correspond to the successive excesses of the versed sines of those arcs.[24] Thus I find no error that cannot be attributed either to the imperfections of my observations or to some defect in the estimation of the speed of their motion; indeed, these [errors] are imperceptible rather than anything else.[25]

Galileo himself had passed from direct observation to the projection method a few days earlier. On 3 May, three hours after sunrise, around 9:30 a.m., he made a drawing about 6 cm in diameter, of the sunspots as seen through the telescope, and later in the day he drew traced sunspots on the projected image of the Sun, 12.5 cm in diameter (fig. 6.3).[26] The difference between these two sketches shows the transition from opportunity-driven qualitative observation

23. No longer extant.

24. The versed sine or versine is defined as the quantity obtained by subtracting the cosine from unity. Castelli divided the (small) circles on which the spots moved into fifteen intervals and projected these onto the diameters of these circles. He found that the sequences of distances thus created exactly matched the distances traveled by the spots on the face of the Sun on successive days.

25. Castelli to Galileo, 8 May 1612, *OGG*, 11: 294

26. In *Istoria e dimostrazioni intorno alle macchie solari*, the observations at the end of Galileo's second letter are all between 123 and 126 mm in diameter. We thank Owen Gingerich for this information.

to systematic (and quantitative) research on sunspots. From this date forward, there is a sequence of daily sunspot observations until well into July of that year, with only a few interruptions caused by cloudy skies.

The day after he made this sunspot tracing, Galileo finished his first letter on sunspots to Marc Welser, and sent copies of the work to the banker in Augsburg and to Federico Cesi in Rome. At the end of the letter, Galileo promised Apelles and Welser that he would send "some observations and drawings of the solar spots, ones of absolute precision, in their shapes as well as in their daily changes in position, without a hairsbreadth of error, all made by a most exquisite method discovered by one of my students" (see pp. 126–127, below).

Within a few days, he moved from Villa delle Selve back to Florence, where, somewhat improved in health, he undertook a sustained program of observations of sunspots, presumably with Castelli. He sent the first results of this program to Maffeo Cardinal Barberini in Rome, whom he had met in Florence at a dinner arranged by Grand Duke Cosimo II in September of the previous year. At this dinner, Galileo had debated the Bolognese Aristotelian Flaminio Papazzoni on the subject of floating bodies, and Barberini had obligingly taken Galileo's side.[27] Along with the nine drawings, an uninterrupted series from 3 to 11 May, Galileo explained his conclusions about the sunspots to Barberini, and then asked,

> If you happen to discuss my solution with learned men in that city, I would like to hear something about their opinions, and in particular of the Peripatetic philosophers, for this novelty appears to be the final judgment of their philosophy, because there have already been signs in the Moon, the stars, and the Sun [*iam fuerunt signa in luna, stellis et sole*].[28] Whence, together with the mutability, corruption, and generation even of the most excellent substance of the heavens, this doctrine itself shows signs of corruption and mutation, but not without the hope of regenerating something better.[29]

While he was engaged in this observational program and beginning to draft his second letter, Galileo also turned to Carlo Cardinal Conti, who had strong

27. On Papazzoni see Michele Camerota, "Flaminio Papazzoni: un aristotelico bolognese maestro di Federico Borromeo e corrispondente di Galileo," in Daniel A. Di Liscia, Eckhard Kessler, and Charlotte Methuen, ed., *Method and Order in Renaissance Philosophy of Nature* (Aldershot: Ashgate, 1997), 271–300.

28. As if in response to Scheiner's appropriation of the Virgilian verse *Sol quoque . . . signa dabit*, Galileo adapted Luke 21: 25, "Et erunt signa in sole, et luna, et stellis," "And there shall be signs in the Sun, and in the Moon, and in the stars," to reflect his own recent celestial discoveries.

29. Galileo to Barberini, 2 June 1612, *OGG*, 11: 311. See appendix 2.

ties to the Jesuits,[30] asking whether Sacred Scripture supported the Aristotelian belief in the incorruptibility of the heavens. Conti replied,

> There is no doubt whatsoever that Scripture does not support Aristotle, but rather, it even supports the contrary judgment, such that it was commonly believed by the Church Fathers [i.e., the Patristic writers] that the heavens were corruptible.[31]

Conti urged caution, however, noting that such matters required both lengthy consideration and a causal account based on repeated observation.[32] Some two months later, after he had received a series of sunspot observations from Galileo, the cardinal repeated his opinion, one that linked the spots to the appearances of the satellites of Jupiter, as Scheiner's *Accuratior Disquisitio* was to do:

> Your observations are very diligent and beautiful; and whatever these spots are, they are matters beyond what has hitherto been believed. But because this is an issue of great consequence, and in places so distant from us, observations made over a lengthy period are needed, especially since someone seizing the opportunity of the Medicean Stars that you observed could maintain that these spots had been generated by stars, ones that are so minute as to be invisible when separate, but whose aggregation causes these spots. They could also say that they are so numerous, and have such a variety of motions around the Sun, that in coming together in different patterns, they cause this diversity of spots. To convince such people, long observations are needed, and even more study to determine what else the spots might be, and when they might render the heavens subject to change, from which we have it that these spots are not on the body of the Sun itself, but rather in another part of the heavens.[33]

For Conti, who had probably learned Apelles' identity and affiliation from one of his contacts at the Collegio Romano, settling the question about the

30. As a very young man, Conti had studied at the Collegium Germanicum in Rome, and after his death in 1615, he was commemorated with a plaque in the Collegio Romano. On this figure, see S. Andretta, "Conti, Carlo," in *Dizionario biografico degli italiani*, 28: 376–378.

31. Conti to Galileo, 7 July 1612, *OGG*, 11: 354. See appendix 3.

32. Conti was referring here to the Aristotelian notion of a phenomenon that could be adduced in a scientific demonstration, for which multiple repetition of the experience was necessary. See Peter Dear, *Discipline and Experience: The Mathematical Way in the Scientific Revolution* (Chicago: University of Chicago Press, 1995), 32–62.

33. Conti to Galileo, 18 August 1612, *OGG*, 11: 376. It is significant that Conti had not yet seen Galileo's second sunspot letter; Cesi did not report receiving it until 8 September.

corruptibility or perfection of the heavens was a matter of knowing the essence of sunspots, an undertaking he judged difficult because of the extraordinary remoteness of the phenomena. Galileo was to address this point specifically in his third letter, where he argued that questions about the essences of substances, whether nearby or far away, were inevitably matters of infinite regression, never leading to satisfactory answers. Questions about accidents—positions, motions, color, and other such qualities—could, however, be answered, and thus the discoveries made in the heavens had a bearing on natural philosophy (see p. 254, below).[34]

While Galileo was composing his second letter, Scheiner was preparing an expanded version of his own work, the *Accuratior Disquisitio*, for publication. When Welser received Galileo's first letter around the end of May 1612, he had responded by claiming that it was "dear beyond measure" to him, and by adding rather optimistically that the letter was so reasonable and modest in tenor "that Apelles, although you contradict his opinion for the most part, must consider himself much honored." Welser went on to acknowledge that Scheiner's ignorance of Italian, and the relative scarcity of translators conversant with astronomy, would delay the Jesuit's acquaintance with the text.[35] Galileo explained a few weeks later to Paolo Gualdo in Padua that he had chosen to write in Italian in order to appeal to readers gifted with a natural intelligence but nonetheless unschooled in Latin, who tended to regard works in that language as devoted to lofty matters entirely beyond them, or as he put it, "as big books full of great novelties in logic and philosophy."[36] He nonetheless asked Martino Sandelli to translate his letter into Latin for the benefit of foreigners like Scheiner, but the aging scholar's eyesight proved too poor for this task.[37]

34. Dear, *Discipline and Experience*, 101.

35. Welser to Galileo, 1 June 1612, *OGG*, 11: 304–305.

36. Galileo to Paolo Gualdo, 16 June 1612, *OGG*, 11: 327. On the question of the vernacular in scientific works, see Michele Camerota, "Adattar la volgar lingua ai filosofici discorsi: una inedita orazione di Niccolo Aggiunti contro Aristotele e per l'uso della lingua italiana nelle dissertazioni scientifiche," *Nuncius* 13:2 (1998): 595–623.

37. Sandelli to Galileo, 28 September 1612, *OGG*, 11: 401–402.

7

Galileo's first letter

First letter from Welser to Galileo about the solar novelties

Virtus, recludens immeritis mori
Coelum, negata tentat ire via[1]

Most Illustrious and Excellent Sir,

Human reason is already launching a serious assault on heaven, and the most vigorous are going to conquer it.[2] You were the first to scale the walls, and have brought back the mural crown.[3] Now others are following you, their courage increased with the knowledge that it would be an open display of faintheartedness not to support so auspicious and honored an undertaking, now that you have shown the way. Take a look at what this friend of mine has ventured, and if it does not seem to you something entirely new, as I believe, I nevertheless hope that it will be to your liking, seeing that on this side of the [Alps] as well there is no lack of men who follow in your footsteps. You will do me a favor by freely telling me your opinion about these solar spots, whether you judge these substances to be stars or something else, where you believe they are situated, and what their motion is. I kiss your hands with wishes for a happy New Year, and I

1. Horace, *Odes* 3:ii: 21–22: Virtue, opening heaven to those who do not deserve to die, makes her course by paths untried. On Welser, see p. 107, note[1], below.

2. For the change in the first sentence demanded by the censor, see p. 239, below.

3. The reward given in Antiquity to the man who first scaled the enemy's walls.

Figure 7.1. Portrait of Marc Welser. Welser, *Opera Historica et Philologica* (Nuremberg, 1682), frontispiece.

ask that in making your new observations known, you do not neglect to share them with me.

From Augsburg, 6 January 1612.
Your Very Illustrious and Excellent Lordship's
affectionate servant
Marc Welser

First letter from Galileo to Welser about the solar spots, in reply to the foregoing

Most Illustrious Lordship and Most Honorable Patron,

To the courteous letter from Your Most Illustrious Lordship, written to me three months ago, I reply belatedly, having been almost forced to adopt such silence by various unforeseen incidents, and in particular by many a long indisposition or, more precisely, by many and long indispositions, which, preventing all other pursuits and occupations, deprived me mainly of my ability to write. And to a great extent they keep me from it even now, but not so much that I cannot at least reply to some of the letters, of which there are not a few, from friends and patrons, all of which await responses. I have also kept silent in the hope of being able to give some satisfaction to the query from Your Most Illustrious Lordship about the solar spots, concerning which you have sent me these brief discourses of the masked Apelles.[4] But the difficulty of the matter and my inability to make many sustained observations[5] have kept me and still keep me uncertain and undecided. The farther the recently observed phenomena [are] from common and popular opinions, the more it behooves me, rather than others, to proceed cautiously and circumspectly in proclaiming any novelty. As Your Most Illustrious Lordship well knows, these new observations have been violently contradicted and impugned, and they make it necessary for me to conceal and to be silent about any new concept whatsoever until I have a more than certain and palpable demonstration. For the enemies of novelty, who are infinite in number, would attribute every error, even if venial, as a capital crime to me, now that it has become customary to prefer to err with the entire world than to be the only one to argue correctly. Add to this the fact that I am content to be the last to produce some true concept rather than to precede others and later be forced to retract what was offered with greater haste and with less reflection. These considerations have made me slow to respond to the questions of Your Most Illustrious Lordship, and they still render me timid in producing anything other than a few negative propositions, since it seems to me that I know better what the solar spots are not than what they truly are, and also since it is much more difficult for me to find a truth than to expose a falsehood. But in order to satisfy the desire of Your Most Illustrious Lordship at least in some

4. "Finto Apelle." While the term refers to the author's disguise, it would not exclude the less kind suggestion of fraudulence, as if Scheiner were an unworthy successor to the ancient artist.

5. On the difficulty of observing sunspots by looking directly at the Sun through a telescope, see p. 27, above.

measure, I will consider those things that appear to me worthy of attention in the three letters of the masked Apelles because you command it, and because these letters contain what has been thus far imagined about determining the essence, the place, and the motion of these spots.

And first, there is no doubt whatsoever that these are real things and not simple appearances or illusions of the eye or the lenses, as the friend of Your Most Illustrious Lordship demonstrates well in the first letter; I have observed them for eighteen months now,[6] having shown them to several of my intimates, and likewise this time last year I made it possible for many prelates and other gentlemen to observe them in Rome. It is also true that they do not remain fixed on the body of the Sun but appear to move with respect to it, and also with regular motions, as the author has noted in that same letter. It is true that it seems to me that the motion is in the opposite direction from that which Apelles asserts, that is, from west to east, drifting [also] from south to north, and not from east to west and from north toward south. This [fact] is also seen very clearly in the observations that he himself described, and which in this particular agree with mine and with what I have seen of those of others. [Thus] one notes that the spots observed on the setting Sun change from one evening to the next, descending from its upper toward its lower limb; and in the morning they ascend from the lower to the upper, showing themselves upon the first appearance in the southernmost part of the solar body, and hiding or separating themselves from it in the most northern parts—in short, describing lines on the face of the Sun in exactly the same way as Venus or Mercury would do if, in their passage under the Sun, they were to interpose themselves between it and our eye. Therefore, the motion of the spots with respect to the Sun appears to be similar to those of Venus and Mercury and also to those of the other planets around the same Sun, which is from west to east, and which appears to run from south to north because of the obliquity of the horizon. Had Apelles not supposed that the spots were turning about the Sun but rather that they were only passing below it, it is true that their motion would have to be designated to be from east to west, but supposing that they describe circles about it and that they are now higher, now lower on it, such revolutions should be termed

6. Galileo is claiming here that he has been observing sunspots since October or November 1610, when he still lived in Padua. On this claim, see Antonio Favaro, "Sulla priorità della scoperta e della osservazione delle macchie solari," *Memorie del Reale Istituto Veneto di Scienze, Lettere ed Arti* 22 (1882): 729–783; J. D. North, "Thomas Harriot and the First Telescopic Observations of Sunspots," 129–136; B. Dame, "Galilée et les Taches Solaires (1610–1613)," *Revue d'Histoire des Sciences* 19 (1966): 306–370, reprinted in *Galilée: Aspects de sa Vie et de son Oeuvre* (Paris: Centre Internationale de Synthèse, 1968), 186–251; Michele Camerota, *Galileo Galilei e la cultura scientifica nell'Età della Controriforma* (Rome: Salerno, 2004), 242–243.

"from west to east" because they move in this direction when they are on the higher parts of their circles.[7]

The author, having established that the observed spots are not illusions of the lens or defects of the eye, seeks to determine something general about their location, showing that they are neither in the air nor on the solar body. As to the first, the absence of sensible parallax shows that it must necessarily be concluded that the spots are not in the air, that is, near the Earth, within the space that is commonly assigned to the element of air.[8] But the hypothesis that they cannot be on the solar body does not appear to me to have been fully and necessarily demonstrated. For it is not conclusive to say, as he does in the first argument, that because the solar body is very bright it is not credible that there are dark spots on it, because as long as no cloud or impurity whatsoever has been seen on it we have to designate it as most pure and most bright, but when it reveals itself to be partly impure and spotted, why shouldn't we call it both spotted and impure? Names and attributes must accommodate themselves to the essence of the things, and not the essence to the names, because things come first and names afterwards. The second reason would follow necessarily if these spots were permanent and immutable, but of this I will say more below.[9]

What Apelles said in that passage, namely, that the spots appearing on the Sun are much darker than any which have ever been seen on the Moon, I believe to be absolutely false. On the contrary, I believe that the spots one sees on the Sun are not only less dark than the dark spots one sees on the Moon, but [also] that they are no less bright than the most luminous parts of the Moon even when it is most directly illuminated by the Sun. And my reasoning is as follows. Even though at its evening rising Venus is filled with such great splendor, it is not visible until it is many degrees from the Sun, especially when both are above the horizon. This is because those regions of the aether, encompassed by the Sun, are no less bright than Venus, and from this it can be argued that if we could put the Moon, shining with the same light that it has when full, beside the Sun, it would in fact remain invisible, as would something located in a field no less shining and bright than its own face. Now consider [the fact that] when we gaze at the very bright solar disc with the telescope—that is, with the spyglass—how much brighter it appears than the surrounding field, and when we then compare the blackness of the solar spots both with the light of the Sun

7. Just as for an epicycle, the direction of motion was specified at its apogee, where the Sun rotates from west to east.

8. If the spots were a significant distance "below" the Sun, as for example at the distance of the Moon, then if they happened to appear on the Sun in the morning, they would, because of parallax, no longer appear on the Sun in the evening.

9. See p. 94, below.

and with the darkness of the adjacent setting, we will find from the one and the other comparison that the spots of the Sun are not darker than the surrounding field. If, therefore, the darkness of the solar spots is not greater than that of the field that surrounds the Sun, and also if the splendor of the Moon would remain imperceptible in the brightness of the same surroundings, then it follows as a necessary consequence that the solar spots are by no means less bright than the brightest parts of the Moon, even though against the most dazzling background of the solar disc they manifest themselves as dark and black. And if these spots yield nothing in brightness to the most luminous parts of the Moon, how bright will they be in comparison with the darkest spots of the Moon, especially if we were referring to those caused by the projected shadows of the lunar mountains, which compared to the illuminated parts are no less black than the ink compared to this paper? And I intend this not so much to contradict Apelles as to show that it is not necessary to suppose that the material of these spots is very opaque and dense, as one must reasonably assume the matter of the Moon and the other planets to be, for density and opacity similar to that of a cloud interposed between the Sun and us is sufficient to cause such a darkness and blackness.

As to that which Apelles next points out in this passage, and which he treats more fully in the second letter—that by this route we can ascertain whether Venus and Mercury make their revolutions below or rather around the Sun—I am a bit surprised that it has not come to his ears, or if it has, that he has not capitalized upon the most exquisitely clear method [of doing so], one that is perceptible and can be used with some frequency, which I discovered almost two years ago and communicated to so many that by now it has become famous; and this is that Venus changes its shapes in the same way as the Moon.[10] And at present Apelles can observe it with the telescope and he will see it with a perfectly circular shape and very small, although it appeared somewhat smaller still at its evening rising. He can then continue to observe it and he will see it in the shape of a semicircle near its maximum digression [from the Sun]. From this shape, it will proceed to the horned shape, very gradually becoming thinner as it approaches the Sun. Near its conjunction with the Sun it will look as thin as the two- or three-day-old Moon, and the size of its visible circle will be augmented such that it will be recognized that its apparent diameter at evening rising is less than the sixth part of the diameter that it will have at evening occultation or at morning rising, and consequently its disc appears about forty

10. Galileo verified the fact that Venus goes through phases like the Moon in November and December 1610. See Galileo to Christoph Clavius and Benedetto Castelli, 30 December 1610, *OGG*, 10: 500, 503. See also *SN*, 106–109.

times greater in the latter position than in the former. These things will not leave room for anyone to be in doubt about the revolution of Venus; they will lead with absolute necessity to the conclusion—one consistent with the positions of the Pythagoreans and of Copernicus—that its revolution is about the Sun, around which, as the center of their revolutions, all the other planets turn. It is therefore necessary neither to wait for physical conjunctions in order to assure oneself of such a manifest conclusion, nor to produce reasons subject to any retort, however weak, in order to earn the assent of those whose philosophy is strangely perturbed by this new constitution of the universe. For these people, when something else does not constrain them, will say that Venus either shines by itself or is of a substance that can be penetrated by the solar rays, so that it is luminous not only on its surface but also throughout its entire depth. And they can shield themselves with this reply all the more boldly because there is no lack of philosophers and mathematicians who have believed this—*pace* Apelles, who maintains otherwise—and it suited Copernicus himself to consider one of these positions possible, or rather, necessary, because he could not [otherwise] explain how Venus, when it is below the Sun, does not appear horned.[11] And indeed, nothing else could be said before the telescope came to show us how Venus itself really is dark like the Moon, and that like the Moon it changes shape. But beyond this, I can cast great doubt on Apelles' inquiry because he tries to discern Venus on the disc of the Sun during the conjunction observed by him, supposing that it must appear there in the form of a spot very much larger than any of those that have been seen, its apparent diameter being three minutes, and in consequence its surface more than a 130th part of that of the Sun. But, by his leave, this is not true, and the apparent diameter of Venus was then not even the 6th part of a minute, and its surface was less than one 40,000th part of the surface of the Sun, as I know from sensory experience and in due time will make known to everyone. Your Most Illustrious Lordship may therefore see that this would leave a lot of room for argument for those who, with Ptolemy, would still retain Venus below the Sun, for they could say that one would seek in vain to observe such a small mole on its immense and very bright face.[12] And I add finally that such experience will not necessarily convince those who would deny the revolution of Venus about the Sun, for they could always resort to saying that it was above the Sun, fortify-

11. Nicholas Copernicus, *De Revolutionibus*, I, 10; *Nicholas Copernicus on the Revolutions*, 18. On whether Copernicus predicted the phases of Venus, see Neil Thomason, "1543: The Year that Copernicus Didn't Predict the Phases of Venus," in Guy Freeland, ed., *1543 and All That: Image and Word, Change and Continuity in the Proto-Scientific Revolution* (Dordrecht: Kluwer, 2000), 291–332.

12. See Bernard R. Goldstein, *The Arabic Version of Ptolemy's "Planetary Hypotheses,"* American Philosophical Society, *Transactions* 57 (part 4), 1967, pp. 6–7; see also *Ptolemy's Almagest*, 419.

ing themselves with the authority of Aristotle, who considered it such.[13] It is thus not enough that Apelles has demonstrated that during its corporeal morning conjunction[14] Venus does not pass below the Sun if he has not also demonstrated that during its evening conjunction it does pass below it. But such evening conjunctions that are corporeal happen very seldom, and we will not succeed in seeing them. Apelles' argument is therefore insufficient to prove his point.[15]

I come now to the third letter, in which Apelles more decisively determines the place, the motion, and the substance of these spots, concluding that they are stars that turn at no great distance around the Sun's body as do Mercury and Venus.

In order to determine the place, he begins by demonstrating that they are not on the body of the Sun itself, whose revolution on its axis would present them to us as moving spots. The visible hemisphere passing [from sight] in fifteen days, they would have to return unaltered every month, which does not happen.

The argument would, however, be conclusive if he had first established that these spots were permanent, that is, that they were not arising anew and likewise fading away and vanishing. But whoever says that some come into being and others pass away will also be able to maintain that the Sun, turning on itself, carried them with itself without the necessity of ever presenting them to us unaltered, either arranged in the same order or with the same shape. Now I hold it to be a difficult and even impossible matter, one against which sense rebels, to prove that [the sunspots] are permanent. Apelles himself will have seen that some show themselves at their first appearance far from the circumference of the Sun, and others vanish and are lost before they finish traversing the Sun, for I, too, have seen many of these. I neither affirm nor deny, however, that they are on the Sun; I only say that it has not been sufficiently demonstrated that they are not on it.

13. The order of the planets followed by Aristotle and Ptolemy was Moon—Mercury—Venus—Sun—Mars—Jupiter—Saturn. It was Plato who placed Mercury and Venus "above" the Sun.

14. Galileo is referring here not to common conjunctions occurring between planetary bodies when their ecliptic longitudes are equal, but to conjunctions that happen when both Venus and the Earth are on or very near the line of intersection between their orbital planes and Venus is observed to transit the Sun's disc. Such transits are indeed very rare. In his *Admonitio*, published in 1629, and reissued the following year, Kepler was the first to predict actual transits, of Mercury for 1631 and for Venus in 1639. See Albert Van Helden, "The Importance of the Transit of Mercury of 1631," *Journal for the History of Astronomy* 7 (1976): 1–10.

15. Galileo had originally included here the wry remark, "Let Apelles not remain so well hidden behind his canvas that he misses those who are passing to and fro." See *OGG*, 5: 100 n. 1.

And then, in the rest [of the argument], which the author adds in order to demonstrate that they are not in the air or in any of the orbs below the Sun, it appears to me that there is some confusion and, in some fashion, inconsistency, for he takes as still true the old and familiar system of Ptolemy, a system whose falsity he had already recognized a bit earlier, when he concluded that Venus does not have its sphere below the Sun but goes around it and is now above and now below it. And he affirmed the same for Mercury, whose digressions [from the Sun], being somewhat smaller than those of Venus, necessitate putting it closer to the Sun [than Venus]. But here, almost refuting what he had believed a little earlier and which is in effect the true constitution [of the universe], he introduces the false one, making Mercury follow the Moon, and Venus Mercury. I wanted to excuse this bit of contradiction by saying that he had not thought it important after the Moon to name Mercury before Venus, or the latter before the former, since it matters little if they are designated in reversed order in words, provided that they remain in fact in the true order. But seeing him then prove by means of parallax that solar spots are not in the sphere of Mercury, and then add that this method would perhaps not be efficacious in the case of Venus because it, like the Sun, has a small parallax, renders my excuse void because Venus will have parallaxes rather greater than those of Mercury and the Sun.

Nevertheless, it seems to me that Apelles, being of a free and not servile mind and quite capable of sound knowledge, moved by the force of such novelty, begins to lend his ear and approval to the true and good philosophy, and especially in the matter concerning the constitution of the universe, but that he cannot yet totally free himself from those fancies previously impressed on him, fancies to which his intellect still returns from time to time, habituated to assent by long custom. And this is also apparent, moreover, in the same passage when he tries to demonstrate that the spots are not in any of the orbs of the Moon, of Venus, or of Mercury, where he maintains as true and real, actually distinct from each other, and mobile those eccentrics taken altogether, or those piecemeal deferents, equants, epicycles, and so forth posited by pure astronomers [*puri astronomi*] to facilitate their calculations but not, of course, considered as such by the philosophical astronomers [*astronomi filosofi*]. The latter, besides the task of saving the appearances in whatever way necessary, try to investigate, as the greatest and most marvelous problem, the true constitution of the universe, because this constitution exists, and it exists in a way that is unique, true, real, and impossible to be otherwise; because of its greatness and nobility, this problem takes precedence over every other intelligible question of speculative minds. I do not, of course, deny circular motions around the Earth and around centers other than its center, and still less do I deny other circular motions separated entirely from the Earth, that is, that do not go around it and enclose it in their orbits. For Mars, Jupiter, and Saturn, with their approaches and retreats, assure me of those former configurations, and Venus and Mercury, and also the four Medicean Planets prove the validity of the latter to me. Consequently, I am most certain that there are circular motions that describe eccentric circles and epicycles, but that Nature really uses that medley of spheres and orbs imagined by the astronomers [*astronomi*] so that she can describe such motions, *that* I consider not so much a necessary belief as a requirement to facilitate astronomical calculations. And I am of an opinion halfway between those astronomers [*astronomi*] who admit not only the eccentric movement of the stars [i.e., planets], but the eccentric orbs and spheres that conduct them as well, and those philosophers [*filosofi*] who with equal force deny both the orbs and the motions about any center other than that of the Earth.

However, when it is a question of investigating the place of the solar spots, I would wish that Apelles had not expelled them from a real place [*luogo*] that is among the immense spaces [*spazii*] in which the little bodies of the Moon, of Venus, and of Mercury go around—expelled, I say, by virtue of an imaginary supposition that these spaces [*spazii*] are entirely filled with orbs, eccentrics, epicycles, and deferents, and that they are disposed and even obliged to carry with them every other body placed in them, so that [such a body] cannot wander on its own toward any place [*banda*] other than where the surrounding heaven, with its too harsh shackles, compels it to go. And I would wish this all the less the more I see Apelles himself very close to conceding what he had earlier denied. He had said that the spots cannot be in any of the orbs of the Moon, of Venus, or of Mercury because if they were, they would follow the motion [of these planets]. Thus he supposes that the spots cannot have any motion of their own, and then, concluding that they are in the orb of the Sun, he allows that there they can move with their own revolutions, such that they are capable of wandering over the solar sphere. But if it is conceded to me that they can move through the heaven of the Sun, then it should not be denied to me that they can likewise course through that of Venus; and if it is conceded to me that they move a little and are not entirely obedient to the whirling movement of the sphere that contains them, then I will not consider it inappropriate that they move a lot and do not obey such spheres at all.

I do not want to pass over another small point that occurs to me about this same passage, at the end of which Apelles makes his last inference. There it appears that he ascertains that the spots are, finally, in the heaven of the Sun (and they absolutely have to be placed there, because in his opinion they revolve about it in very narrow circles). He then adds that they can neither be in the eccentric of the Sun, nor in the eccentrics *secundum quid*,[16] nor in another orb, were there any other. Now, here I cannot understand how they can be in the heaven of the Sun and revolve around the solar body without being in any of the orbs of which the heaven of the Sun is composed.

The three arguments that Apelles puts forward next as convincing proof that the spots move circularly about the Sun appear very probable, but they are not without some cause for doubt. As for the first, it is that the decrease in the width of the spots near the limb of the Sun would be a sign that they are stars that, turning in circles slightly more ample than the solar body, were beginning to show their illuminated parts in the manner of the Moon or of Venus, whence the dark part starts to diminish. And yet it appears to some who have very diligently observed the sunspots that the diminution of the dark part happens contrary to the way it would have to, that is, not in the part that faces the center of the Sun, but on the opposite side. And it seems to me to be nothing other than the fact that they are becoming thinner. As to the second argument—the division of what had appeared to be a single spot near the circumference into many—it has this impediment, that vast changes are perceived among these spots also in the middle regions of the Sun: they increase, diminish, join, and separate, and below I shall include some [illustrations of] changes observed by me. As regards the third argument, regarding the difference that one then notes between the speed of their motion around the middle regions and the slowness at the extremes—a difference that seems very pronounced—it would appear that it suggests rather that those spots must be on the solar body itself, and that they move with its own motion, rather than turning about it in other orbits. For a similar difference in speed would remain almost invisible to the unaided eye whenever these circles expanded beyond the surface of the Sun by a perceptible but not very great space, as one understands in the very figure included by Apelles. And here it seems that a bit of contradiction arises in him: for in this passage it is necessary to put the circles of rotation of the spots very close to the solar globe, because otherwise there would have been no increase in the speed of the motion and the [mutual] separation of the spots toward the middle of the disc, which near the circumference seemed to touch each other. And on the contrary, from the argument with which a little bit earlier he proved that the spots are not contiguous to the Sun, he would necessarily have to conclude that these circles were rather far from the same [body], because only the fifth part of their circumference, at most, could remain interposed between the solar disc and our eye, given that they cross the visible hemisphere in fifteen days,

16. A common term in scholastic writing meaning "in a certain sense" or "specially defined," that is, a restricted meaning as compared with a general meaning.

and that they had not yet returned and shown themselves in two months. It is necessary, therefore, to observe diligently, from when a spot first appears until it is finally hidden, the measure by which its speed increases and then decreases, for from this proportion one can then estimate whether its motion is on the surface itself of the solar body or rather in some circle separated from it, under the assumption, however, that such variation in the spots' velocity depends on a simple circular motion.

It remains for us to consider what Apelles decides about the essence and substance of these spots, which is, in sum, that they must be neither clouds nor comets, but rather stars revolving about the Sun. I confess to Your Most Illustrious Lordship, in regard to such a determination, that I do not yet have enough confidence to dare to establish and affirm any conclusions as certain, for I am very sure that the substance of the spots could be a thousand things unknown and unimaginable to us, and that the accidents that we observe in them—their shape, opacity, and motion—being very common, can provide us with either no knowledge at all, or little but of the most general sort. Therefore I do not believe that the philosopher who was to acknowledge that he does not and cannot know the composition of sunspots would deserve any blame whatsoever. But if by way of analogy with materials familiar and known to us we desired to proffer something that they might be, in truth I would be of an opinion entirely opposed to that of Apelles. For it does not appear to me that any condition of the essentials that belong to the stars [i.e., planets] fits these [spots], and, on the contrary, I do not find a single characteristic in them that is not likewise seen in our clouds. This we will discover if we reason in the following manner:

The solar spots appear and vanish at shorter and longer intervals; some of them come together and draw apart greatly from one day to the next; their shapes change, and most are very irregular, with greater and less darkness here and there; and because they are either on the solar body or very close to it, they must be absolutely immense bodies; because of their varying opacity, they are capable of impeding the light from the Sun here more, here less; and sometimes many appear, at other times few, and then again none at all. Now, of very large and immense masses, which appear and disappear in brief periods, which sometimes last longer and sometimes less, which expand [*distraghino*] and contract [*condensino*], which easily change their shapes, and which are here denser and more opaque, and there less so, nothing like that is found near us except for clouds; all other materials, on the contrary, differ greatly from the combination of these characteristics. And there is no doubt whatsoever that if the Earth were luminous in itself and if the illumination of the Sun did not come upon it from outside, then to someone who was able to observe it from a very great distance it would truly offer a similar appearance. For as now this and now that region were obstructed by the clouds, it would appear strewn with dark spots by which, according to the greater or lesser density of their parts, the Earth's splendor would be more or less impeded, and therefore [the spots] would appear here more and there less dark; they would appear sometimes many, at other times few, and sometimes expanding and at other times shrinking; and if the Earth rotated on its axis, those [spots] would also follow its motion; and because they are not very thick with respect to the width to which they commonly extend, those which in the middle of the visible hemisphere would appear very wide would, in approaching the extremity, seem to contract. In sum, I don't believe that a single feature would emerge that doesn't have its analogue in the solar spots. But because the Earth is dark and the illumination comes from the external light of the Sun, if the globe could be viewed now from a most distant place, one would absolutely not see any blackness or any spot caused by the scattering of clouds over it, because these, too, would receive and reflect the light from the Sun. Of the change in shape, of their irregularity, and of the unequal densities, please accept these two examples, Your Lordship.

Spot A, which on the 5th of this past April, at sunset, looked very thin and not very dark, the following day appeared once again at sunset, [but,] like spot B, increased in darkness and changed in shape, and on the seventh day [of the month] it was similar to figure C, and their position was always far from the circumference of the Sun.

On the 26th day of the same month, at sunset, on the upper part of the Sun's circumference, a spot similar to D began to appear, which on the 28th day was like E, on the 29th like F, on the 30th like G, on the 1st of May like H, and on the 3rd like L; and the changes of spots F, G, H, and L occurred rather far from the circumference of the Sun, so that such a change in appearance could not have been caused by the fact that they were seen under different angles [*diversamente*] (which near the circumference, because of the receding of the curved surface, causes a great difference).[17]

From these and other observations, and from those that can be made from day to day, one can clearly gather that among our [terrestrial] substances there is none that better imitate the characteristics of these spots than clouds, and the reasons that Apelles adduces to demonstrate that they cannot be clouds appear to me to be extremely feeble. To his question "Who would ever posit that there are clouds about the Sun?" I would respond, "He who saw these spots and wanted to say something plausible about their essence, for he would not find anything known to us that resembles them more." To the question he asks about how large they are, I would say: "As large as we see them in comparison

17. The original observations are reproduced at *OGG*, 5: 253–254.

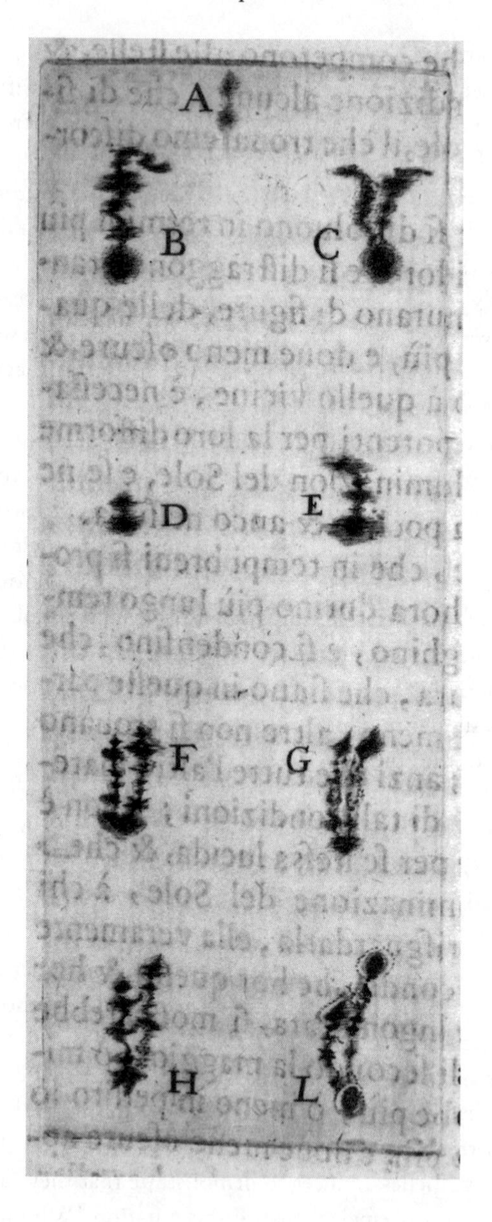

to the Sun, as large as those that sometimes cover a large territory of the Earth." And were this not large enough, I would say two, three, four, and ten times as great. And finally, to the third impediment that he produces, how these [spots] could cast such a shadow, I would respond that their blackness is less than that which our densest clouds would present to us when interposed between our eye and the Sun. This can be easily seen whenever one of the darkest clouds covers a part of the Sun and there are some spots in the uncovered portion, for no small difference will be observed between the blackness of the former and the latter, even though the edge of the cloud crossing the Sun cannot be of great thickness. And from this we can infer that a very thick cloud could cause much greater darkness than that of the darkest spots. But if, however, that were not the case, who would prohibit us from believing and saying that some of the solar clouds are denser and deeper than the terrestrial ones?

By this I am not asserting that these spots are clouds of the same substance as ours, consisting of aqueous vapors ascended from the Earth and attracted by the Sun; but I am merely saying that we have no knowledge of anything that resembles them more. Whether they are vapors or exhalations or clouds or smoke produced by the solar body or attracted to it from elsewhere is uncertain to me, because they could be a thousand other things imperceptible to us.

From what has been said, it can be gathered that the term "stars" is poorly suited to these spots, for stars, whether they are fixed or wandering, are always observed to maintain their shape, and that shape is circular; and it is not observed that some [stars] dissolve and others appear again anew, but rather that they always remain the same. And they have periodical motions such that they return after some fixed time, but these spots are not observed returning unchanged; on the contrary, some can be seen to dissolve on the face of the Sun. And I believe that one would wait in vain for the return of those that Apelles believes capable of moving around the Sun in very small circles. The principal conditions that belong to these natural bodies to which we have given the name "stars" are thus lacking. That they should be called stars, then, because they are opaque bodies and denser than the substance of the heavens, and therefore better withstand the Sun['s light], and are illuminated strongly by it on the side struck by its rays and on the opposite side produce a heavy and deep shadow, and so forth, these are conditions that belong to every stone, to wood, to the denser clouds, and in sum to all opaque bodies. And a ball of marble, because of its opacity, resists the light of the Sun, is illuminated by it, as is the Moon or Venus, and on the opposite part produces a shadow, so that in this respect it could be called a star. But because they lack the other, more essential, conditions—conditions of which the solar spots are also likewise deprived—it therefore appears that the name "star" should not be attributed to them.

Indeed, I would wish that Apelles had not counted the companions of Jupiter—by which I believe he means the four Medicean Planets—in this group, because they appear most constant, like all the other stars, [and] always bright except when they enter the shadow of Jupiter, for then they are eclipsed like the Moon in the shadow of the Earth. They have their fixed periods, which are different for each of them, and already precisely determined by me.[18] Nor do they move in a single circle, as Apelles seems to have believed or at least thought that others had believed, but rather they each have their distinct circles of different sizes around Jupiter as center, the sizes of which I have likewise determined. I have also found the causes of when and why now one, now another of them declines either to the north or to the south in relation to Jupiter; and once he has specified them, perhaps I shall have responses to the objections that Apelles suggests arise in this matter. But that there are more of these Medicean Planets than the four observed up to now, as Apelles says he holds for certain, this could perhaps be true, and such a resolute affirmation by a person who is, as far as I know, very knowledgeable makes me believe that he might have some grand hypothesis about it, one that I surely lack. But I would not be so bold as to affirm anything, because I would fear that I would have to retract it with time. And in this respect, I would not decide to propose anything around Saturn except what I have already observed and discovered, that is, two small stars that touch it, one toward the east and the other toward the west, in which no change whatsoever has yet been seen; nor surely will any [alteration] be seen in the future, if not perhaps some very strange property not only most remote from the other motions known to us, but also far removed from our every fantasy. But what Apelles proposes, that Saturn appears sometimes oblong and sometimes accompanied by two stars on its sides, let Your Lordship understand that this is caused by imperfection in the instrument or in the eye of the observer, for the shape of Saturn being thus, oOo —as perfect instruments reveal to perfect eyes—where this flawlessness is lacking it appears like this: ⬭, the separation between the three stars and their shapes not being perfectly distinguished. But I, who have observed it a thousand times under different conditions with an excellent instrument, can assure Your Lordship that no change whatsoever is perceived in it [Saturn], and reason itself, based on the experience that we have of all the other motions of the stars, can render us certain that likewise none will take place. For if there were any motion in these stars similar to the motions of the Medicean or of other stars, they would already have to be either separated from or completely conjoined to the principal star of Saturn, even if their motion were a thousand times slower than

18. Galileo published the periods of Jupiter's satellites for the first time in his *Discorso intorno alle cose, che stanno in sù l'acqua, ò che in quella si muovono* (Florence 1612). See *OGG*, 4: 63; Stillman Drake, ed., Thomas Salusbury, tr., *Discourse on Bodies in Water* (Urbana: University of Illinois Press, 1960), 1.

any other such movement of any other star that might go wandering through the heavens.

In response to what Apelles has put forward as his final conclusion, that is that these spots are more likely wandering than fixed stars, and that between the Sun and Mercury and Venus there are rather many of them, of which only those that interpose themselves between the Sun and us are manifest to us, I say, as for the first part, that I believe that they are neither wandering, nor fixed, nor stars, nor that they move about the Sun in circles distinct and distant from it. And were I obliged to offer my opinion in confidence to a friend and patron, I would say that the solar spots were produced and dissolved near the surface of the Sun, and that they were contiguous to it, and that the Sun itself, rotating around its axis in about a lunar month, carried them with itself, and perhaps sometimes brought back some of them, those of a duration longer than one of its revolutions, but so changed in shape and company that we cannot easily recognize them. And however far my conjecture now extends, I have great hopes that with what I have pointed out to you, Your Lordship will consider this matter finished. That there could then be some other planet between the Sun and Mercury that moves about the Sun, and remains invisible to us because of its small digressions [from the Sun], and can make itself visible to us only when it passes linearly below the solar disc, is for me not at all improbable. And it appears to me equally credible that [such stars] exist as that they don't exist. But I would not believe in a vast multitude [of these stars], for if they existed in great numbers, then it stands to reason that frequently one would have to be seen below the Sun, and this has thus far not befallen me, nor have I seen anything [below the Sun] other than these spots. And it is not likely that a star of this kind could have passed among those [spots], even if this star were to manifest itself, in terms of its appearance, as a black spot. I say it is not likely, for its movement would have to appear uniform and very rapid with respect to that of the spots: very rapid because moving in a smaller circle than that of Mercury, it is probable, in analogy with the motions of all the other planets, that its period would be shorter and its motions more rapid than the motion and the period of Mercury. In its passage below the Sun, Mercury traverses its disk in about six hours, such that another planet whose motion is yet more rapid should not have to remain in conjunction with it for a longer interval, excluding the possibility that one wanted to make this hypothetical body move in a circle so small that it almost touched the solar body, an arrangement that would be entirely too chimerical. But even in circular orbits whose diameters were two or even three times as great as the diameter of the Sun, it would take place as I have said: yet in fact the spots remain in conjunction with the Sun for many days, and therefore it is not likely that any planets whatsoever pass among them or in their guise. Such a planet, besides its velocity, would also have to move almost uniformly, being some considerable space away from the Sun, because only a small part of its circle would be under the Sun, and this small part would face the rays from our eyes directly and not obliquely. Thus, equal parts of it would be seen under angles of negligible inequality, that is, almost equal, and therefore its motion would appear uniform. This does not happen with the motion of the sunspots, which traverse the middle region [of the Sun] with great rapidity, while the closer they are to the circumference, the more sluggishly they proceed. Thus, the stars that go wandering about between the Sun and Mercury can in all likelihood not be many in number, and [there would be fewer still] between Mercury and Venus, for since [these hypothetical stars'] greatest digressions [from the Sun] are necessarily greater than that of Mercury, they would, like Venus and Mercury itself, have to be visible as brilliant [stars], and especially since they are not very far from the Sun and from the Earth. And so, because of the small distance from us and because of the Sun's strong illumination they would be visible because of the vividness of the light, even if they were very small in size.

I know that I have excessively wearied Your Most Illustrious Lordship with my great prolixity and few conclusions. Please recognize in the length [of this letter] the pleasure I take in conversing with you, and the desire to obey and serve you as much as my strength will allow me. And in these respects, please pardon my excessive loquacity, and receive with pleasure the readiness of my affection. May my uncertainty be excused by the novelty and difficulty of the material, where the various ideas and different opinions that have passed through my imagination, sometimes finding assent and sometimes rejection and contradiction, have rendered me bashful and perplexed, for I hardly dare open my mouth to affirm anything. I do not want on this account to despair and to abandon the enterprise; on the contrary, I would hope that these novelties might serve me wonderfully to adjust a few pipes of this grand [but] discordant organ of our philosophy, which, in my view, many organists labor in vain to tune to perfection. And this is because they go about leaving and preserving three or four of the principal pipes out of tune, such that it is impossible for the others to respond in complete harmony.

As a servant of Your Lordship, I desire to take part in the friendship that you have with Apelles, because I consider him a person of sublime skill and a lover of truth. I ask you, therefore, to greet him cordially in my name, giving him to understand that in a few days I will send him some observations and drawings of the solar spots, ones of absolute precision, in their shapes as well as in their daily changes in position, without a hairsbreadth of error, all made by a most exquisite method discovered by one of my students. These [observations] will perhaps be of benefit to him in philosophizing about their essence. It is time to cease troubling you. Therefore, kissing your hands with every reverence, I recommend myself to your good graces, and I pray to God for the greatest happiness for you.

From the Villa delle Selve, 4 May 1612
Your Most Illustrious Lordship's
Most Devoted Servant
Galileo Galilei L[inceo]

8

Galileo's second letter

Second Letter from Welser to Galileo

Most Illustrious, Excellent, and Worthy Lordship,

Your Lordship pays high interest for a slight delay, sending me such a copious and lengthy discourse in reply to a letter of a few lines. I read it, and I can even say that I devoured it with a gusto matched by my appetite and eagerness. And I tell you that it offered me some relief from a long and painful indisposition that has tormented me enormously in the left thigh, because the physicians have thus far not been able to find an effective remedy.[1] In fact, one of the principal ones told me in very clear terms that the earliest practitioners of the profession wrote of this malady: *Some are cured with difficulty; others not at all*, for which reason it is necessary to submit oneself to the

1. On Welser, see especially Christoph Arnold, "Vita, genus, et mors auctoris nobilissimi," in Marc Welser, *Opera historica et philologica, sacra et profana* (Nuremberg: Wolfgang Moritz, 1682), 3–68; Pierre Bayle, "Velsérus," in *Dictionnaire historique et critique*, 16 vols. (Paris: Desoer, 1820), 14: 342–349; Antonio Favaro, "Dal carteggio di Marco Velser con Giovanni Faber," *Memorie dell'Istituto Veneto* 24 (1891): 79–100; Giuseppe Gabrieli, "Marco Welser Linceo Augustano," *Rendiconti della Reale Accademia Nazionale dei Lincei*, 6th ser., 13 (1937): 74–99; R. J. W. Evans, "Rantzau and Welser: Aspects of Later German Humanism," *History of European Ideas* 5, no. 3 (1984): 257–272; Bernd Roeck, "Geschichte, Finsternis und Unkultur: Zu Leben und Werk des Marcus Welser (1558–1614)," *Archiv für Kulturgeschichte* 72 (1990): 115–141' and Jean Papy, "Lipsius and Marcus Welser: The Antiquarian's Life as *via media*," *Bulletin de l'Institut Historique Belge de Rome* 68 (1998): 173–190.

fatherly disposition of God's goodness: *He is the Lord; He must do what is good in His eyes.*[2] But I have made too much of melancholy matters.

As I was saying, your discourse was precious to me beyond measure, and, from the little that I can judge in this area, it appears to me written with such good and well-founded reasons, and ones advanced with such modesty, that even though it contradicts the greater part of his views, Apelles must feel himself highly honored. It will take time to acquaint him fully with the content, because he does not understand Italian, and translators familiar with the profession [i.e., astronomy], as the occasion requires, are not always available, but we will attempt to overcome this difficulty as well. I have written to the most honorable Signor Sagredo, and I repeat to you, that were I in a city where there were Italian printers, I would hope to procure through your kindness the possibility of publishing this labor quickly. And I believe that I might manage it easily, for it proceeds in such judicious and circumspect fashion that if in the future something is discovered in this connection that at present we do not even imagine, you will never be accused of having been precipitous, or of having asserted doubtful matters as certain. And it would be a public benefit if brief treatises about these new discoveries were to be published one after the other, both to keep memory [of these novelties] alive, and to encourage others to greater diligence, it being impossible that a machine of such immensity burden the shoulders of one single individual, however robust.[3] On your Lordship's word, I will promise Apelles observations and drawings of the solar spots of absolute precision, which I know he will value like a treasure. For now, I cannot write more, and close by kissing your hand and wishing for your every benefit,

From Augsburg, 1 June 1612
Your Illustrious and Most Excellent Lordship's
Most Affectionate Servant
Marc Welser

Second letter from Galileo to Welser about the solar spots

Most Illustrious Sir and Most Honorable Patron,

Many days ago, I sent you a very long letter, written a propos of the matters contained in the three letters of the masked Apelles,[4] where I raised those diffi-

2. I Samuel 3:18.
3. The image invoked is that of Heracles and then Atlas with the *machina mundi* on their shoulders.
4. See chapter 7, note 4, p. 89, above.

culties that prevented me from agreeing with the opinions of that author, and I indicated in part to him, moreover, the position to which I then inclined. Since then, I have not retreated from this inclination, but rather have become wholly convinced of it, because continuous daily observations made at every possible opportunity, and without any contradiction whatsoever, have shown me that my opinion agrees with the truth. Thus it seemed good to me to give Your Lordship some account of these [observations], on the occasion of sending you a few illustrations of these spots, drawn with accuracy, and also of the method of drawing them, together with a copy of a little treatise of mine about bodies that float on water or sink in it, which has just lately emerged from the press.[5]

Therefore, I reply to Your Most Illustrious Lordship, and reaffirm more resolutely, that the dark spots that are seen on the solar disc by means of the telescope are not far from its surface, but rather adjoin it; or they are separated from it by a distance so small as to remain entirely invisible. Moreover, they are not stars or other solid and long-lasting bodies, but rather some arise and others disappear constantly, certain of them being of brief duration—one, two, or three days—and others longer, such as ten, fifteen, and, in my opinion, also of thirty and forty and more [days], as I will relate below. They are for the most part of very irregular shapes, shapes that change constantly, some with rapid and substantial alterations, and others more slowly and with smaller variations. They also change by increasing and decreasing their darkness, for they show that some times they condense, and on other occasions they spread out and rarefy. Besides changing into many diverse shapes, some of them are frequently observed splitting into three or four, and often many unite into one, and this is not as common close to the circumference of the solar disc as it is around the middle regions. Besides these disordered and particular motions of aggregating and disaggregating, condensing and rarefying, and changing shapes, the sunspots have a single great, common and orderly motion with which they travel across the body of the Sun in uniform fashion and in parallel lines. We are informed by the particular characteristics of this motion, first, that the body of the Sun is absolutely spherical, and second, that it moves of itself and about its own center, carrying the said spots with it in parallel circles, and finishing an entire turn in about one lunar month, with a revolution similar to that of the orbs of the planets, that is, from west to east. Further, it is worth noting how the greater part of these spots always appear to fall in a band, or rather a zone of the solar body that is contained between two circles that correspond [i.e., are similar] to those that limit the declination of the planets. Beyond these limits I do not think I have thus far observed any spot, all being within that

5. *Discourse on Bodies in Water.*

limit, such that they seem to decline neither toward the north nor the south from the greatest circle of the revolution [i.e., equator] of the Sun by more than around 28 or 29 degrees.[6]

Their different densities and blackness, their changes in shape, and their mingling and separation are in themselves manifest to the eye without the need of further discussion, and therefore a few simple comparisons of those details, which we will make below, will suffice in the drawings that I am sending you. But that they are contiguous to the Sun and that they are carried around by its revolution, this must be deduced and concluded by reasoning from certain particular accidents provided to us by sensory observations.

To begin with, seeing them always move with a general motion shared by all, even though often they numbered more than twenty and even thirty, was a strong argument that there was but a single cause of this apparent change, rather than that each spot went wandering by itself around the solar body in the manner of the planets, much less in different circles at different distances from the same Sun. From this it necessarily had to be concluded either that they were in a single orb that carried them, like the fixed stars, around the Sun, or else that they were on the solar body itself, which led them with it as it rotated about itself. Of these two positions, this second one, in my opinion, is true and the other false, just as any other position that one might want to adopt will be found to be false and impossible, as I will attempt to demonstrate by means of [their] manifest inconsistencies and contradictions.

The hypothesis that they are contiguous to the surface of the Sun and that because of its rotation they are carried around it corresponds entirely to all the appearances; no problems or difficulties are encountered here. In order to explain this, it is useful that we define the poles and the circles of longitude and latitude, like those that we imagine on the celestial sphere, on the globe of the Sun. And therefore, as the Sun rotates on itself, and has a spherical surface, the two stationary points will be called its poles, and all the other points marked on its surface will describe circles parallel to each other and larger or smaller according to the greater or lesser distance from the poles; the largest will be the circle of the middle, equally distant from both poles. The longitude or length of the solar surface will be the dimension that is observed along the length of the circumference of these circles, but the latitude or breadth will be the extension in the other direction, that is, from the greatest circle to the poles. Whence the dimension taken along a line parallel to the said circles, that is, taken in the direction along which the rotation of the Sun occurs, will be called the

6. Galileo is comparing the zone extending on either side of the Sun's equator in which sunspots are contained to the band of the zodiac in which all the planets are contained.

"longitude" of the spots, and the "latitude" will be understood to be that which extends toward the poles and which is determined by a line perpendicular to the lines of longitude.

Having defined these terms, let us start by considering all the particular accidents that are observed in the solar spots, from which their place and motion can be learned. To begin with, [there is] the fact that at their first appearance and final disappearance near the circumference of the Sun the spots generally show themselves [to be] very scant in [solar] longitude, but with as much [solar] latitude as they have when they are on the most interior parts of the solar disc. To those who understand, because of [the laws of] perspective, what the receding of the spherical surface near the extremities of the visible hemisphere means, this will be a clear argument for both the sphericity of the Sun and the proximity of the spots to the solar surface, and also for [the fact that] they are carried on the same surface toward the middle regions, showing themselves ever increasing in [solar] longitude and maintaining the same [solar] latitude. And although when they are very close to the circumference, not all of them appear equally attenuated and reduced to the thinness of a thread—for the oval shape of some is more slender, others less so—this comes from the fact that they are not simple superficial spots, but rather, that they also have thickness, or we should say "depth," and some have more of it, and others less. This is also the case with our clouds, which spread themselves in length and breadth tens and sometimes hundreds of miles, but when it comes to their thickness they are sometimes more and sometimes less deep, but we do not see this depth exceed many hundreds, or at most thousands, of *braccia*.[7] Therefore, because the thickness [i.e., depth] of solar spots, although small in comparison to the other two dimensions, can be greater in one spot and smaller in another, it will happen that near the circumference of the Sun, where they are seen edge-on, the thinnest spots will appear very slender, especially because the inner half of this edge is illuminated by the light of the nearby Sun, and others of greater depth will appear thicker. But that many of them are reduced to the thinness of a thread, as experience teaches us, this could on no account occur if the motion with which they seem to traverse the disc of the Sun were along circles removed by ever so small an interval from the solar globe. For the great diminution of the length occurs under the maximum foreshortening, that is, near the edge of the disc, but it would be seen to occur beyond the body of the Sun, if the

7. The Florentine *braccio* ("arm," pl. *braccia*) = 58.4 cm. In their well-known translation of Galileo's *Discorsi* (1638), Henry Crew and Alfonso de Salvio translate *braccia* as "cubits" (see *Dialogues Concerning Two New Sciences* [New York: MacMillan, 1914], e.g., 178). We believe this term confuses rather than clarifies the issue.

spots were carried in circles removed by any appreciable space from the Sun's surface.

One notes, second, the size of the apparent spaces through which the same spots seem to move from day to day, and one observes that the spaces crossed in equal intervals by the same spot appear to become smaller, as it approaches the circumference of the Sun, and one sees through careful observation that these decreases and increases, recorded at equal time intervals, correspond closely to the versed sines and their excesses proportional to equal arcs.[8] This phenomenon does not take place in any motion other than a circular one contiguous to the Sun, for in circles, even in ones not very far from the solar globe, the spaces traversed in equal intervals would appear to differ very little among themselves in comparison to the surface of the Sun.

The third detail, which wonderfully confirms this conclusion, derives from the distances between one spot and another, some of which always remain the same, while others increase greatly toward the middle regions of the solar disc. Near the Sun's limb, before approaching the middle and after moving beyond it, these distances are very small, and almost regular; and other distances also change, but at very different rates. Nevertheless, they are such that similar ones could not be found in any other motion than a circular one made by different points variously situated on a rotating globe. The spots that have the same declination, that is, those that are located on the same parallel, appear almost to touch each other at the beginning, if their actual distance is small. If this distance is somewhat greater, they will appear well separated, but much closer than when they are toward the middle of the solar disc. And as they move away from the circumference, they come to separate and distance themselves increasingly from each other until they are equidistant from the center of the disc, and there is their greatest separation. Moving away from there, they begin again to approach each other more and more the closer they come to the circumference. And if the proportions of these approaches and retreats are observed with accuracy, it will be seen that these likewise cannot take place except in motions made on the surface itself of the solar globe. And because this reason is very powerful—such that it alone would suffice to demonstrate the essence of this point—I wish to give Your Lordship a practical method that will explain more clearly what I have in mind and at the same time may show you the truth of this.

First, Your Lordship should note that because the distance between the Sun and us is very great in proportion to the diameter of its body, the angle subtended by the rays produced from our eye to the endpoints of this diameter will

8. The versed sine or versine is defined as the quantity obtained by subtracting the cosine from unity. See p. 82.

be so acute that we can, without appreciable error, take these rays to be parallel lines. Further, because not any two spots taken indiscriminately are suitable for the experiment I have in mind, but only those that are carried on the same parallel, we have to choose two that are disposed in this way. We will know them to be so disposed whenever they both pass, in their motion, through the center of the solar disc itself, or at any rate equally distant from it, and toward the same pole. Such conditions come about sometimes, as in the case of the two spots A and B, in the drawing of the first of July.[9] B passed the second day [of the month] close to the center, and A passed at a similar distance on the seventh day, and both with a northern declination, and because their distance from the center was very small, the parallel described by them is almost imperceptibly smaller than the greatest circle. Therefore, let Your Lordship first of all imagine the line GZ, which may represent to us the distance of the Sun, and let Z be our eye and G the center of the Sun around which let there be described the semicircle CDE of semidiameter equal to or slightly less than the semidiameter of the circles on which I marked the spots, so that the circumference CDE will represent that [path] described by the spots A and B. [The circumference CDE] will appear as a straight line to the eye at Z, because the eye is very distant and is in the same plane of the circle CLE, and [this circumference] will also appear identical to the diameter CGE. (And I say this because from the observations I have been able to make up until now I do not judge that the revolution of the spots is oblique to the plane of the ecliptic, in which the Earth lies.)[10] Next, take the distance of the spot A from the circumference closest to it and assign to it [the distance] CF, and through point F let there be drawn a line perpendicular to CG, which will be FH, and which will be parallel to GDZ. It will be the visual ray which goes from the eye to spot A, which spot, while it appears at point F of the diameter of the Sun, CE, will come to point H [on its surface]. After this, take the interval between the two spots, A and B, and project it on the diameter CE, from F to I, and likewise erect the perpendicular IL, which will be the visual ray of [to] spot B, and the line FI will be the apparent distance between spots A and B. The true interval, however, will be determined by the [chord] HL subtending the arc HL, but because this distance is bounded by

9. In the pages that follow, Galileo is asking Welser to compare the apparent intervals between specific sunspots, or groups of sunspots, on particular days in the observations made between 2 June and 8 July 1612, with various distances indicated on this geometrical diagram. The observational images in question are on pp. 131–165.

10. In fact, the axis of rotation of the Sun is 7° 15′ inclined to the ecliptic, and thus the revolution of the spots is oblique to the ecliptic. This inclination, not noticed during the period of Galileo's detailed sunspot observations (May to July 1612), was later to provide him with an argument for the motion of the Earth. See pp. 325–327, below.

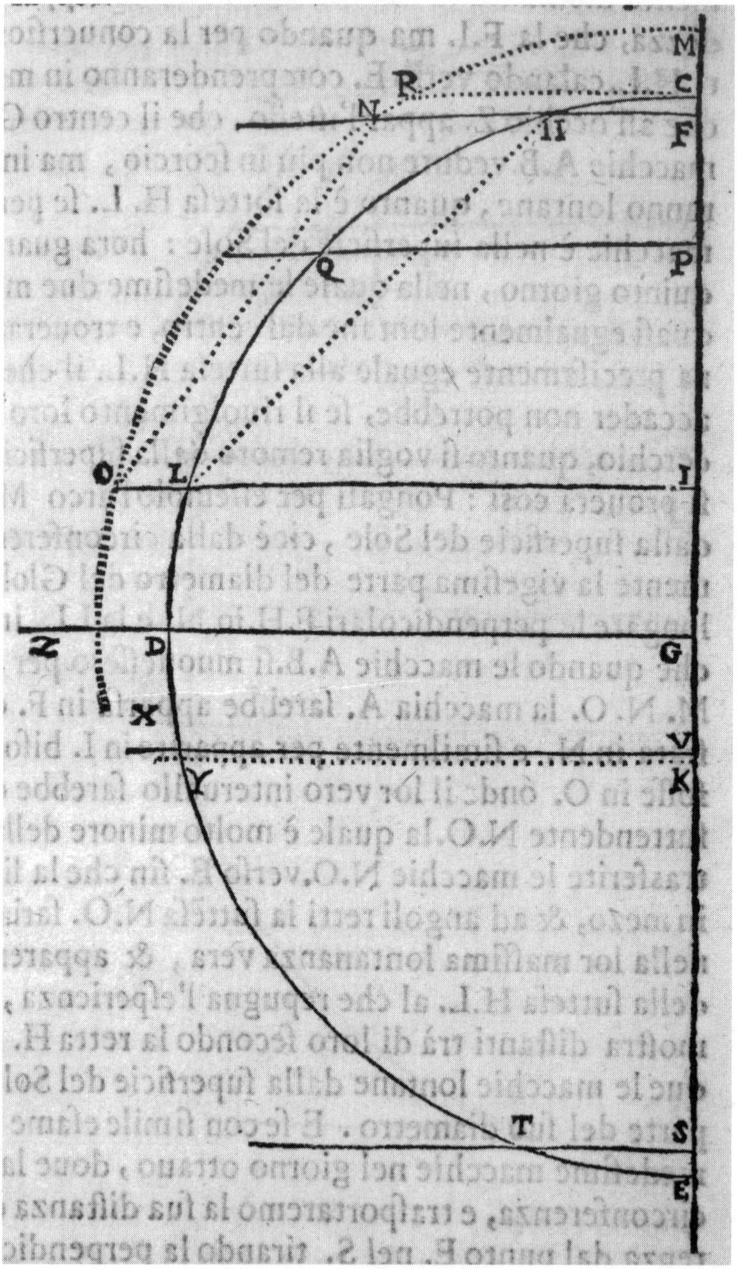

nza, che la F.L. ma quando per la conuertio
la L. caschde [illegible] E. comprenderanno in m
co alt [illegible] apparit [illegible] che il centro G
macchie A.B. [illegible] non più in scorcio, ma in
ranno lontane, quanto è la suttesa H. L. se pe
macchie è [illegible] superficie del Sole: hora guar
quinto giorno, nella quale le medesime due m
ciasi egualmente lontane dal centro, e trouera
za precisamente eguale alla suttesa H. L. il che
accader non potrebbe, se il mouergimento loro
cerchio, quanto si voglia remoto dalla superfic
si prouerà così: Pongasi per esempio l'arco M
dalla superficie del Sole, cioè dalla circonfere
mente la vigesima parte del diametro del Glo
lungare le perpendicolari F.H. in N. e I.L. in
che quando le macchie A.B. si muouessero per
M. N. O. la macchia A. sarebbe apparita in F.
[illegible] in N. e similmente per apparire in L. biso
sole in O. onde il lor vero interuallo sarebbe
suttendente N.O. la quale è molto minore dell
trasferite le macchie N.O. verso E. sin che la li
in mezo, & ad angoli retti la suttesa N.O. saria
nella lor massima lontananza vera, & appare
della suttesa H.L. al che repugna l'esperienza,
mostra distanti trà di loro secondo la retta H.
che le macchie lontane dalla superficie del Sol
parte del suo diametro. E se con simile esame
le istesse macchie nel giorno ottauo, doue la
circonferenza, e trasportaremo la sua distanza
tanza dal punto E. nel S. tirando la perpendic

M
R
C
N
H
F
P
Q
O
L
I
Z
D
G
X
V
Y
K
T
S
E

the rays FH and IL, and is seen obliquely because of its inclination, it does not appear to be of a magnitude other than FI. But when because of the rotation of the Sun, as points H and L decline toward E, they will contain between them point D, which to the eye, Z, appears the same as the center G, then the two spots A and B, no longer seen foreshortened but rather head on, will appear as distant from each other as the chord HL, that is, if the location of these spots is on the surface of the Sun. Now if one looks at the illustration of the fifth day [of July], in which the same two spots, A and B, are almost equidistant from the center, one will find that their distance [from each other] would be exactly equal to the chord HL. This could in no way happen if their revolution took place along a circle at any distance whatsoever from the surface of the Sun. This is proved as follows.

Let it be supposed, for example, that the arc MNO is distant from the surface of the Sun, that is, from the circumference CHL, only by a twentieth part of the diameter of the solar globe, and let the perpendiculars FH and IL be extended respectively to N and O. It is clear that if the spots A and B were to move on the circumference MNO, spot A would be seen at F when in reality it was at N, and likewise spot B would have to be at O in order to appear at I, and hence their true interval would be as long as the subtended straight line NO, which is much shorter than HL. For which reason, if the spots N and O were moved toward E, until the line GZ cut the chord NO in half and was at right angles to it, then the spots would be at their greatest true and apparent distance from each other, and the chord NO much shorter than the chord HL. Our sensory experience, which shows them separated from each other by the straight line HL, is contrary to this. The spots are, therefore, not removed from the surface of the Sun by the twentieth part of its diameter. And if through a similar examination we were to observe the same spots on the eighth day [of July] (see p. 165, below), when B is close to the circumference, and we were to transfer its distance from that circumference from point E to S, drawing the perpendicular ST on the diameter CE, point T will be the place of this spot on the surface of the Sun. And assigning then the distance BA [in the illustration of 8 July] to SV in the diagram, and producing likewise the perpendicular VX, we will find that the interval TX (which is the true distance between the spots A and B) is the same as HL. This phenomenon could not occur in any way if the spots B and A were moving in circles perceptibly removed from the surface of the Sun. And note that if one chose two spots that were less distant from each other and closer to the edge C or E, this phenomenon would be much more noticeable. For if there were two spots, one of which were at point C at its first appearance and the other at F, so that the apparent distance between them were CF, the true interval between them, if they were on the surface of the Sun, would be the chord HC, seven or more times greater than CF; but if these spots had been at R and N, their true distance [from each other] would have been the chord RN, which is less than the third part of CH. Therefore, with these spots moved to the region around point O, when sensory experience would make their distance appear to us equal to CH, that is seven times greater than CF and not equal to RN—which is hardly double CF—there would be no room to doubt that the actual spots were contiguous to the Sun, and not separated from it. But sensory experiences will prove to us that the chord CH (that is, the true distance between the spots when they are near the center of the solar disc) contains not only seven but ten and [even] fifteen times the first apparent distance CF. This will be the case whenever the spots are in reality less and less distant from each other than the chord CH, and this particular case could never arise if the circumference MNZ were distant from the surface of the Sun by even the hundredth part of the solar diameter, as I will demonstrate below. Therefore it follows as a necessary consequence that the distance of the spots from the surface of the Sun is nothing if not imperceptible. And the demonstration of what I have just said will be as follows.

Let, for example, the arc CH be 4°, then the straight line CF will be 24 parts of which the semidiameter CG is 10,000, and from this the chord CH will be 419, that is, seventeen times greater than CF. But if the semidiameter GM were only the hundredth part greater than the semidiameter GC, so that of the parts of which GC is 10,000, GM would be 10,100, the arc MR would be found to be 8°4′, the arc NRM 8°58′, the arc RN 0°54′, and its chord 94, of which CF will be 24, that is, [the chord RN will be] less than four times greater than it [i.e., CF]. Experience disagrees with this no less than it agrees with the other position.[11]

With the same method we can also see from day to day the increases and decreases of the same intervals corresponding to revolutions made only on the surface of the Sun. Let one consider the drawing of the third day of July and suppose the distance PC to be equal to the distance of spot A from the circumference of the solar disc, and then likewise suppose the line PK equal to the interval AB. Then, when the two perpendiculars PQ and KY are produced, we will find the chord QY to be equal to HL—an irrefutable argument that the rotation was made on the surface itself of the Sun.

I say moreover that these spots are not only very close, and perhaps contiguous, to the surface of the Sun, but furthermore that they extend only a little bit above it as regards their thickness or rather their "depth." That is, I am saying that they are very thin in comparison to their length and breadth; I gather this

11. For a complete demonstration of Galileo's argument, see appendix 5.1.

from the appearances of the spaces between them, which are often separate and distinct right up to the limb of the Sun, even when spots are observed not far removed from each other and situated on the same parallel. This happens with the two [spots marked] Y of the 26th day of June, which are beginning to appear; although very close to the extreme circumference of the disc, the one nevertheless does not touch the other, but rather a bright gap can be perceived between them. This would not be the case if they were very high and thick, and especially when they are very close to each other, as is shown in the succeeding drawings of the 27th and 28th days [of the month]. Likewise, the spot M, made up of a vast congeries of very small spots, shows the separation between them right up to the final occultation, even though the entire aggregate is severely foreshortened because of the receding of the globular surface, as is seen in the drawings of the 26th, 27th, and 28th.

But here it could by chance occur to someone that these spots might be purely superficial or at least of an extreme thinness. For when they are close to the circumferences of the disc, the bright spaces between them are not foreshortened more than their lengths are diminished. And it seems that this could not happen if their depth were of any appreciable significance. To that I reply that this consequence is not necessary, and this is because if their depth were noticeable in comparison to their length or to the spaces between one spot and another, the bright gap would still be able to appear up to very near the circumference, and this because of the Sun's splendor, which illuminates the spots themselves from the side.

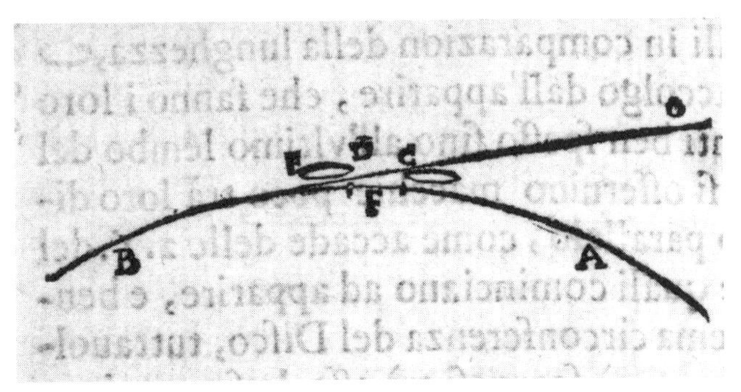

For if Your Lordship will imagine the surface of the Sun along the arc AFB, and above it the two spots C and DE, and the visual ray along the straight line OC, which is so oblique or inclined that it can in no way discern [that part of] the surface of the Sun marked F, that is, the area between the two spots. Nevertheless, they can be perceived as distinct spots, rather than as a single continuous one, because of side D of the spot DE, which is highly illuminated by the neighboring brightness of the surface F. In addition to this, the eye, being so obliquely positioned, sees some part of the surface of the Sun, that is, the surface that is beneath the spot DE, which it did not see while the visual rays were head-on. I also point out that not all the tightly clustered spots look separated right up to the circumference; some, on the contrary, appear united. This can sometimes happen when the spot farther from the circumference is thicker and deeper than the nearer one. Besides this, there is the fact that their own motions are irregular and wandering, which can cause varying appearances in this particular. But I observe generally that the blackness of all [spots] diminishes little by little when they are close to the extreme edge of the disc. In my opinion, this happens because the illuminated edge is exposed, and the dark backsides of the spots are largely hidden: their obscurity thus remains much diffused by the abundance of light. I could adduce many other examples for Your Lordship, but I would be too prolix, and I will reserve writing in more detail for another time. And for now I will content myself with having indicated my opinion, born of a continuous course of observation. In sum, this is that the distance of the spots from the surface of the Sun is either null or so small that it cannot give rise to any phenomenon perceptible by us, and that their depth or thickness is likewise small in comparison to the other two dimensions, imitating in this particular, too, our larger cloud covers.

And these are the findings that we have about the spots that lie on the same parallel. Next, the spots that are located on different parallels but are, so to speak, on the same meridian—that is, the line that joins them cuts the parallels at right angles and not obliquely—these undergo no change in the distances between them, but [rather] the interval separating them at their first appearance is always maintained to the last occultation. Other spots on different parallels and different meridians, however, have these intervals increase and then decrease, but greater differences accompany those that are viewed more obliquely, that is, those spots that are on parallels that are closer together and on meridians that are farther apart. Those that are situated less obliquely with respect to each other, by contrast, manifest less change [in these intervals]. And anyone who measures all these related differences will find that the entirety corresponds to and agrees in perfect proportionality with our hypothesis alone, and disagrees with any other.[12] It must nevertheless be noted that, these spots being neither entirely fixed nor immutable on the face of the Sun, but, on the contrary, for the most part continually changing shape, some joining together and others disintegrating, slight variations in the accurate comparison

12. On Galileo's use of the term *simmetria* to mean "proportionality," see Giora Han and Bernard R. Goldstein, "*Symmetry* in Copernicus and Galileo," *Journal for the History of Astronomy* 35 (2004): 1–20, especially 10–11.

of the observations recounted above can be caused by similar small changes. These divergences, minute in comparison to the great and general rotation of the Sun, should not engender the slightest concern in anyone who goes about judiciously calibrating the uniform and general motion with these minor and accidental fluctuations.

Now, as a result of all these comparisons, as much as the appearances that are observed in the spots correspond exactly to their being contiguous to the surface of the Sun, to this surface being spherical and not of another shape, and to [the spots'] being carried around by the Sun in its revolution on itself, so much do they contradict, through instances of obvious inconsistency, all other positions that one might attempt to assign them.

For if someone wanted to suppose that they were in the air[13]—where it appears that other similar phenomena [*impressioni*] are continuously produced and dissolved, with comparable features of aggregation and disintegration, condensation and rarefaction, and with the most random changes in shape—first of all, because [spots so placed] would block minute areas of the solar disc when they interpose themselves between it and our eye, and would be very close to the Earth, they would have to be masses no larger than the smallest clouds, for a cloud that was not large enough to occult the Sun would have to be very small indeed. And if this were the case, how can there be, in such a small mass, such great density of matter as to resist the force of the solar rays so obstinately that these rays neither penetrate them with their light nor dissolve them over the course of days and days with their force? How, if they are generated in the regions surrounding the Earth, and, if I judge correctly, according to the account of others, perhaps by evaporation from it, how, I ask, do all of them happen to be between the Sun and us, and not in another part of the sky? For none are observed below the illuminated face of the Moon, nor are any of them seen far beyond the Sun, either as dark [masses] or as ones made luminous by its rays, as happens with the dim and bright clouds that we constantly see both near the Sun and in all other parts of the air. Further, recognizing that the material of these spots is by its nature changeable because they come together and separate without any regularity, what force is there, other than a regular movement common to all, that makes them cross the solar disk in about fifteen days, that can convey to them, and in such orderly fashion, the diurnal motion, so that they never fail to accompany the Sun, while the other aerial phenomena not only move across its face but also across much greater spaces in lesser intervals of time?

One cannot respond to such plausible reasons without introducing great

13. That is, the region near the Earth.

implausibilities. But we still have conclusive demonstrations, ones that do not permit any further reply. One of these is that those spots appear arranged in the same order and on the same part of the Sun when viewed at the same time from different places on the Earth, many quite distant from each other, as I have been able to observe from diverse comparisons of drawings received from various regions. This is a conclusive argument for their enormous distance from the Earth. [The fact] that all [spots] fall within that band of the solar globe corresponding to the area of the celestial sphere contained between the tropics or, better, between the two parallels that determine the maximum declinations of the planets agrees marvelously with this measure [of their distance from the Earth]. And I am not forced to believe that this is a particular privilege of the city of Florence, where I myself live, but rather must imagine that they are seen within the same confines from every other place, as southerly or as northerly as you like. Moreover, we are convinced that they are far above the Moon by their failure to undergo any change in place below the solar disc other than the one, general and common to all [spots], that has them move across the Sun in about fifteen days, and those small and random modifications through which at times some come together and others separate. For otherwise, as Apelles rightly notes, it would be necessary that in the time between the rising and setting of the Sun all [spots] leave the solar disc because of parallax. And if anyone still wanted to attribute some motion of their own to them, [one] which would compensate for the difference in aspect, the same spots seen by us today would not be able to appear again tomorrow, which is contrary to experience, because not only do they return to make themselves visible the second day, but the third and fourth, and right up to the fourteenth.

The spots are therefore, by conclusive demonstrations, rather far above the Moon, and being in the celestial region, they cannot without still more contradictions be assigned any place other than the surface of the Sun, nor any motion other than that of its axial rotation. For among all the imaginable hypotheses, the one most suited to satisfy the observational accounts would be to imagine a small sphere between the solar body and us such that our eye and the centers of that sphere and the Sun were in a straight line, and moreover, such that its apparent diameter were equal to that of the solar body; on the surface of this sphere such spots would appear and disappear, and by the rotation of the same sphere on itself they would be carried around. Such a position, I am saying, would be one that would satisfy the aforementioned appearances, provided that it were assigned a place so far above the Moon that it was free of all the difficulties posed by two types of parallax—both the one that depends on the diurnal motion as well as the one that arises from different places on Earth—and this so that at all times and to all observers the centers of the said sphere and the Sun would remain on the same straight line. But with all this an inevitable difficulty convinces us, and that is that we would have to see the spots move below the solar disc with contrary motions, for those that were in the lower hemisphere of the imagined sphere would move in the opposite direction of those placed in the higher hemisphere, and this is not seen to occur. Besides this, just as seeing such spots appear and disappear on the face of the Sun itself has not produced any difficulty for or inconsistency with what is credible to speculative and free minds—who understand perfectly well that it neither has been nor ever can be demonstrated with any effectiveness that the part of the world beyond the concave of the lunar orb is not subject to mutations and alterations—so do I believe that placing spots contiguous to the Sun, rather than in another place, will be only slightly more bothersome to those who would like the celestial substance to be unalterable, when they are obliged, by firm and sensory experiences to imagine them in the heavenly regions.

The introduction of such a sphere between the Sun and us—which alone could satisfy many of the phenomena, but with little profit to those who might desire to remove the spots from the Sun—now having been judged a falsehood, it is not necessary to waste time in re-examining every other conceivable position, for anyone will immediately encounter manifest impossibilities and contradictions himself, as long as he has understood all the phenomena that I have recounted above and are truly observed in these spots at all times. And so that Your Lordship might have examples of all the particulars, I am sending you the drawings of thirty-five days, beginning with 2 June. In these Your Lordship will have, first of all, examples of how these spots appear smaller and thinner in the regions nearest the circumference of the solar disc, by comparing the spots marked A of the 2nd and 3rd days [of June], which are the same;[14] B and C of the 5th day with the same of the 6th; A of the 10th and 11th; likewise B of the 13th, 14th, 15th, and 16th, and C of the 14th, 15th, and 16th; B of the 18th, 19th, and 20th; C of the 22d, 23rd, and 24th; A of the 1st, 2nd, and 3rd of July; C and B of the 7th and 8th;[15] and still others which I omit for the sake of brevity. As for the second observation, which was that the spaces traversed in equal times are smaller and smaller the closer the spot is to the circumference, clear examples of this are given to us by the spots A of the 2nd and 3rd of June; the spots B and C of the 5th, 6th, 7th, and 8th; the spots C and A of the 10th, 11th, 12th, 13th, 14th, 15th, and 16th days; the spots F

14. Note that on the observation of 3 June, p. 132, below, there are two spots marked "A." Galileo was clearly referring to spot on the lower right.

15. B is not marked on the engraving for 15 June, but must be that below H and nearest to the solar limb. C is not marked on the engravings of 7 and 8 July, and perhaps is a misprint for A.

and G of the 16th, 17th, 18th, 19th, 20th, and 21st; the spots C of the 22nd, 23d, 24th, 25th, and 26th; the spots A and B of the 1st, 2nd, 3rd, 4th, 5th, 6th, 7th, and 8th of July; and many others. Moreover, the fact that the transverse distances [i.e., on the meridian] between spots always remain the same, which was the first part of the third observation,[16] is proved by spots B and C from the 5th of June to the 16th, and by spots F and G of the 13th to the 20th, where at the end their distance diminishes a bit because they are not located exactly on the same great circle that passes through the poles of rotation of the Sun. And the same is seen in the intervals between spot A and the center of spot F from the 2nd of July to the 8th, which [intervals] grow somewhat because the said spots face each other obliquely; and spots E and F do the same during the same days, but with smaller differences because they are positioned less obliquely with respect to each other. But that the intervals between the spots that happen to be on the same parallel apparently change, becoming smaller as their distance from the center becomes greater, this is clearly shown by spots B and O from the 5th day of June until the 14th, where their distance increases up to the 8th and 9th day, and then diminishes until the last day. The three spots H of the 17th day were much more widely spaced in the preceding days, and the interval FH from the 14th until the 18th diminishes, and always in increasing proportions.

Next, as to the other phenomena, Your Lordship will see first of all great changes in shape of spot B from the 5th of June until the 14th; you will see more considerable variations in G from day 10 to 20, with a great increase in size and then a decrease. Spot M began to emerge on day 18, and on day 20 it looked extremely large and was a congeries of many together; it then continued to change shape as is seen up to the end. The spots R began to appear, most minute, on day 21, and then with a great increase in size and [taking on] the oddest shapes, they continued changing up to the end. Spot F likewise appeared on day 13, nothing at all having been observed in that region the previous day; it then grew and at the end diminished, while variously changing shape. Spot S began to appear 3 June, and it was then two small little spots, ones that grew and formed another shape, and then they likewise diminished, as can be seen in the drawings. In the group of spots P, which began to appear on 25 June, great changes were subsequently seen, as well as an increase in number and size, and then again a great diminution of the one and the other up to the end. As the drawings show, Spot F, which emerged into view 2 July, underwent enormous and strange changes during the following days. On 8 June the spots E, L, and

16. See above, p. 112. E and F are not marked on the illustrations in question.

N[17] appeared for the first time: spots L quickly dissolved, and N grew in size and number. [Spots] P of the 11th day [of the month], having just then appeared, vanished two days later. [Spot] Q having appeared on the 24th, divided into three the next [day], and then faded away. Spot C likewise emerged on the 25th, splitting into three the following day, and on the same day all the [spots] X were seen for the first time. Spot G of day 27 split into many the next day, and made other divisions and changes of place on the other days, just as great changes in the spots near P were also seen on those same days. The seven spots M and N of 3 July appeared on that day; and the next day N reduced itself to two, having first been five; and M first increased in number and then coalesced, and finally began to split apart again.[18] And from all these and other accidents that Your Lordship will be able to observe in the same drawings, it can be seen to what erratic changes these spots are subject, the sum of which, as I have pointed out to you elsewhere, finds neither example nor analogy among any of our [terrestrial] materials except in clouds.

Next, as for the greatest duration of the larger and denser [spots], although it cannot be determined with certainty if any of the same ones reappear after more than one revolution because the continuous changes in shape take away our ability to recognize them, nonetheless I would be of the opinion that some returned to show themselves to us more than once. And I am inclined to this belief upon seeing a very large one appear and grow continuously while the visible hemisphere turns; since it is credible that it was generated long before its arrival, so it is reasonable to believe that it can last after its departure [from the visible hemisphere], such that its duration will be much longer than the time of half a revolution of the Sun. Therefore some spots can doubtless, or rather necessarily, be seen twice by us. Among these would be those that appear on the visible hemisphere just before disappearing [from view], and then, passing across the other [hemisphere], continue to grow; nor do they dissolve before they have returned once again to show themselves to us. And for this to happen, they have only to last three or four days more than the time of half a revolution. But I believe, further, that there are those that traverse the entire visible hemisphere more than once, such as those that from the first appearance ever increase in size as long as we see them, and become extraordinarily large; these can continue to grow while they are hidden from sight, and it is not credible that then they diminish and dissolve in a briefer interval, because

17. No spots are marked P and N on these dates.

18. Spots M and N do not appear to have been labeled in these engravings, but to judge from Galileo's description of their subsequent appearances, M is the elongated mass beneath B, and the cluster N is slightly higher, and to the left.

none of the largest [spots] has been observed to dissolve suddenly. And I have noticed more than once after the departure of some of the largest [spots], when the time of half a revolution had passed, a spot begin to appear again that was, in my opinion, the same one, and moving along the same parallel.

From what has been said thus far, it seems to me, if I am not mistaken, that it must necessarily be concluded that the solar spots are contiguous or very close to the body of the Sun; that they are of material that is not permanent and fixed, but rather variable in shape and density; that they are also mobile, some less and some more, with some small indefinite and irregular motions; and that generally all appear and disappear, some in shorter and others in longer times. Their rotation about the Sun is also manifest and beyond doubt, but it remains to some extent in doubt whether this happens because the body of the Sun itself turns and goes around on its own axis, carrying them with it, or rather that the solar body remains immobile and the revolution is that of the ambient that contains and carries them with itself, for both hypotheses are possible. Nevertheless, it appears much more probable to me that the motion is that of the solar globe rather than that of the ambient. And I am led to this belief primarily by the certainty I have that that ambient is a very tenuous, fluid, and yielding substance from seeing how easily the spots contained in it change shape and come together and divide, which could not happen in a solid and firm material—a proposition that will appear rather novel in the common philosophy [*comune filosofia*].[19] Now it seems that a constant and regular motion, such as is the general one of all the spots, cannot have its origin and primary basis in a fluid substance composed of parts that do not cohere and is thus subject to the agitations and perturbations of many other accidental motions, but that it can originate in a solid and firm body where the motion of the whole and of the parts is by necessity a single one. And it is credible that such is the solar body in comparison with its ambient. Such a motion communicated to the ambient by contact, and to the spots by the ambient, or rather conferred immediately on the spots by the same contact, can carry them around. Moreover, if someone were to maintain that the circulation of the spots about the Sun proceeded from motion residing in the ambient and not in the Sun, I would believe all the same that it is almost necessary that the same ambient communicate by contact the same motion to the solar body as well.

For I seem to observe that natural bodies have a natural inclination to some motion—heavy ones, for example, tend downward—and they exercise this motion through an intrinsic principle and without need of a particular external mover, as long as they are not impeded by some obstacle. To some

19. Aristotelian philosophy.

other motions they have resistance—those same heavy bodies, for instance, to upward motion—and therefore will never move in this way except when thrust violently by an external mover. Finally, they are indifferent to some movements—as are the same heavy bodies to horizontal motion—to which they have neither inclination, because it is not toward the center of the Earth, nor aversion, because it is not away from the same center. And therefore, with all the external impediments removed, a heavy body on the spherical surface concentric to the Earth will be indifferent to rest and to movement toward any part of the horizon, and it will remain in the state in which it has been put; that is, if it has been put in a state of rest it will remain in it, and if it has been put in motion, toward the west, for example, it will maintain the same state. And thus a ship, for example, having received one single time some impetus, would move continuously through a quiet sea around our globe without ever stopping, and if one were to bring it gently to rest it would perpetually remain at rest, provided that in the first case all the extrinsic impediments could be removed, and in the second one that no external mobile cause came upon it. And if this is true, and it is most certainly true, what would such a mobile body of indefinite nature do when it was continually surrounded by a mobile ambient with a motion to which this natural mobile [body] was by nature indifferent? I do not think there can be any doubt that it would take on the motion of the ambient. Now the Sun, a body of spherical shape, suspended and balanced around its own center, cannot fail to move along with the motion of its ambient because it has neither an intrinsic aversion nor an external impediment to such a rotation. It cannot have an internal aversion, seeing that by such a rotation the whole does not move from its place, nor do the parts change places among themselves or alter in any way their natural constitution, so that as regards the constitution of the whole in relation to its parts such a movement is as though it did not exist. As to the external impediments it does not appear that any obstacle (except, perhaps, magnetic force) can impede [something] without contact [with it], but in our case all that touches the Sun, which is its ambient, does not only not impede the motion that we seek to attribute it, but also moves itself, and in moving it communicates its motion wherever it does not find resistance, of which there can be none on the Sun, so that here all external impediments cease. This can be confirmed even further, because in addition to what has been said, it does not appear that any mobile [body] can have an aversion to one motion without having a natural propensity for its contrary—for there is no aversion in indifference—and therefore whoever might propose a resistance in the Sun to the circular motion of its ambient would likewise suppose that it has a natural propensity to the circular motion opposite to that of the ambient, and this [supposition] is discordant to the well-tuned mind.

Therefore, for the first reason adduced by me, because one must in any case posit the apparent revolution of the spots on the Sun, it is better to suppose it is there naturally, rather than through communication from the ambient.

I could bring up many other considerations for ampler confirmation of my opinion, but I would exceed by too much the limits of a letter. In order, therefore, not to occupy your time further, I will deliver what I promised to Apelles: the method of drawing the spots with the greatest accuracy, which, as I indicated in the other letter,[20] was discovered by one of my students, a Benedictine monk[21] named D. Benedetto dei Castelli, a noble family of Brescia, a man of superior intellect and, as is fitting, free in philosophizing. The method is this. The telescope should be directed to the Sun as if someone wanted to observe it. Having focused and steadied it, one places a flat sheet of white paper opposite the concave glass, about four or five palms[22] away from it, so that upon this the circular image of the solar disc will fall, with all the spots that are on it arranged and disposed with exactly the same proportions as they have on the Sun.[23] The farther the paper is moved from the tube, the larger the image will become, and the better the spots will be depicted; without any injury one will see all of them, including the very small ones that, observed through the tube with great fatigue and damage to the eyesight, could scarcely be perceived. To draw them accurately, I first describe on the paper a circle of a size that suits me and then, moving the paper away from or toward the tube, I find the exact place where the image of the Sun is magnified to the size of the described circle, which also serves me as a guide and a rule for keeping the paper straight and not inclined to the luminous cone of the solar rays that come out of the telescope, for when it is oblique, the section becomes oval and not circular, and therefore does not fit the circumference described on the paper. But by inclining the paper more or less, one easily finds the exact position, which is when the image of the Sun fits the described circle. Having found this position, on the spots themselves one marks their shapes, sizes, and positions with a brush. But one has to proceed deftly, following the motion of the Sun and, frequently moving the telescope, one must keep it accurately directed at the Sun. This is known by looking at the concave lens, where a small luminous circle, concentric to that glass when the telescope is correctly trained upon the Sun, is seen. And in order to see very distinct and sharply delimited spots, it is best to darken the room, shuttering

20. See p. 104, above.

21. Literally, a monk of Monte Cassino; around 530 Saint Benedict established the famous Abbey of Monte Cassino.

22. Galileo is probably referring to the *palmo Romano*, which is about 22 cm long. The total distance was thus about 90 to 110 cm.

23. See Hon and Goldstein, "*Symmetry* in Copernicus and Galileo," 10.

every window, so that no light enters it other than that which comes through the tube, or at least to make it as dim as possible and to fit a rather large piece of pasteboard to the tube such that it throws a shadow on the paper where one has to draw and prevents any sunlight from falling on it, except for what comes through the lenses of the tube. It should be noted next that the spots exit the tube inverted and located opposite to where they are on the Sun: that is, the spots on the right come out on the left side, and the higher ones lower, because the rays intersect each other inside the tube before they emerge from the concave glass. But because we draw them on a surface facing the Sun, when turning back toward the Sun, we hold the drawing up to our eyes, the side on which we drew no longer faces the Sun but is instead turned away from it, and therefore the parts of the drawing on the right-hand side are already in their proper place again, corresponding to the right side of the Sun, and the left ones on the left, such that one only has to invert the upper and lower ones. Therefore, turning the paper over and thus making the top the bottom, and looking through the paper while facing the light, one observes the spots as they should be, as if we were looking directly at the Sun. And in this appearance they must be traced and inscribed on another sheet in order to have them correctly positioned.

I have then recognized the kindness of Nature, which thousands and thousands of years ago put in place the means of having some knowledge of these spots and through them, certain great consequences. For without other instruments, the image of the Sun and the spots is carried over great distances through each small aperture traversed by the solar rays, and imprinted on any surface held up to it. It is true that [such images] are by no means as well defined as those of the telescope; nevertheless, the larger ones are perceived distinctly enough. And when in a church Your Lordship sees the light of the Sun fall on the floor through some faraway broken pane of glass, hasten there with a large unfolded sheet of white paper, because you will discern the spots on it.[24] But I say further that Nature was likewise so benevolent that for our instruction she has from time to time speckled the Sun with a spot so large and dark that it was

24. Many of Galileo's Florentine readers would have recognized this hypothetical remark as an actual description of that city's Dominican church, Santa Maria Novella. In the 1570s, Egnazio Danti was allowed to make one small hole in a large circular window in the upper façade of that church, and later a hole in that same wall, in order to use the projected solar images to measure the length of the tropical year. See J. L. Heilbron, *The Sun in the Church: Cathedrals as Solar Observatories* (Cambridge, MA: Harvard University Press, 1999), 68; Simone Bartolini, *I fori gnomonici di Egnazio Danti in Santa Maria Novella* (Florence: Edizioni Polistampa, 2006); Thomas B. Settle, "Egnazio Danti: le meridiane in Santa Maria Novella a Firenze e gli strumenti collegati," in *Atti del Convegno Il Sole nella Chiesa: Cassini e le grandi meridiane come strumenti di indagine scientifica: Bologna, Archiginnasio, 22–23 settembre 2005*, ed. Fabrizio Bònoli, Gianluigi Parmeggiani, and Francesco Poppi, *Giornale di astronomia* 32 (2006), 91–98.

seen with the naked eye alone by a great number of people. But a false and inveterate idea, that the celestial bodies were exempt from all alterations and mutations, made them believe that such a spot was Mercury interposed between the Sun and us, and [this happened] not without shame of the astronomers of that age. Without any doubt, such was the one mentioned in the *Annals and Histories of the Franks [taken] from the Library of the Jurisconsult Pierre Pithou*, printed in Paris in 1588, where, in the *Life of Charlemagne*,[25] on folio 62, one reads that a black spot whose entrance and exit could not be observed because of the impediment of the clouds was seen by the people of France for eight straight days on the solar disc, and it was believed to be Mercury, at that time in conjunction with the Sun. But this is too great an error because Mercury cannot remain in conjunction with the Sun even for the space of seven hours: such is its motion when it comes between us and the Sun. Therefore, this phenomenon was absolutely one of the largest and darkest spots, and similar ones can also be encountered in the future, and perhaps, if careful observations are undertaken, we will be able to see some shortly. If this discovery [of sunspots] had been made some years earlier, it would have relieved Kepler of the task of interpreting and salvaging this passage with alterations of the text and other emendations of time. I will not at present trouble myself about the matter, for I am certain that this author, as a true philosopher and as one hardly opposed to evident matters, will no sooner hear of these observations and discourses of mine than he will give his total assent to them.

Now, in order to reap some fruit from the unexpected wonders that have remained hidden until our age, it will be well that in the future we go back to lending an ear to those wise philosophers who judged differently from Aristotle about the celestial substance, and from whom even Aristotle would not have distanced himself had he had knowledge of the present sensory observations, for he not only admitted manifest [sensory] experiences as one means of drawing conclusions about natural questions, but he even gave them pride of place. Hence, if he argued for the immutability of the heavens because in times past no alteration whatsoever had been seen in them, it is entirely credible that if vision had demonstrated to him the things that it makes manifest to us, he would have arrived at the opposite opinion, [the one] to which we are led by such wonderful discoveries. And I will further say that I think that I contradict Aristotle's doctrine much less—these observations being truthful ones—with the supposition of mutable celestial material, than do those who would prefer to treat it as inalterable, because I am sure that he was never as certain of the conclusion of inalterability as he was of the notion that all human discourse

25. See pp. 11–13, 19–20, of the introduction.

must defer to evident experience. And therefore one will philosophize better by giving assent to those conclusions depending on manifest observations than by persisting in opinions contrary to sense perception itself, and confirmed only by probable or apparent reasons. It is not difficult to understand, then, of what kind and how many are the observed phenomena that invite us to more certain conclusions. Behold, someone[26] was inspired by a greater power with the means necessary to free us from all doubt, from which it was learned that the generation of comets occurs in the celestial region. The majority of those who instruct others resist this person, who like a witness, quickly passes through and is gone. New flames of very long duration are then sent to us, in the form of very bright stars,[27] ones that are produced and then disappear in the most remote parts of the heaven, but neither does this suffice to sway those whose minds are untouched by the necessity of geometrical demonstrations. Finally, it is detected in that part of the heaven that rightly must be considered the cleanest and purest, I mean upon the face of the Sun itself, that an innumerable multitude of dark, dense, and cloudy matter is constantly produced and quickly dissolved. Here we have a series of productions and disintegrations, one that will not draw to an end any time soon[28] but, lasting through all the future ages, will give human minds as much time to observe as they desire, and to learn the doctrines that will render them certain about their own location. However, in this, too, we must recognize divine grace, for the means that suffice for such understanding are very easy and quickly learned, and he who is not capable of anything more, let him arrange to have drawings made in far-flung regions and let him compare them with the ones made by himself on the same days, because he will find that they agree entirely with his own. And I have just received some made in Brussels by Signor Daniello Antonini[29] on 11, 12, 13, 14, 20, and 21 July, which fit exactly with mine and with others sent to me from Rome by Signor Lodovico Cigoli, a very famous painter and architect, an argument that alone should suffice to persuade everyone that these spots are a great distance above the Moon.

And with this I want to stop occupying Your Lordship further. Please do me the favor of sending the drawings to Apelles at your convenience, accompanied

26. Tycho Brahe, whose judgment about the superlunary location of comets was well known. See *De Mundi Aetherei Recentioribus Phaenomenis* (1588) and *Astronomiae Instauratae Progymnasmata* (1602), in *Tychonis Brahe Dani Opera Omnia*, vols. 1–2, 4.

27. The novae of 1572 and 1604.

28. In fact, sunspots virtually disappeared by the middle of the seventeenth century. See John Eddy, "The Case of the Missing Sunspots," *Scientific American*, 236, no. 5 (1977): 80–95.

29. Daniello Antonini (1588–1616) studied mathematical subjects under Galileo at the University of Padua. He became a professional soldier, and in 1612 was stationed in Flanders.

by my singular affection toward his person.[30] And I reverently kiss Your Lordship's hand and I pray to God for your happiness.

From Florence, 14 August 1612
Your Most Illustrious Lordship's
Most devoted servant
Galileo Galilei

P.S. In conformity with what I had supposed and written, six days later the effect followed. For the 19th, 20th, and 21st days of the present month I and many other gentleman friends of mine saw, with the naked eye, a dark spot near the middle of the solar disc at sunrise, one which was the largest among the many others visible with the telescope, and I am sending Your Lordship the drawings of this spot as well.

30. *Persona*, or mask: a pun on the disguise assumed by the *finto Apelle*.

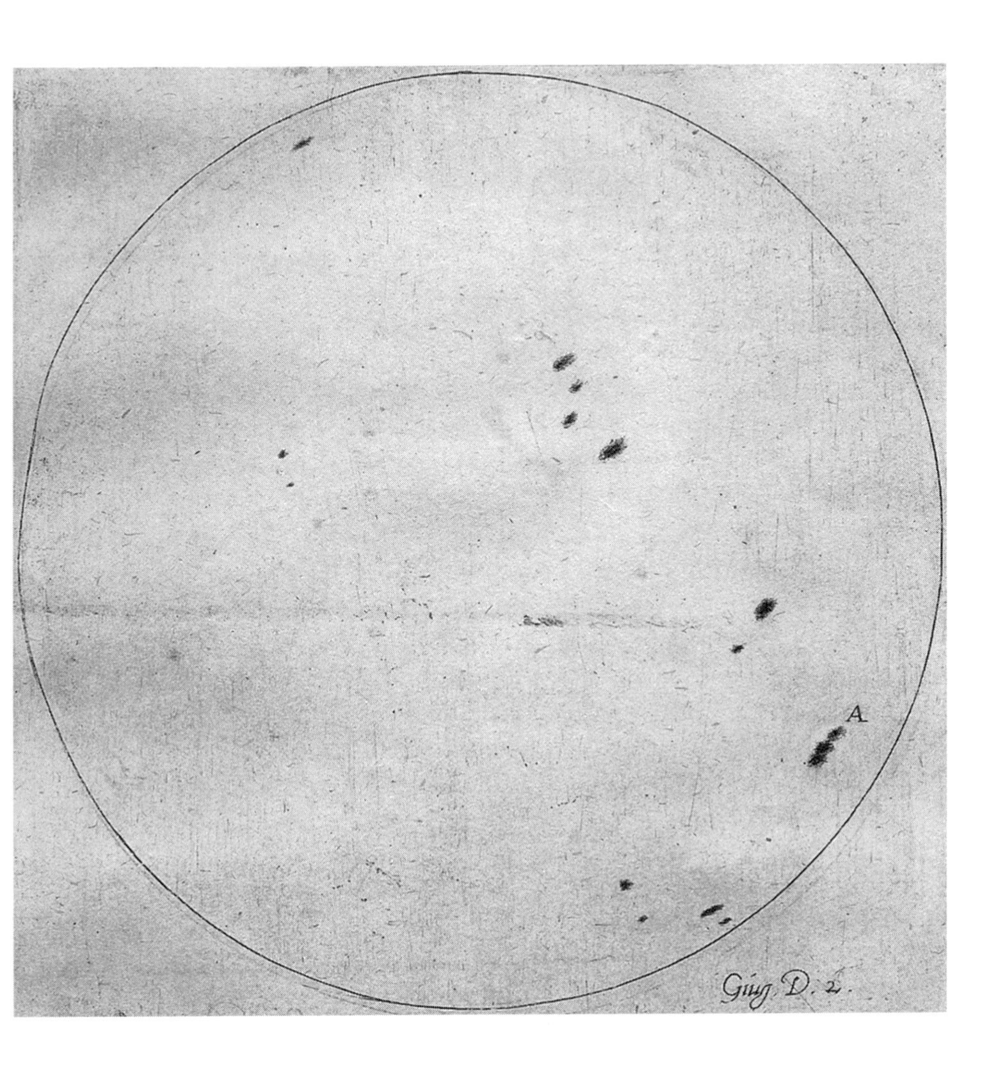
A
Giug. D. 2.

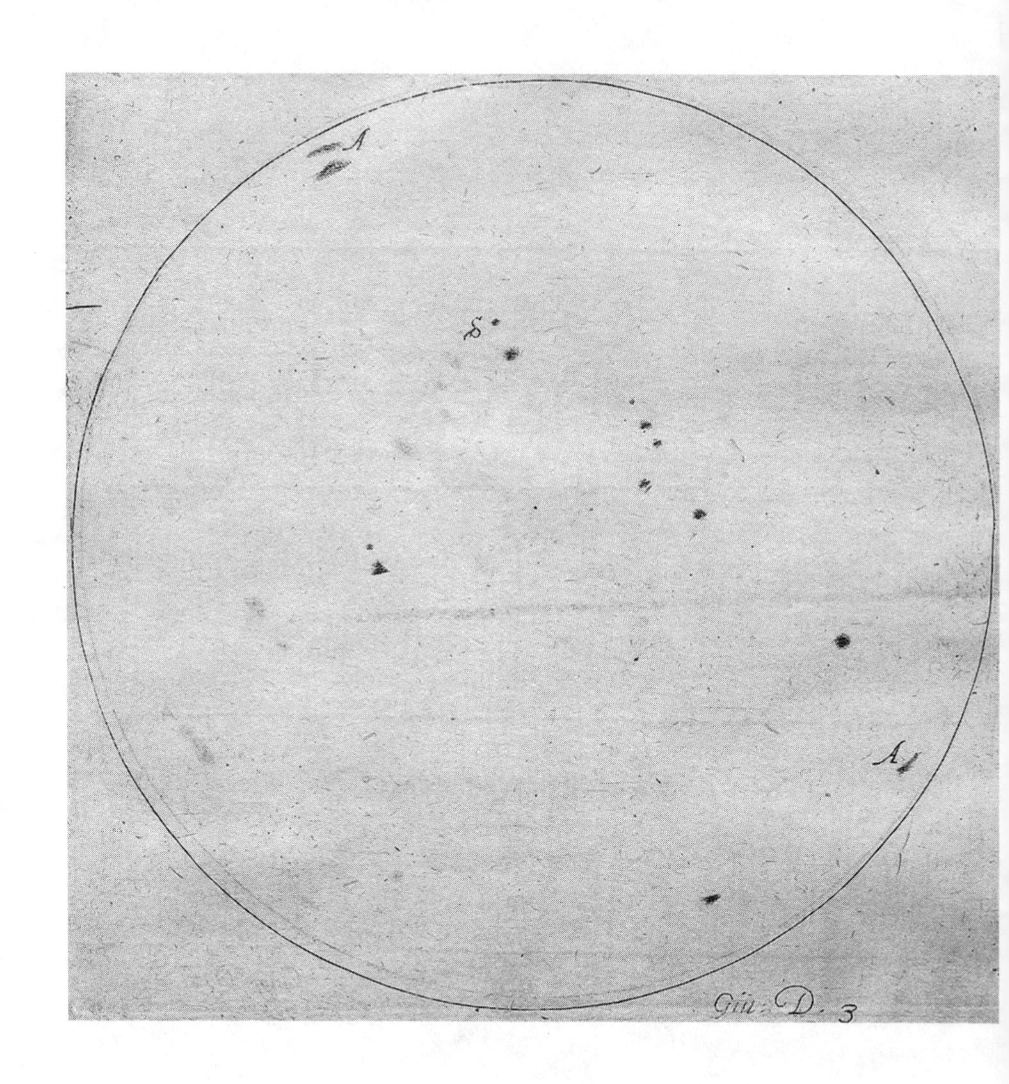
A
S
A
Giu: D. 3

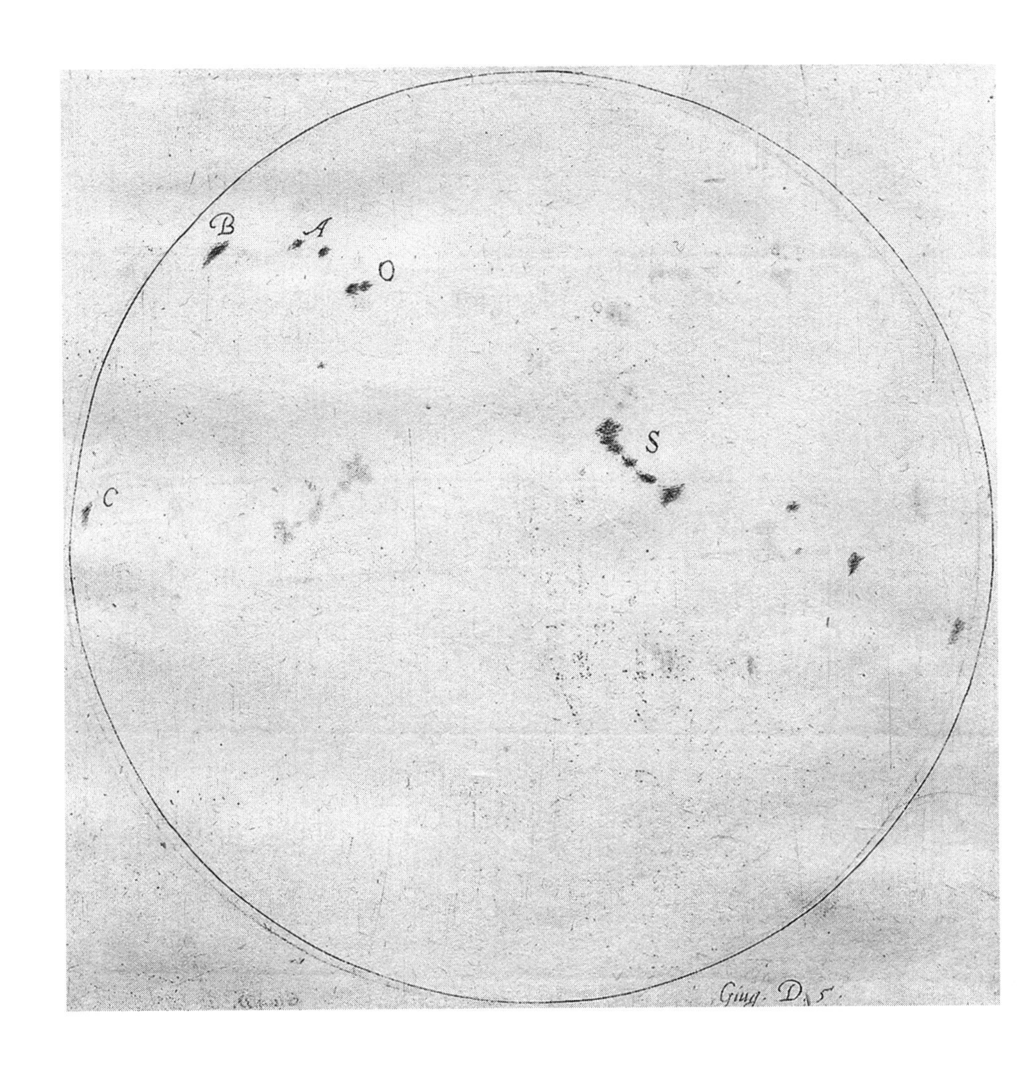
B
A
O
S
C

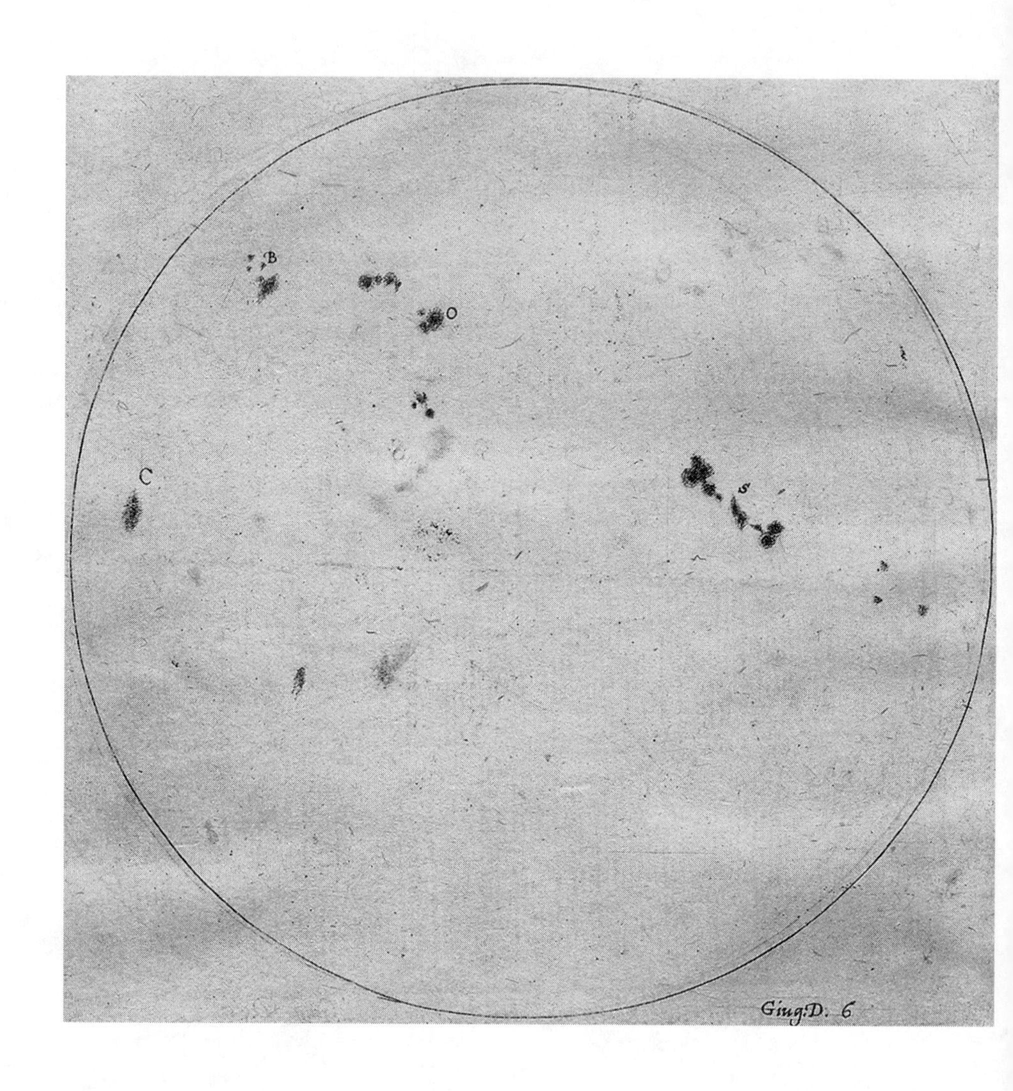
B
O
C
S
Ging:D. 6

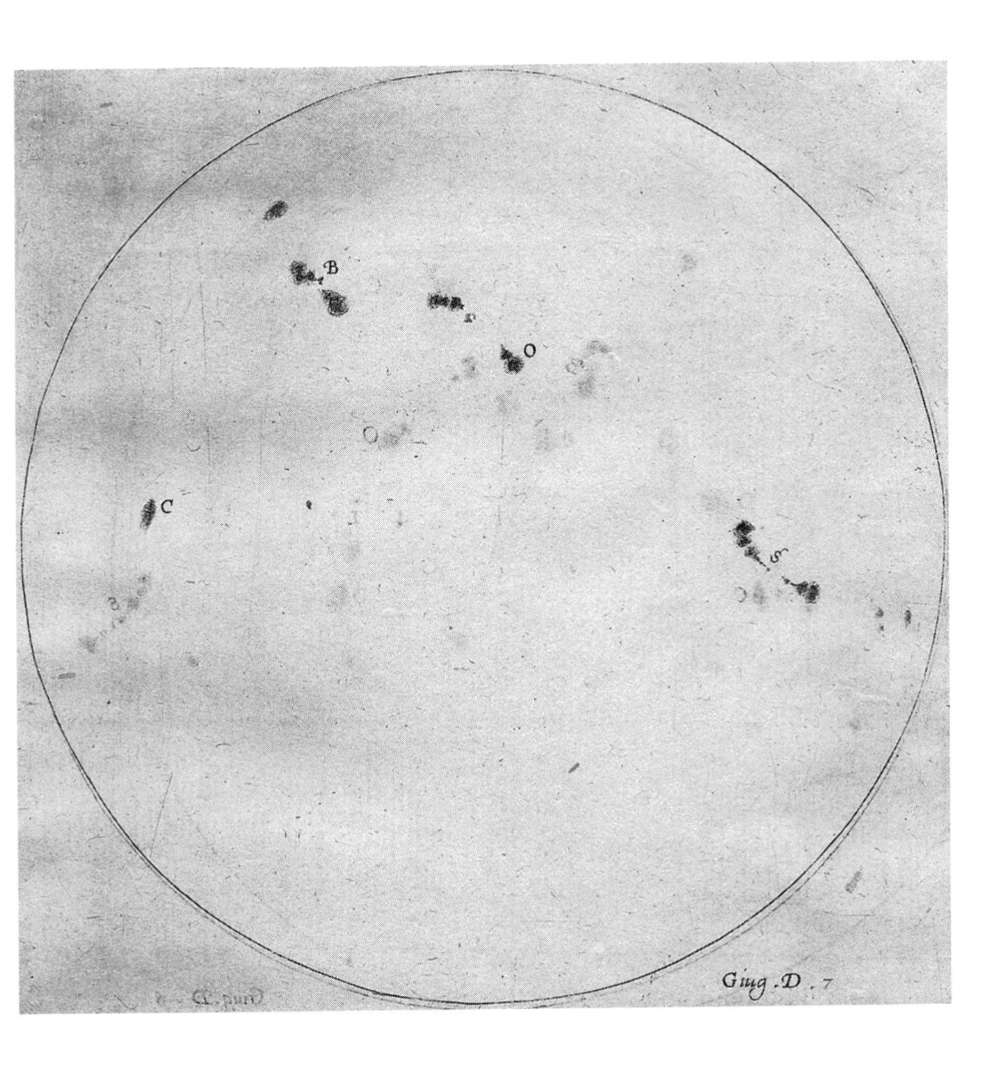
B
O
C
Giug .D . 7

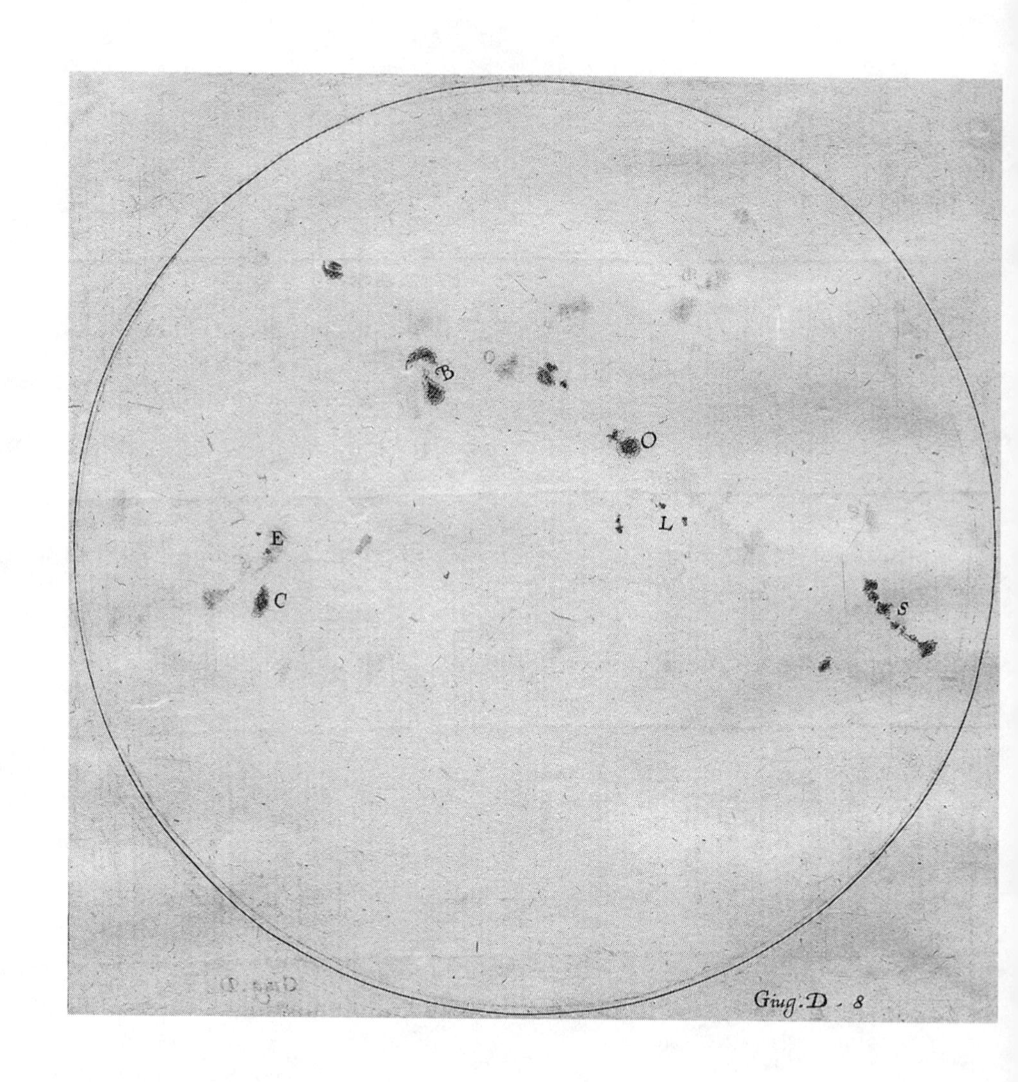
B
O
L
E
C
S
Giug. D - 8

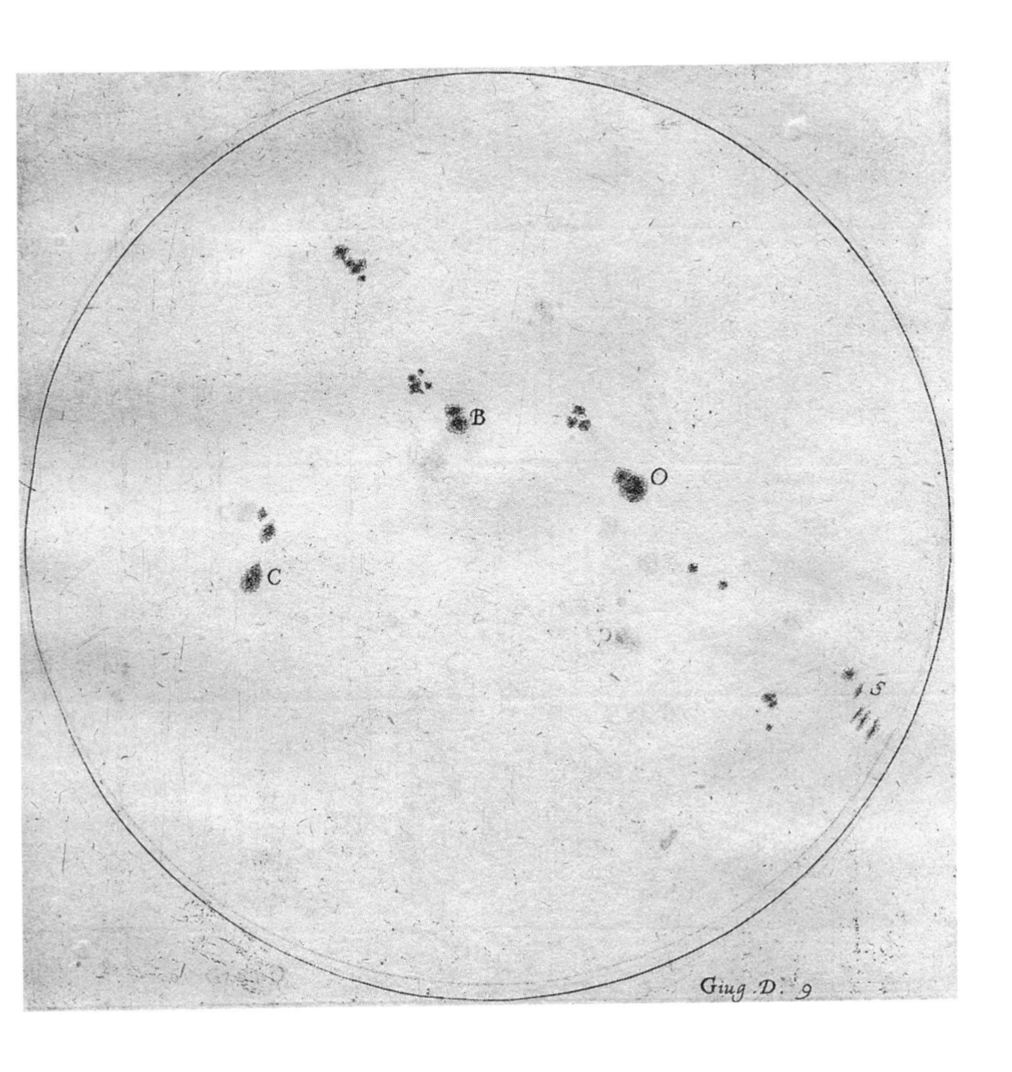
B
O
C
S
Giug. D. 9

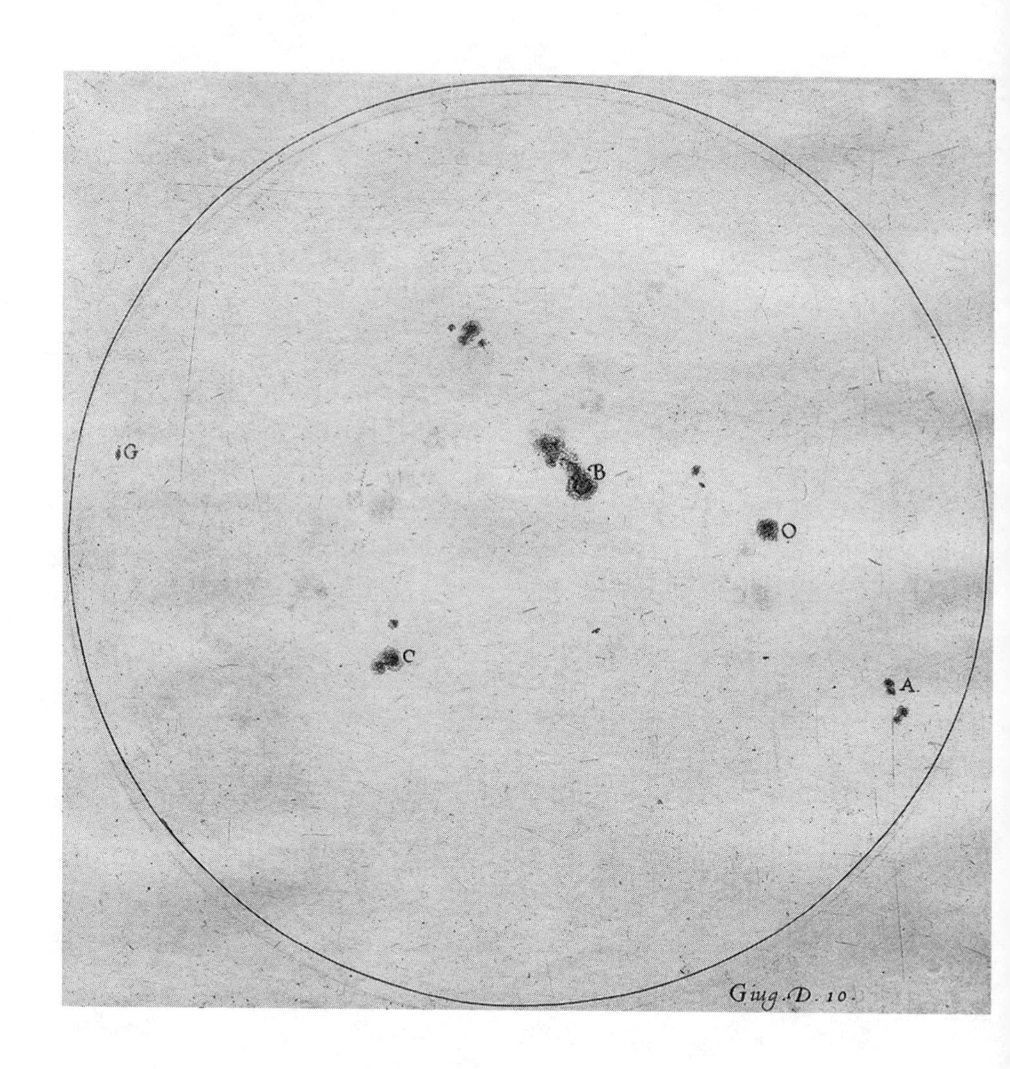
G
B
O
C
A.
Giug. D. 10.

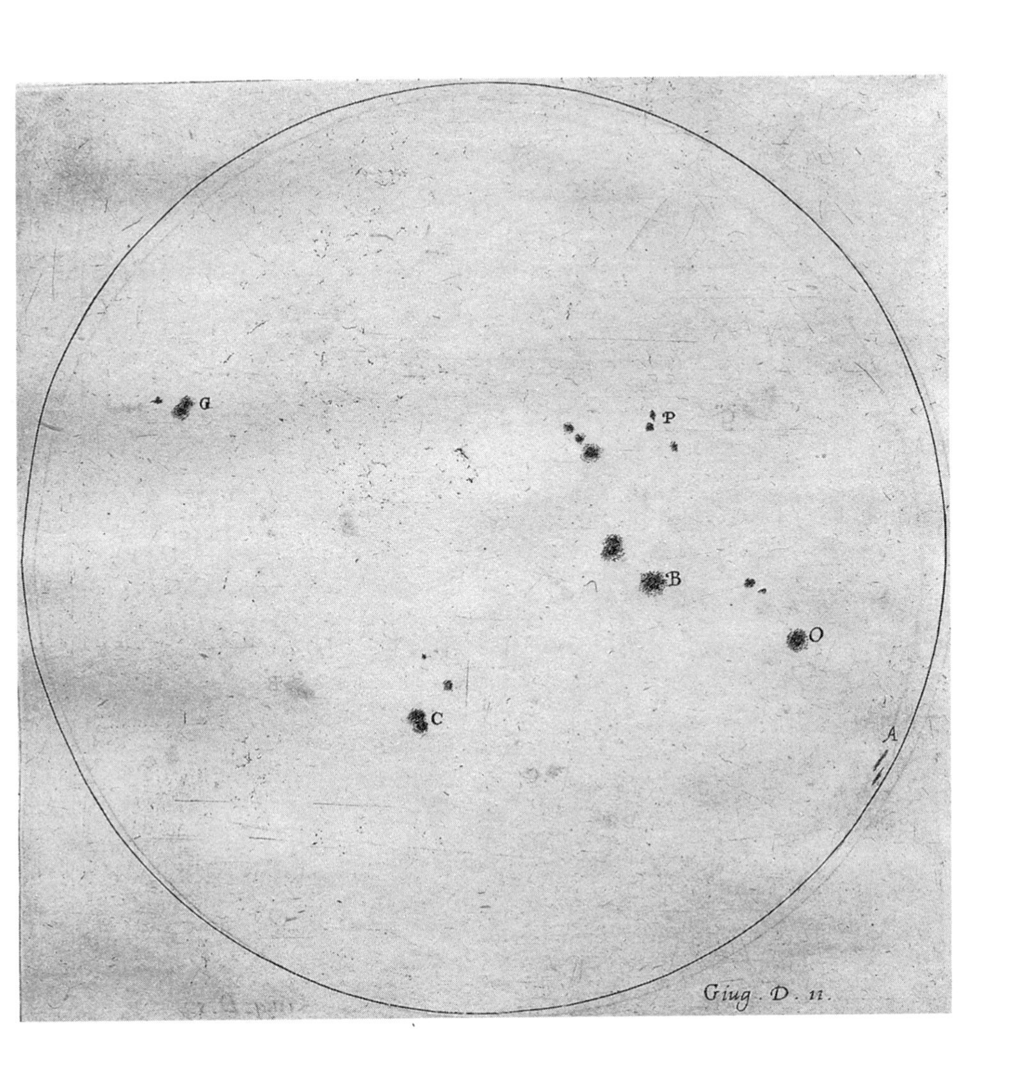
G
P
B
O
C
A
Giug. D. 11.

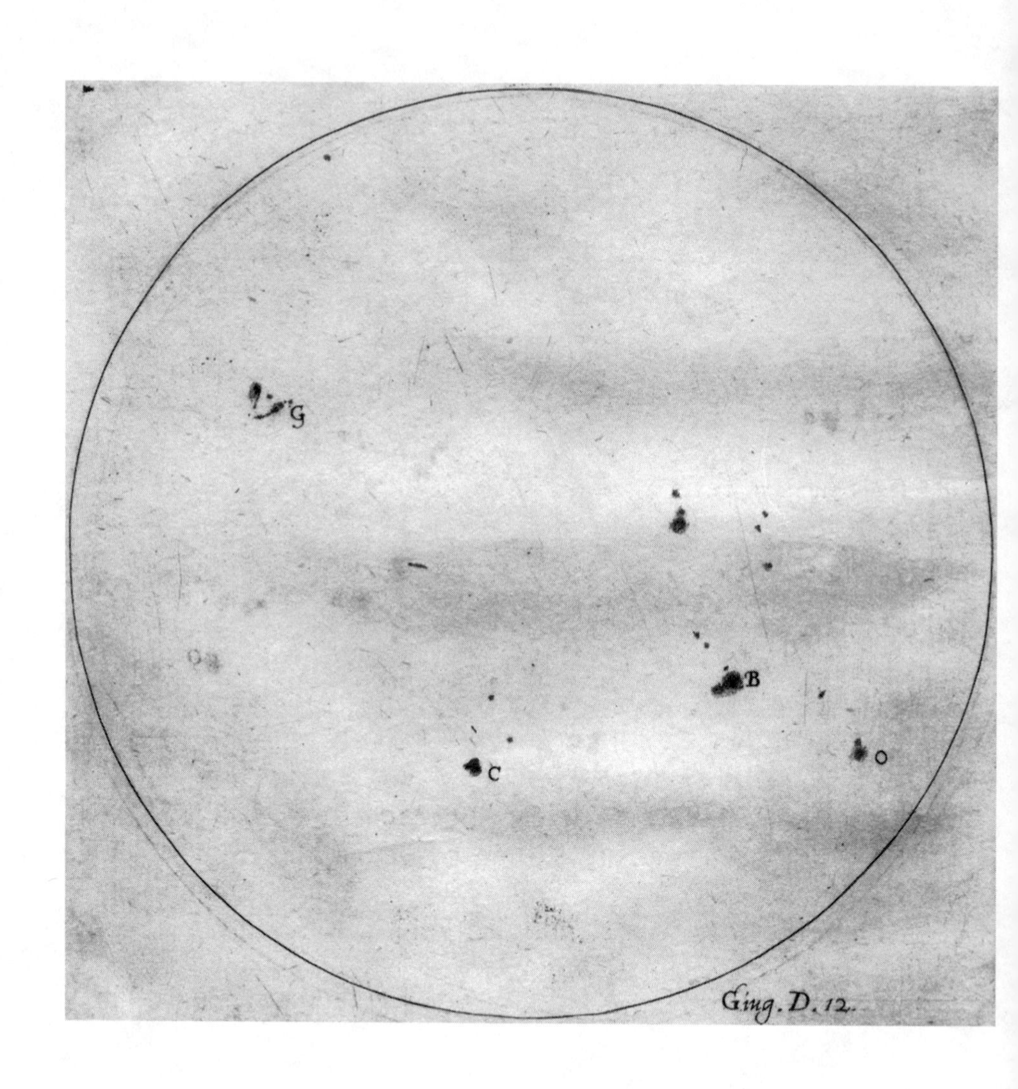
G
B
C
O
Ging. D. 12.

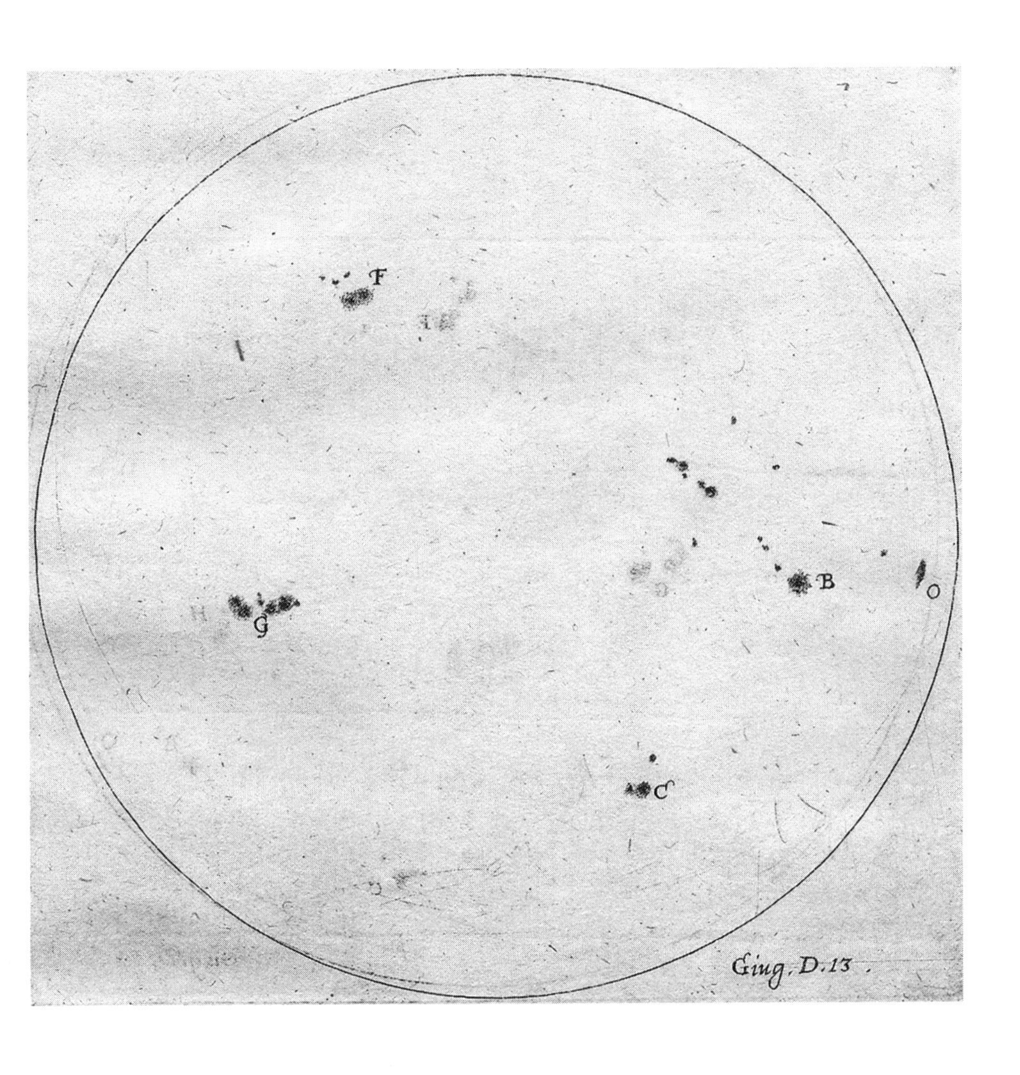
F
G
B
O
C
Giug. D.13.

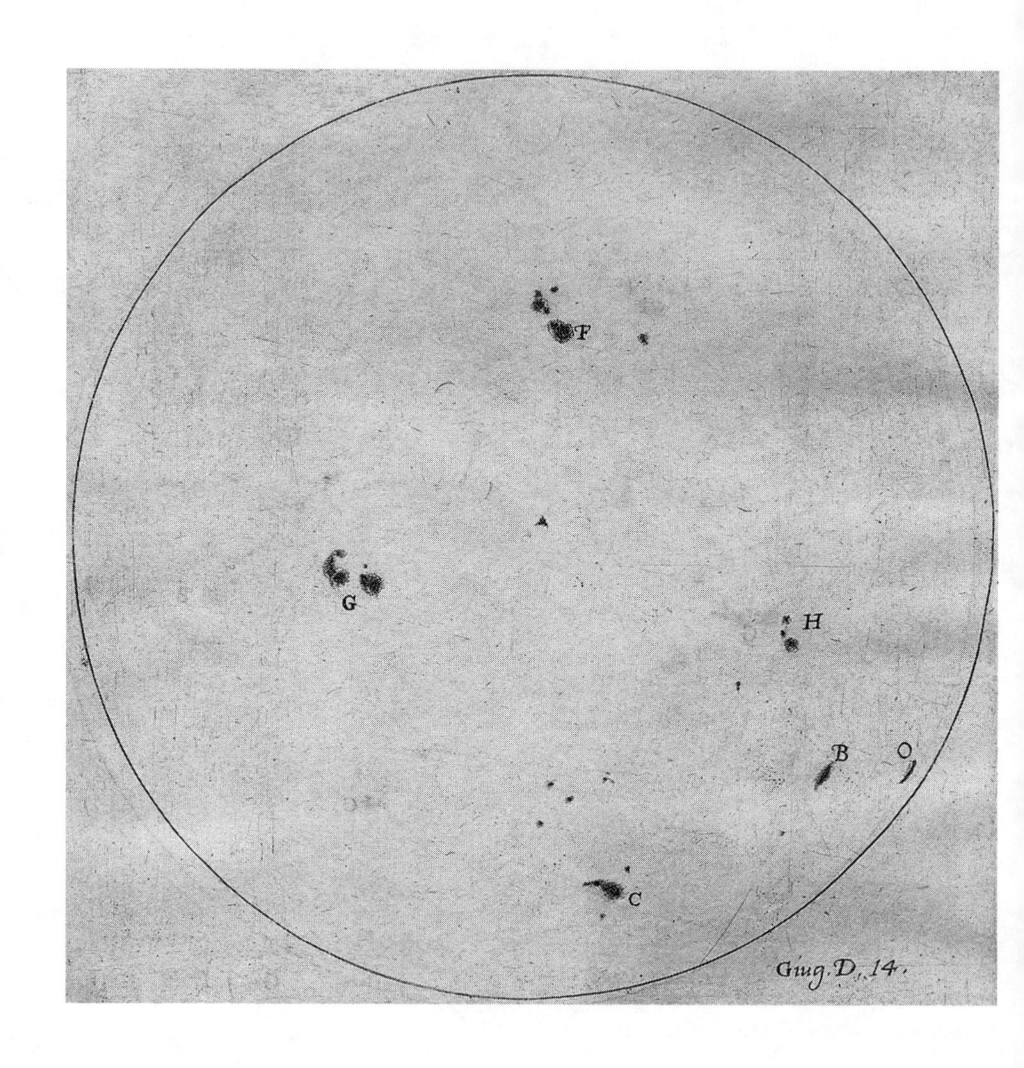
F
G
H
B
O
C
Giug. D., 14.

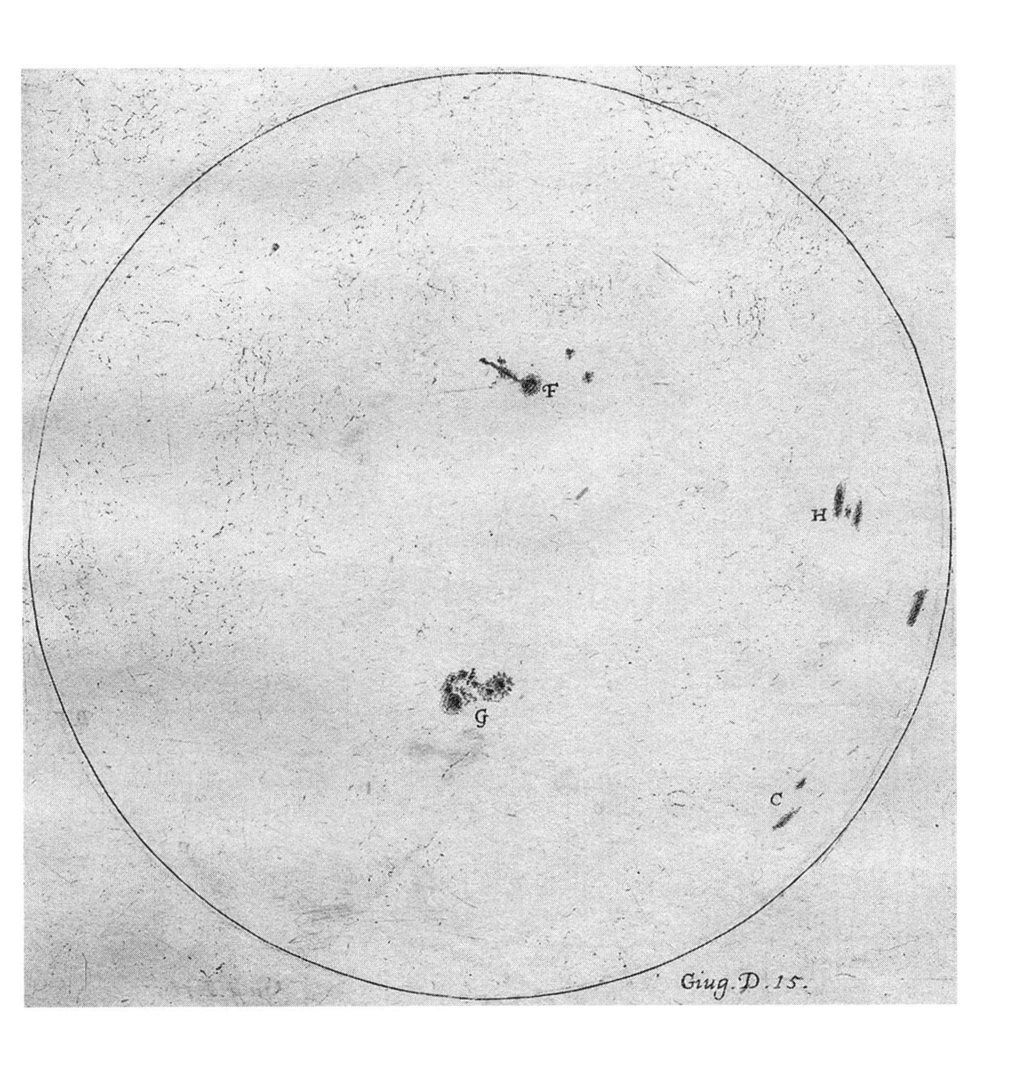
F
H
G
C
Giug. D. 15.

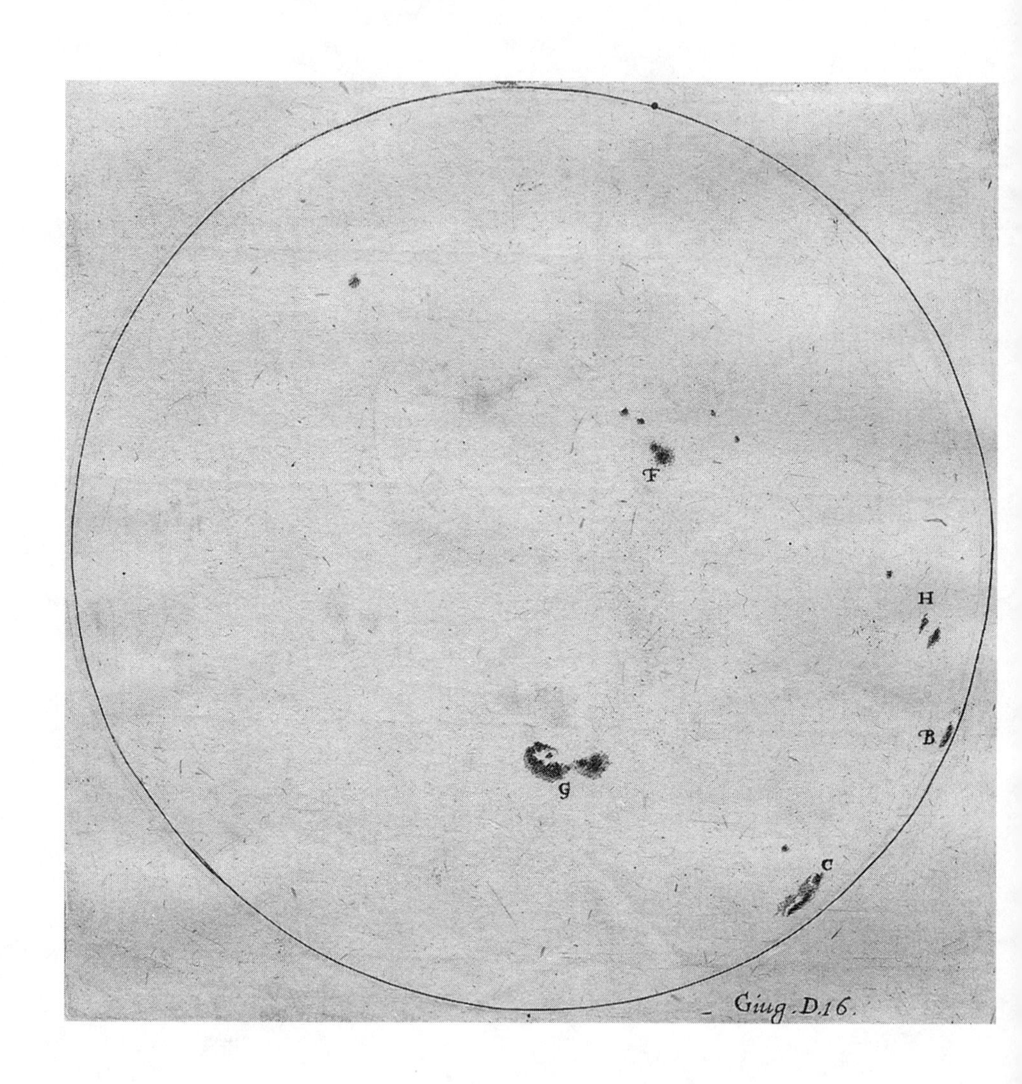
F
H
B
G
C
Giug. D.16.

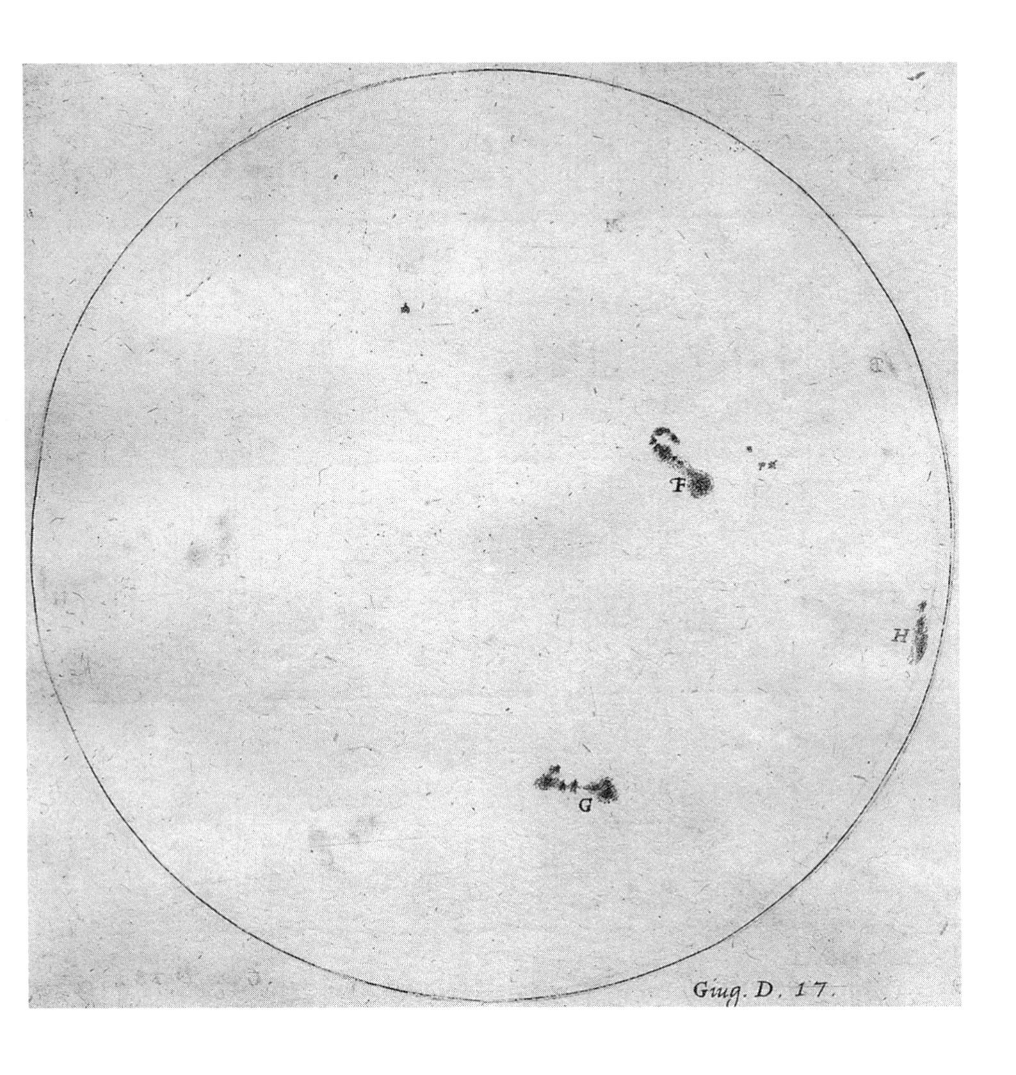
F
G
H
Giug. D. 17.

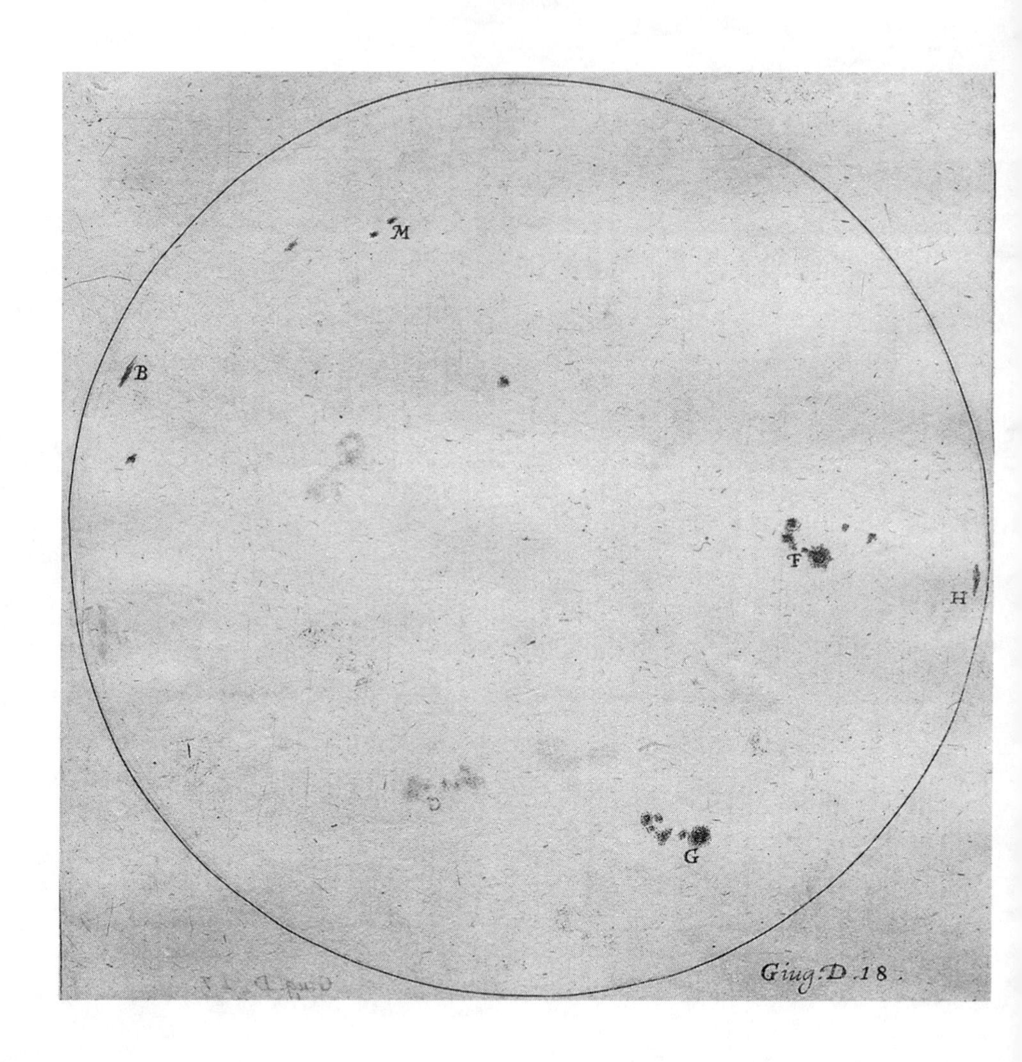
M
B
F
H
G
Giug: D. 18.

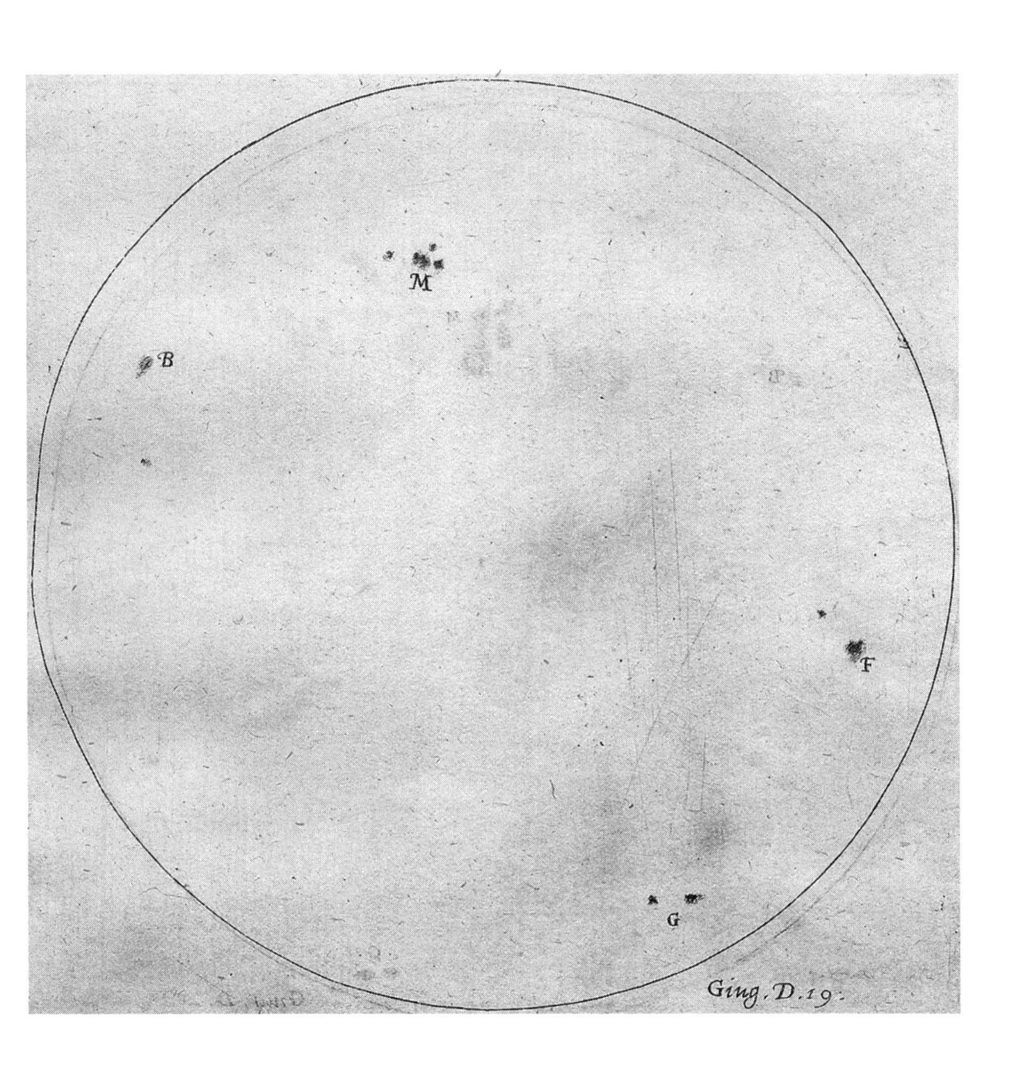
M
B
F
G
Ging. D.19.

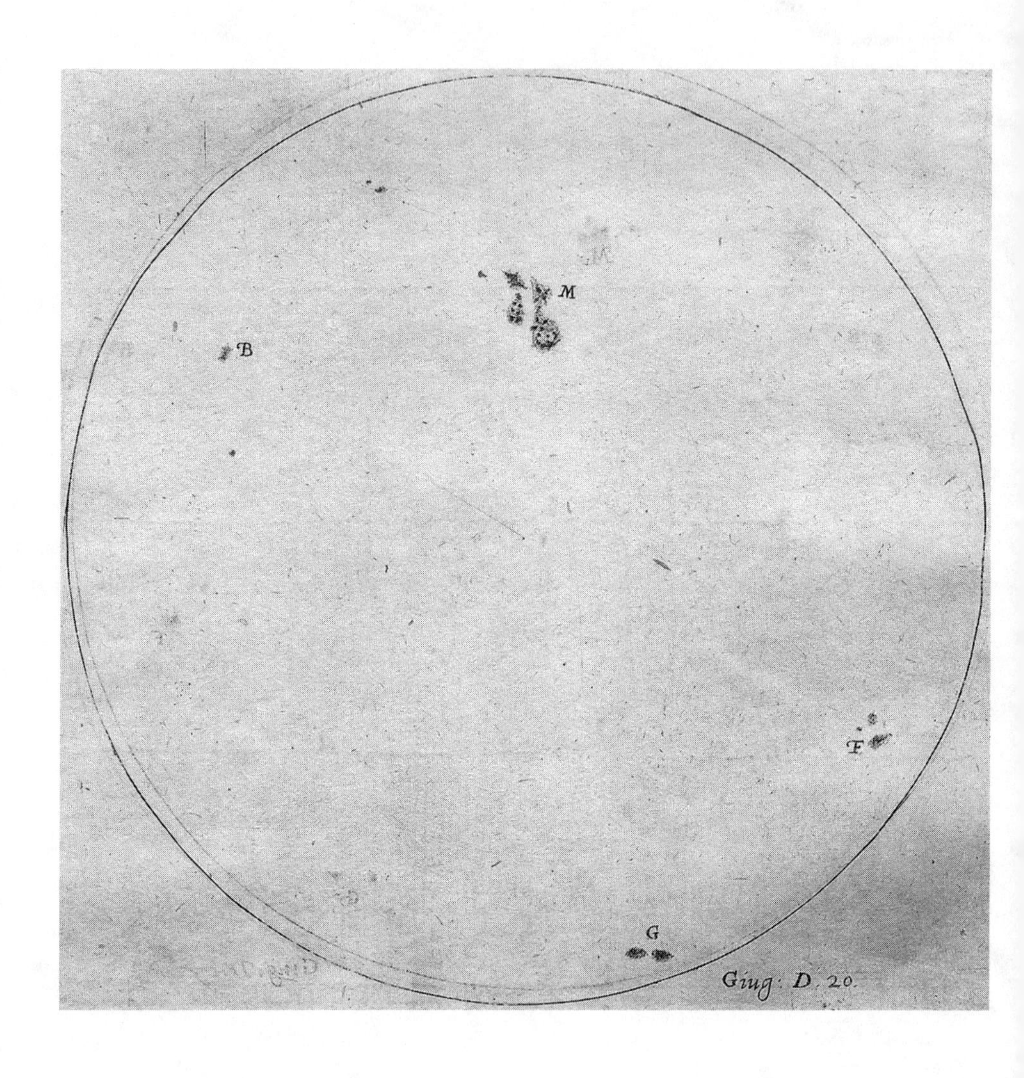
M
B
F
G
Giug: D. 20.

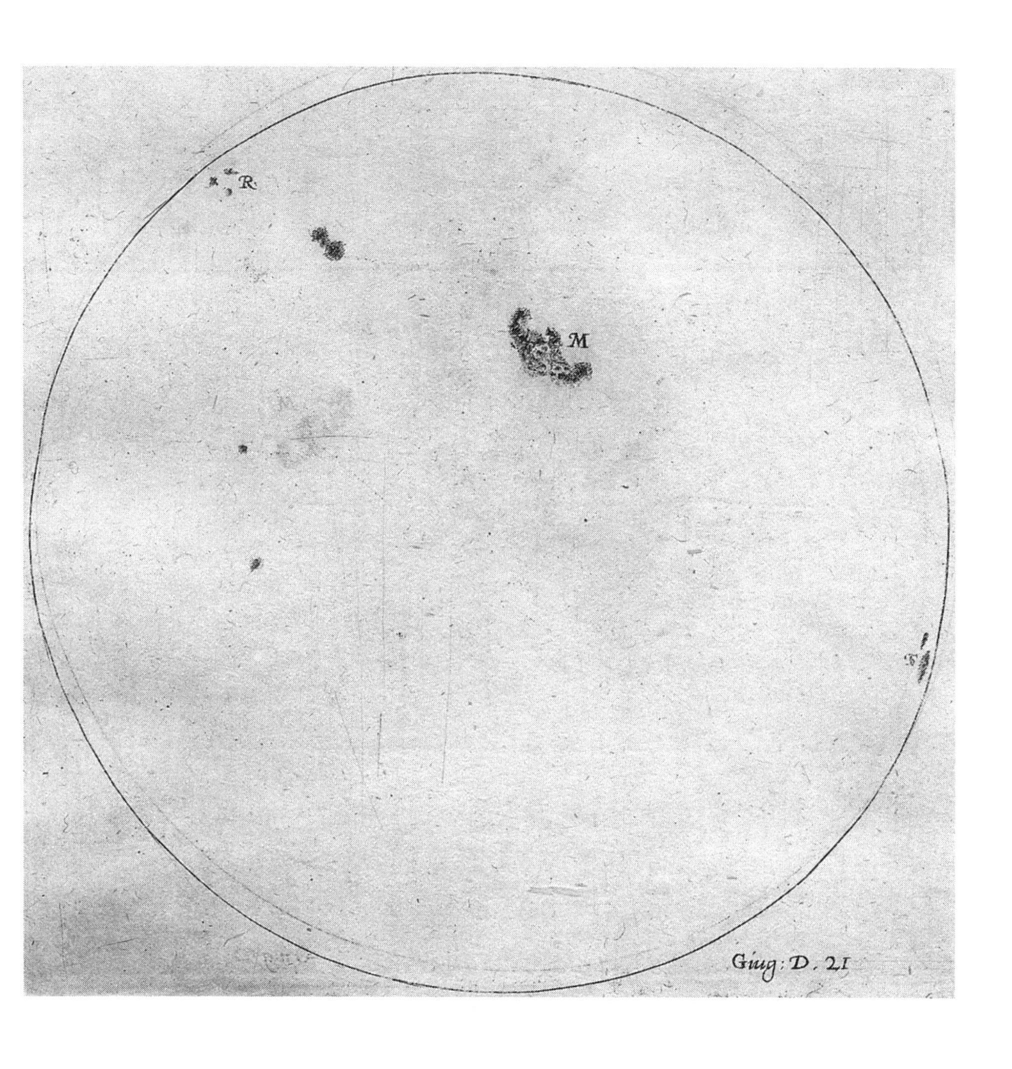
R.
M
Giug: D. 21

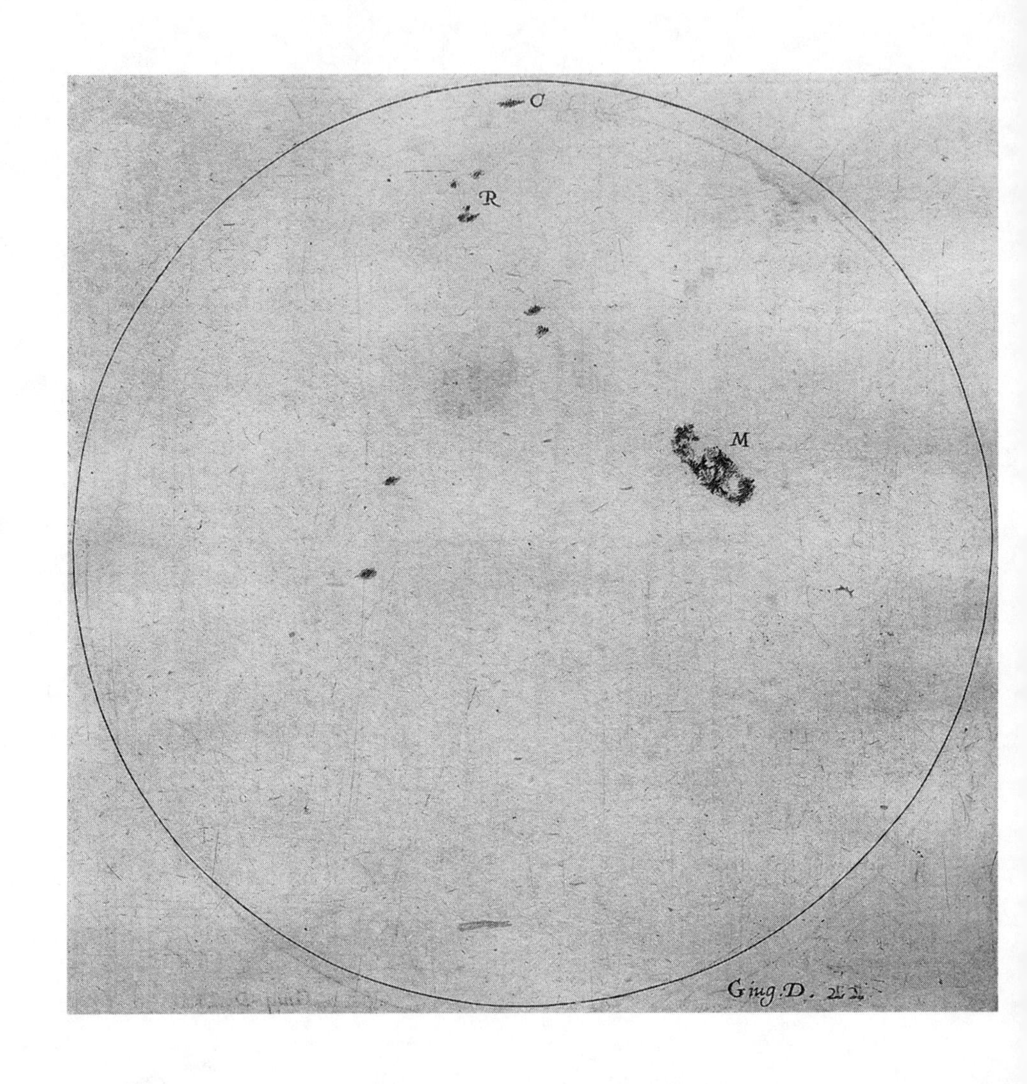
C
R
M

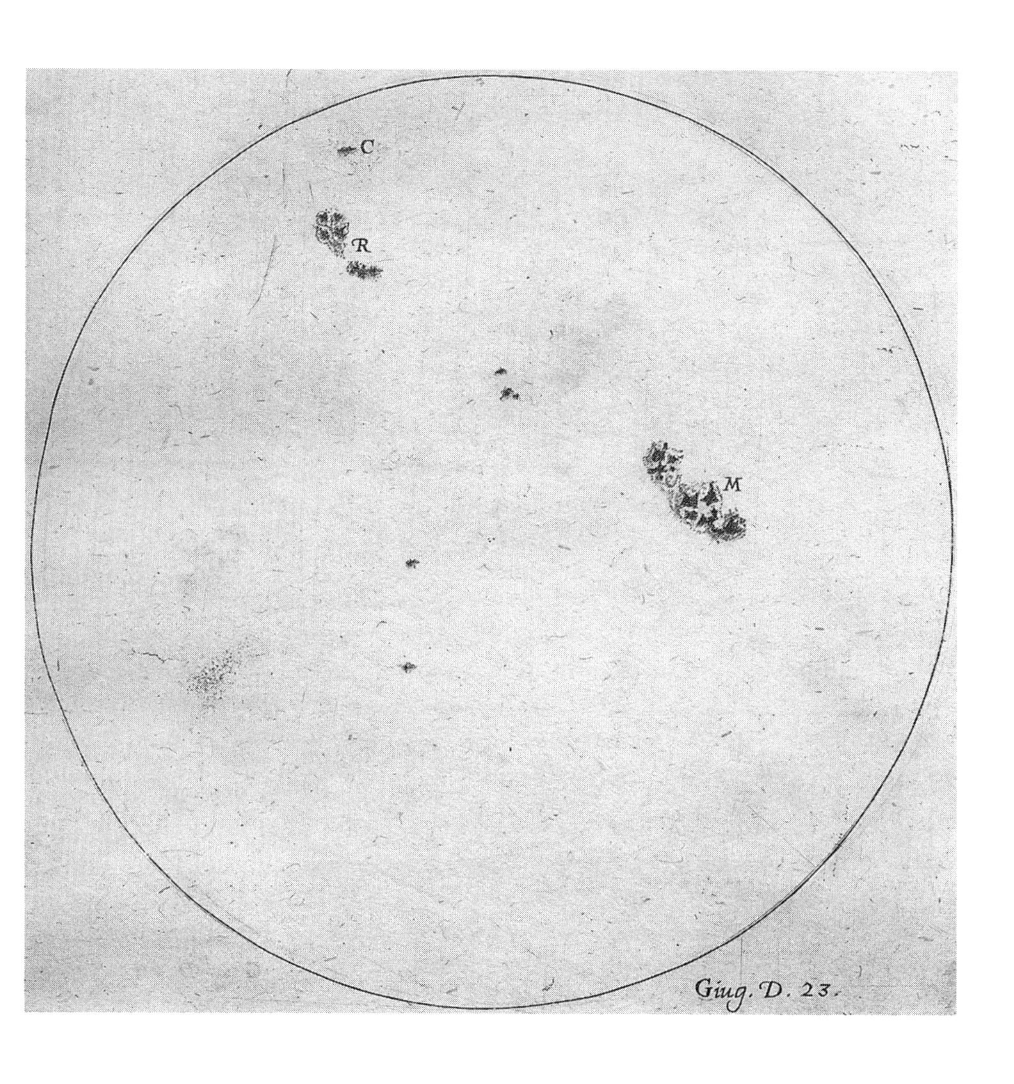
C
R
M
Giug. D. 23.

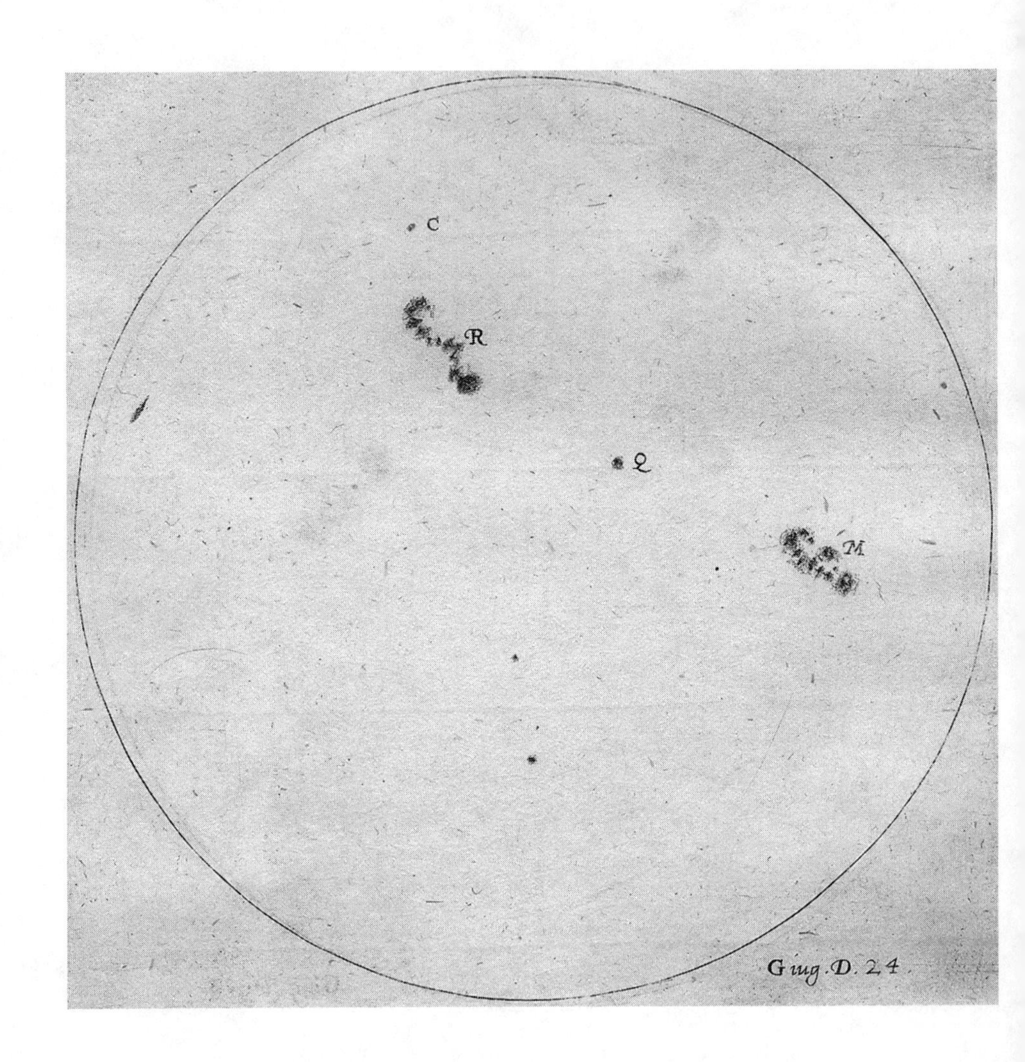
C
R
Q
M
Giug. D. 24

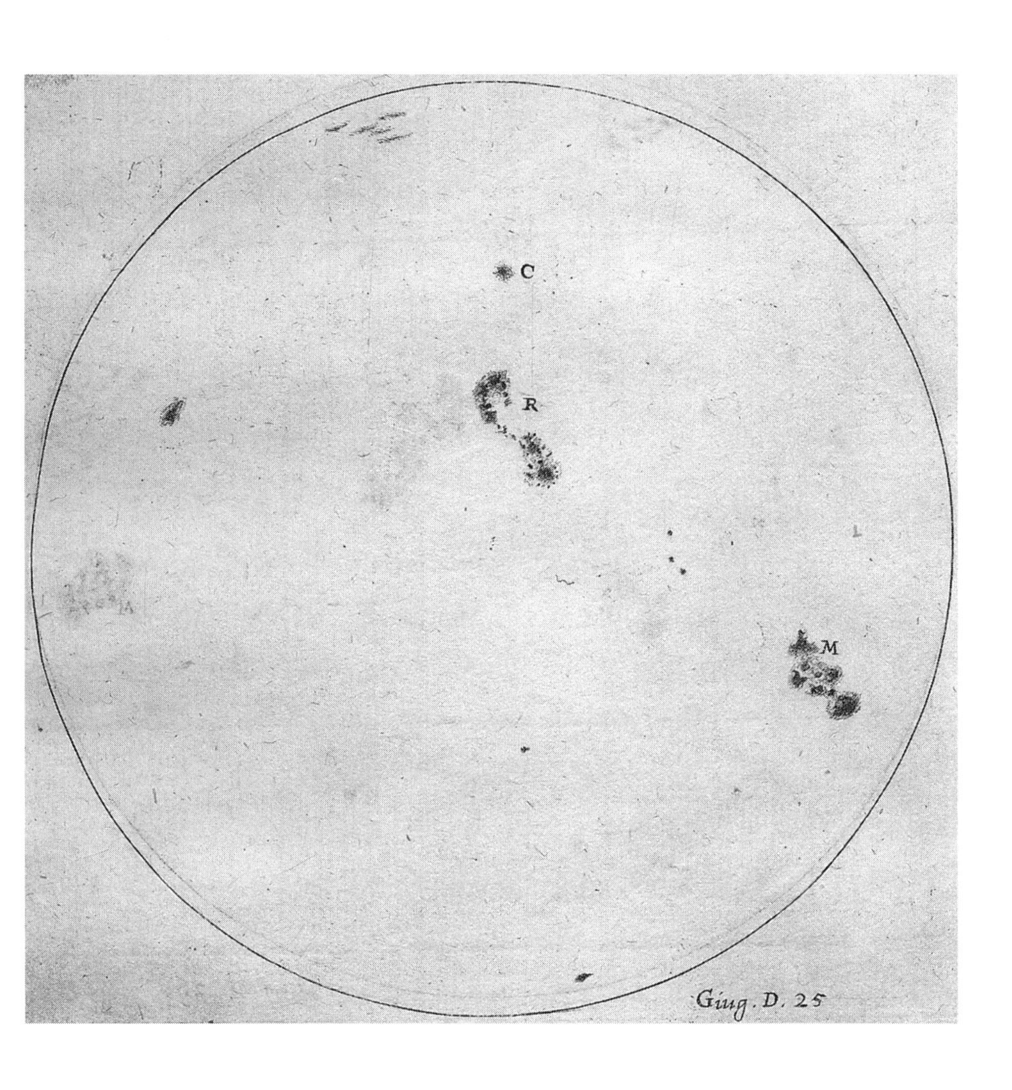
C
R
M
A
Giug. D. 25

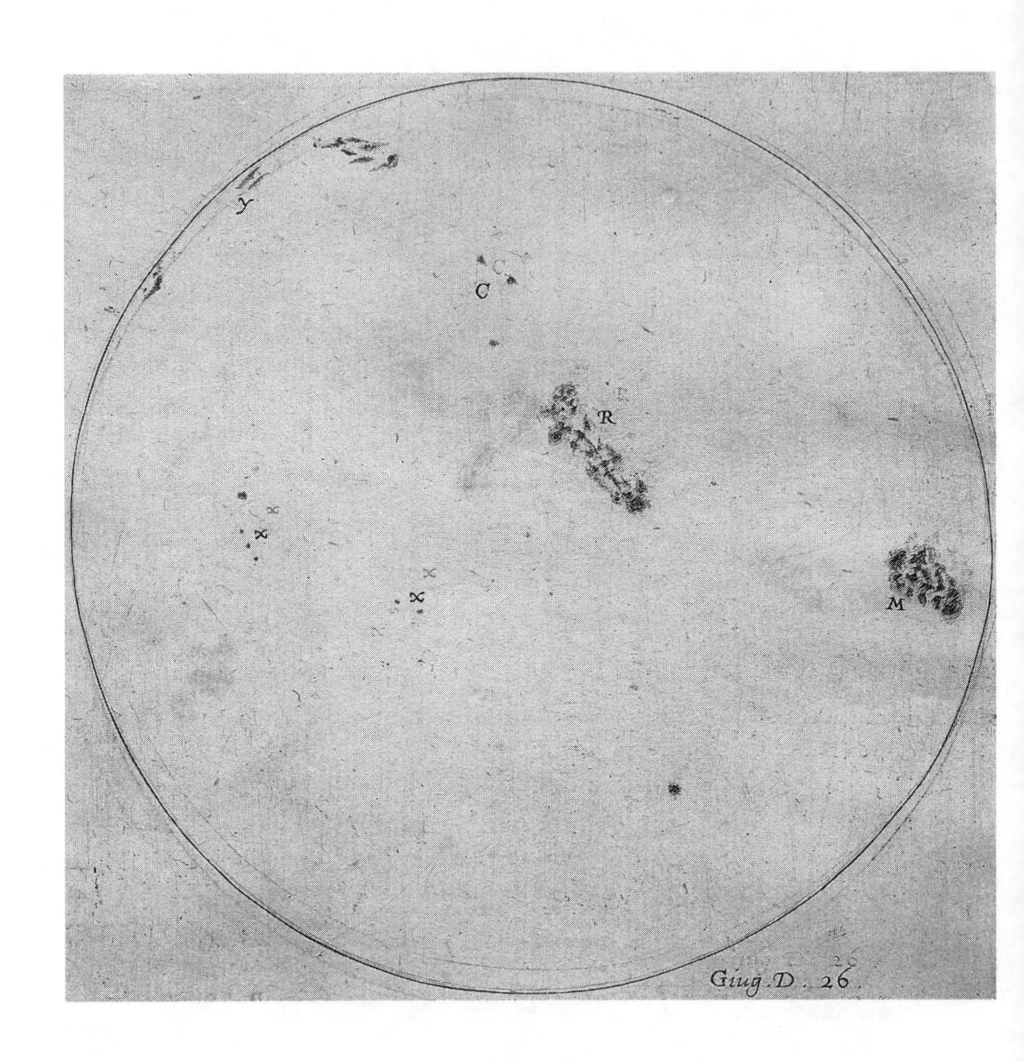
y
C
R
x
x
M
Giug. D. 26.

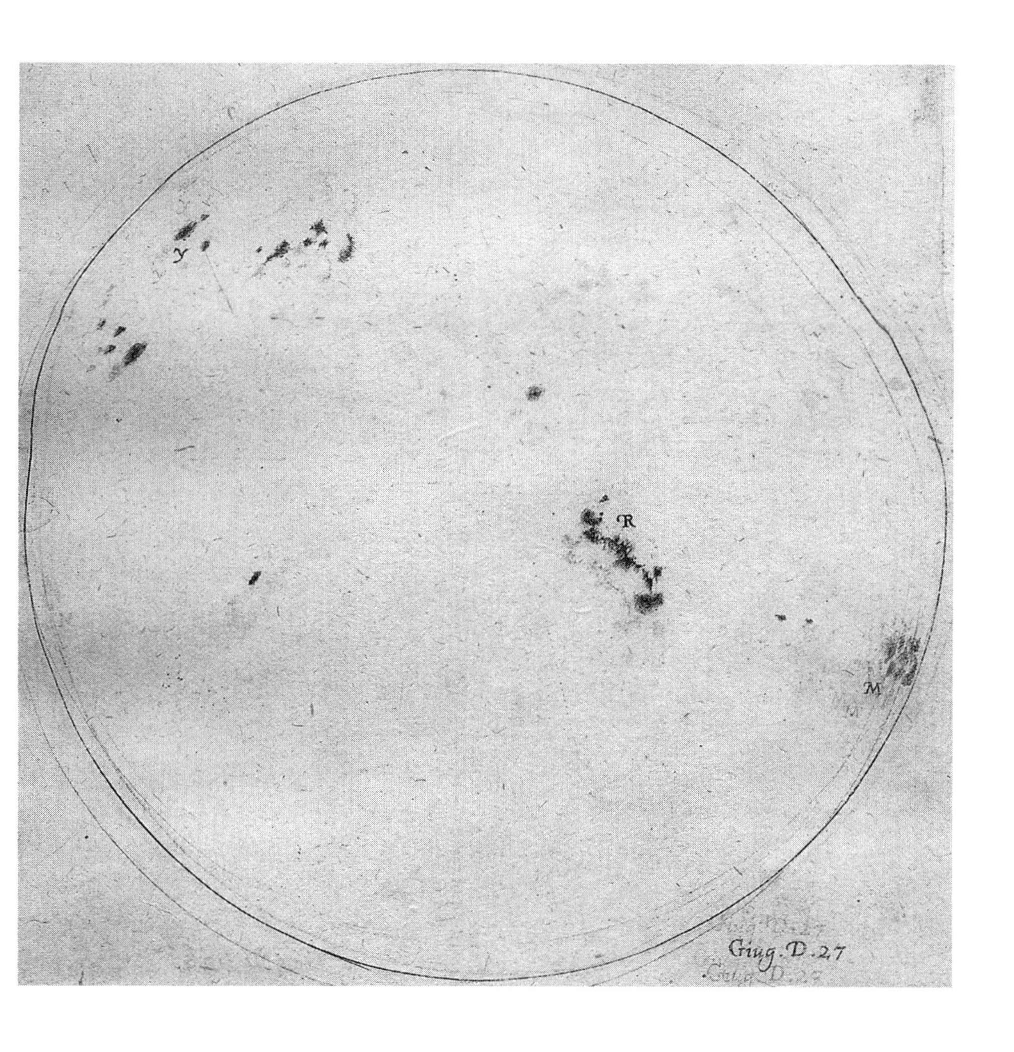
Y
R
M
Giug. D. 27

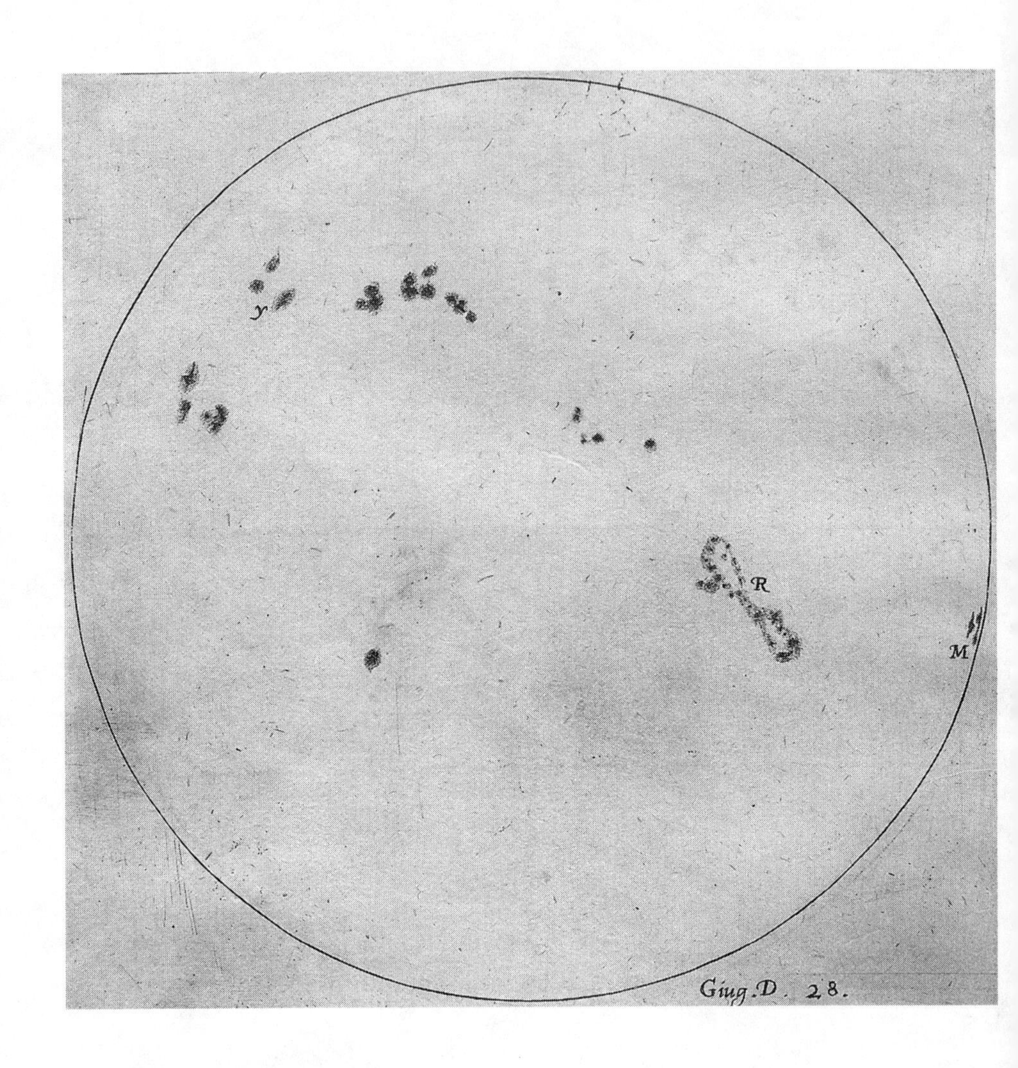
y
R
M
Giug. D. 28.

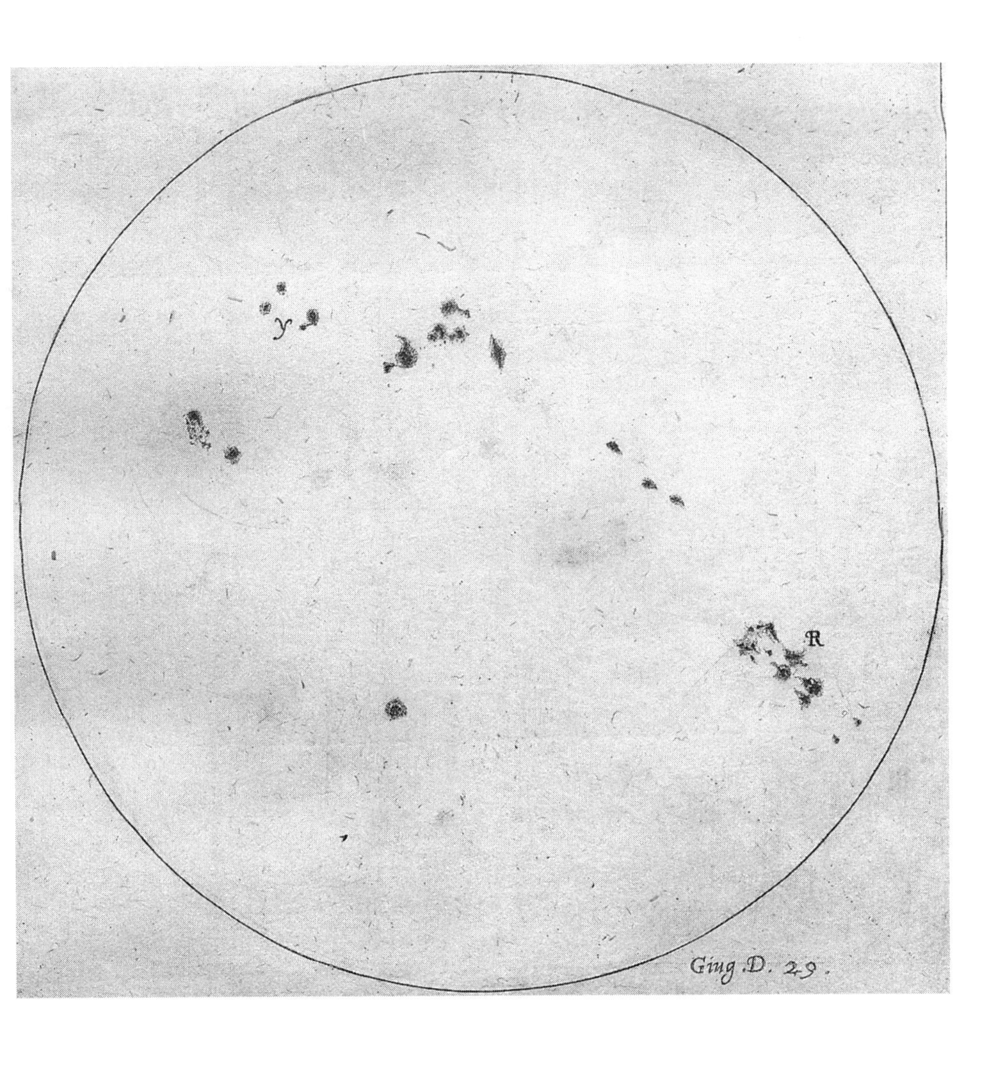
y
R
Giug. D. 29.

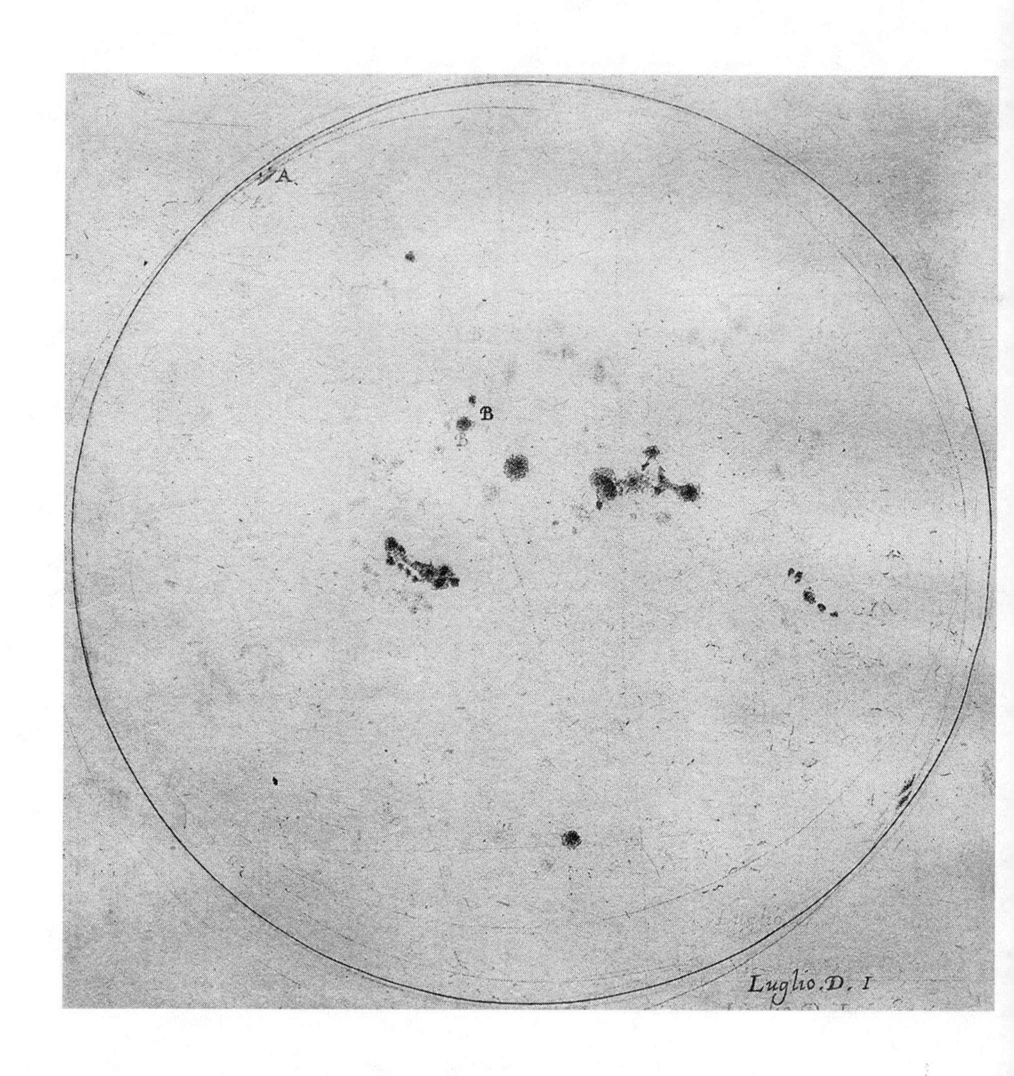
A
B
Luglio.D. I

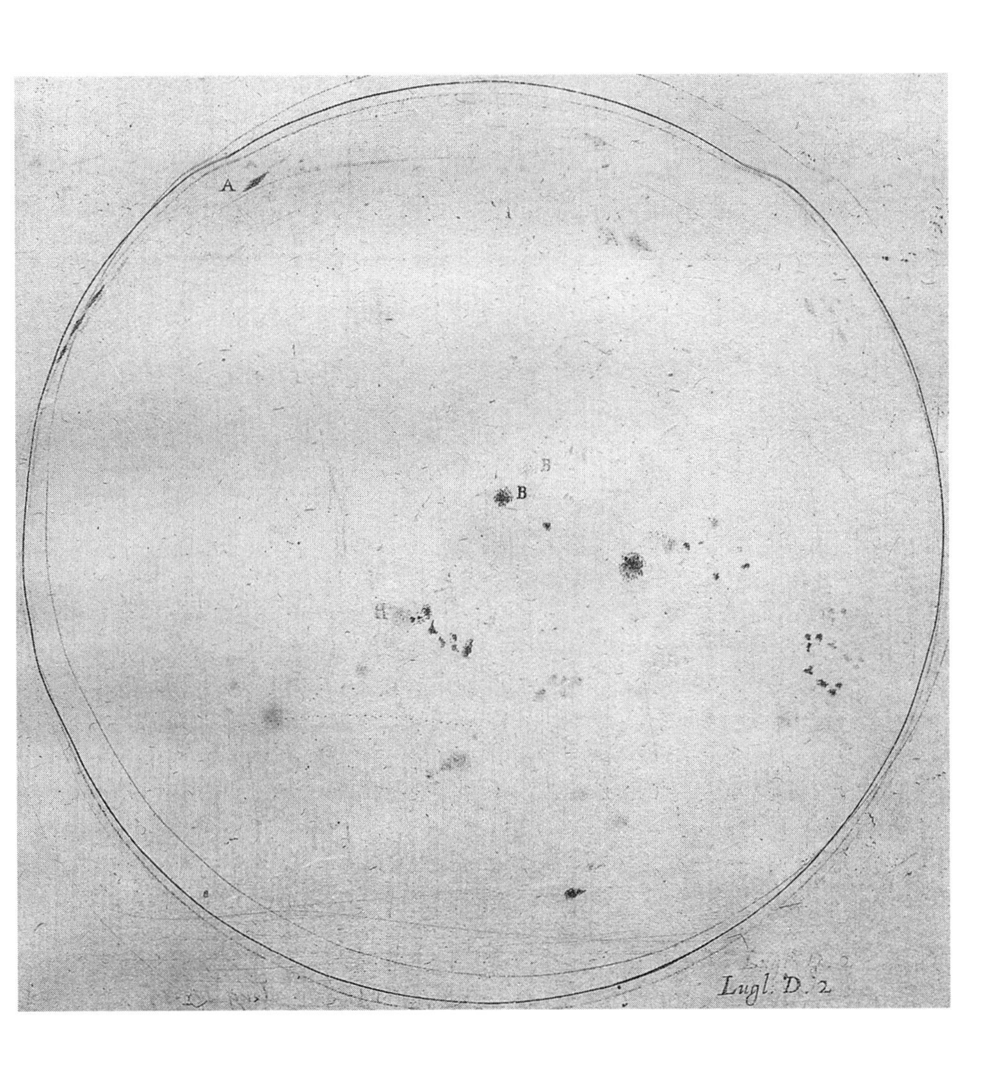
A
B
B
Lugl. D. 2

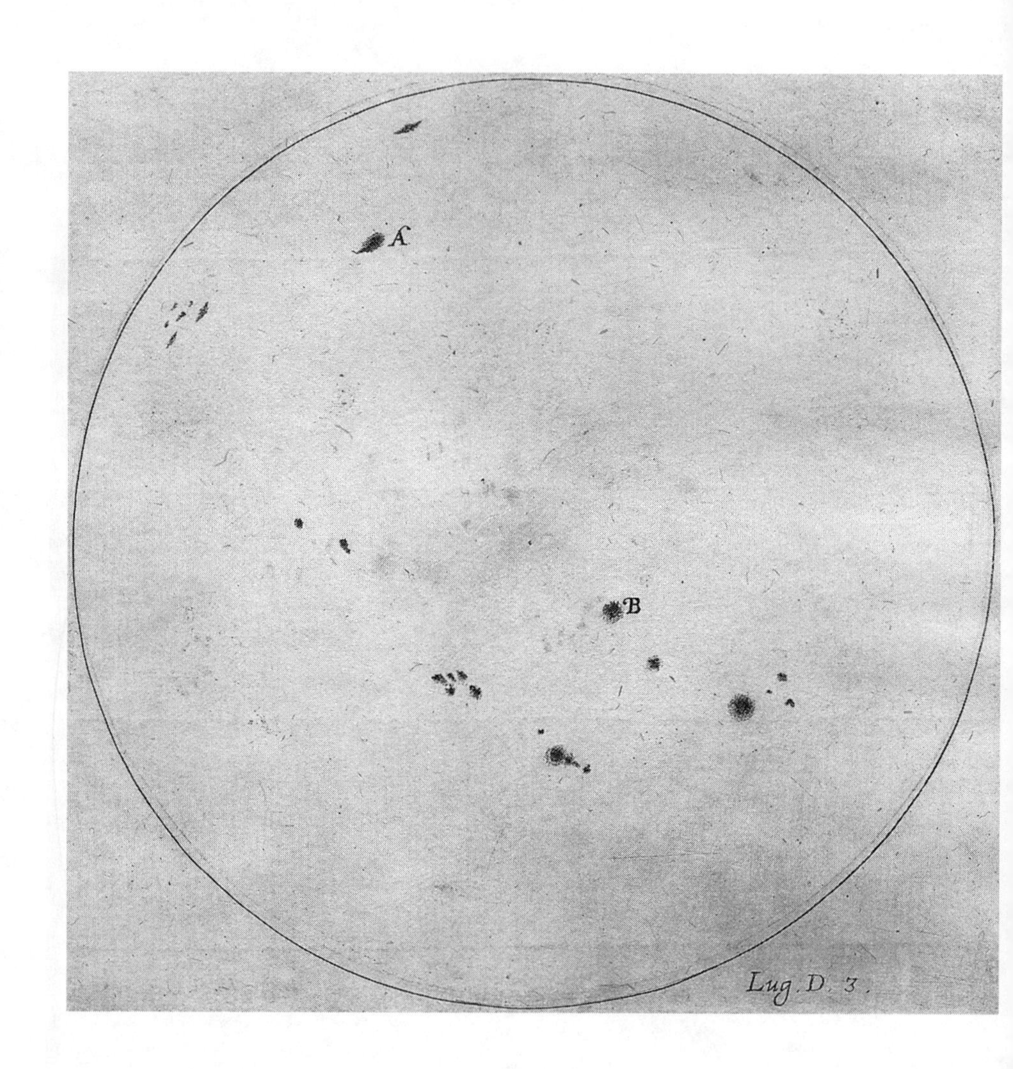
A
B
Lug. D. 3.

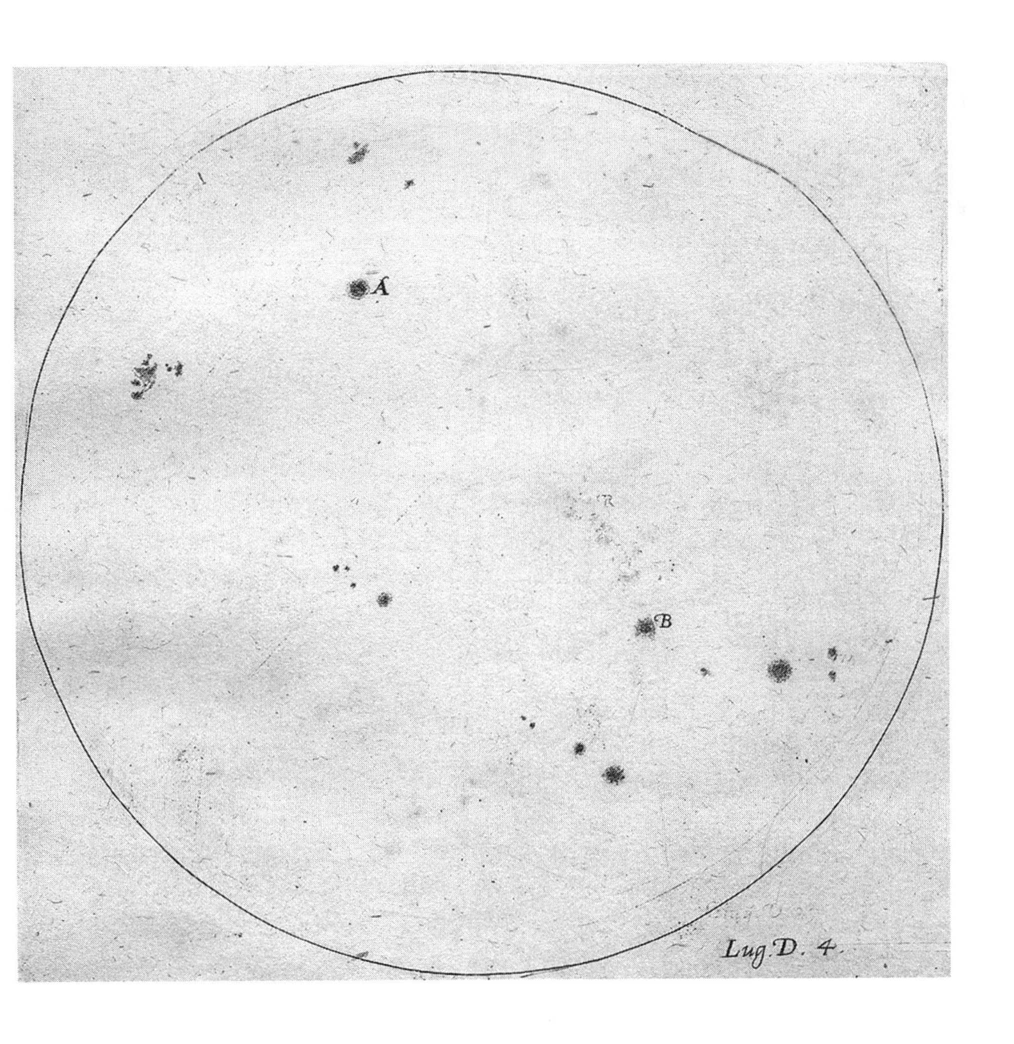
A
B
Lug. D. 4.

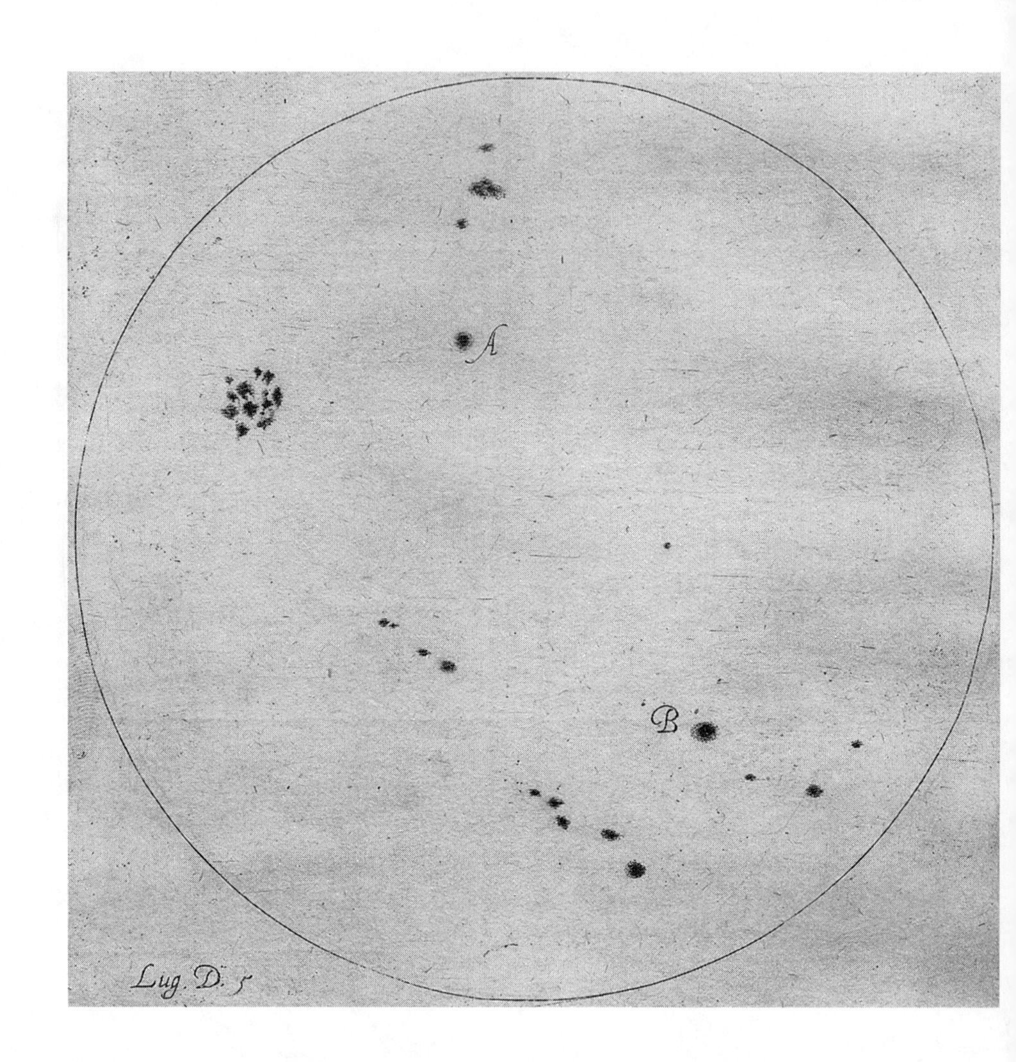
A
B
Lug. D. 5

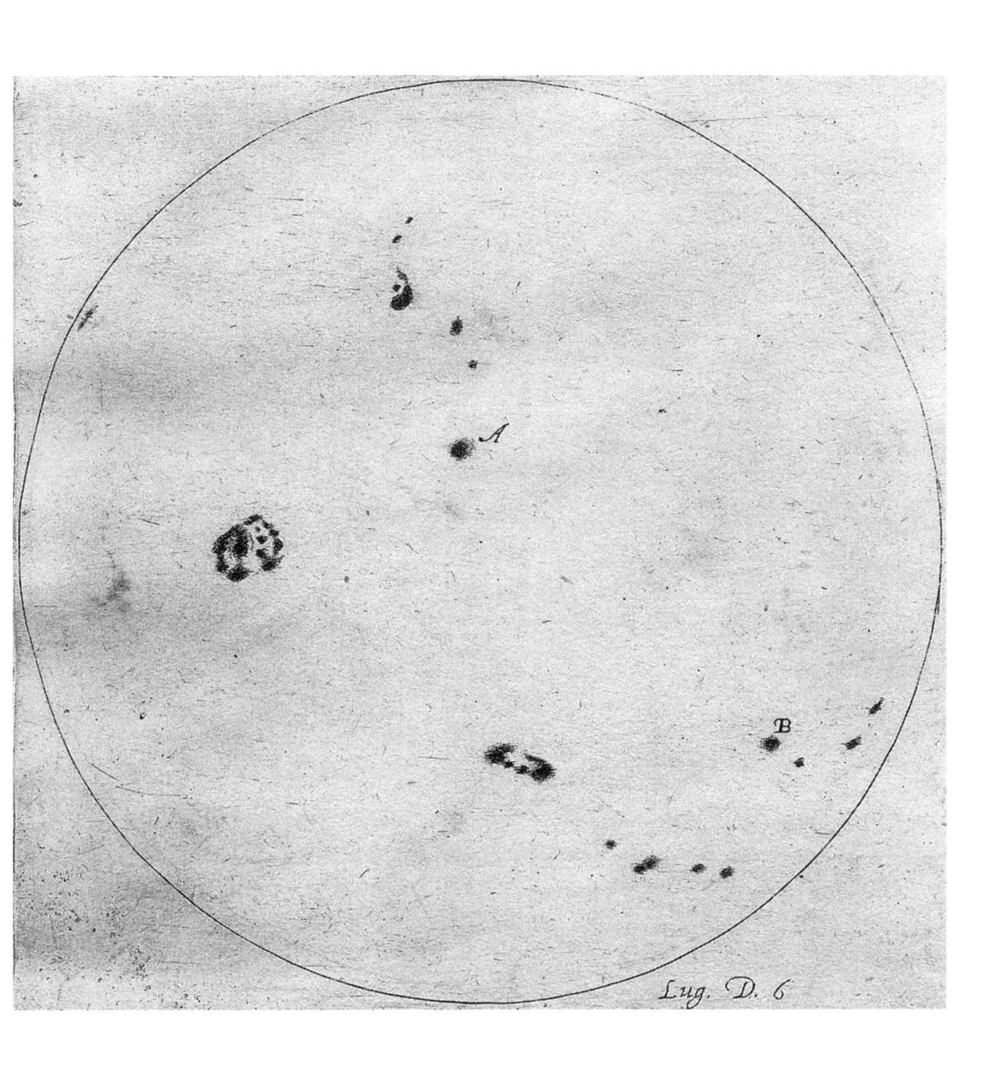
A
B
Lug. D. 6

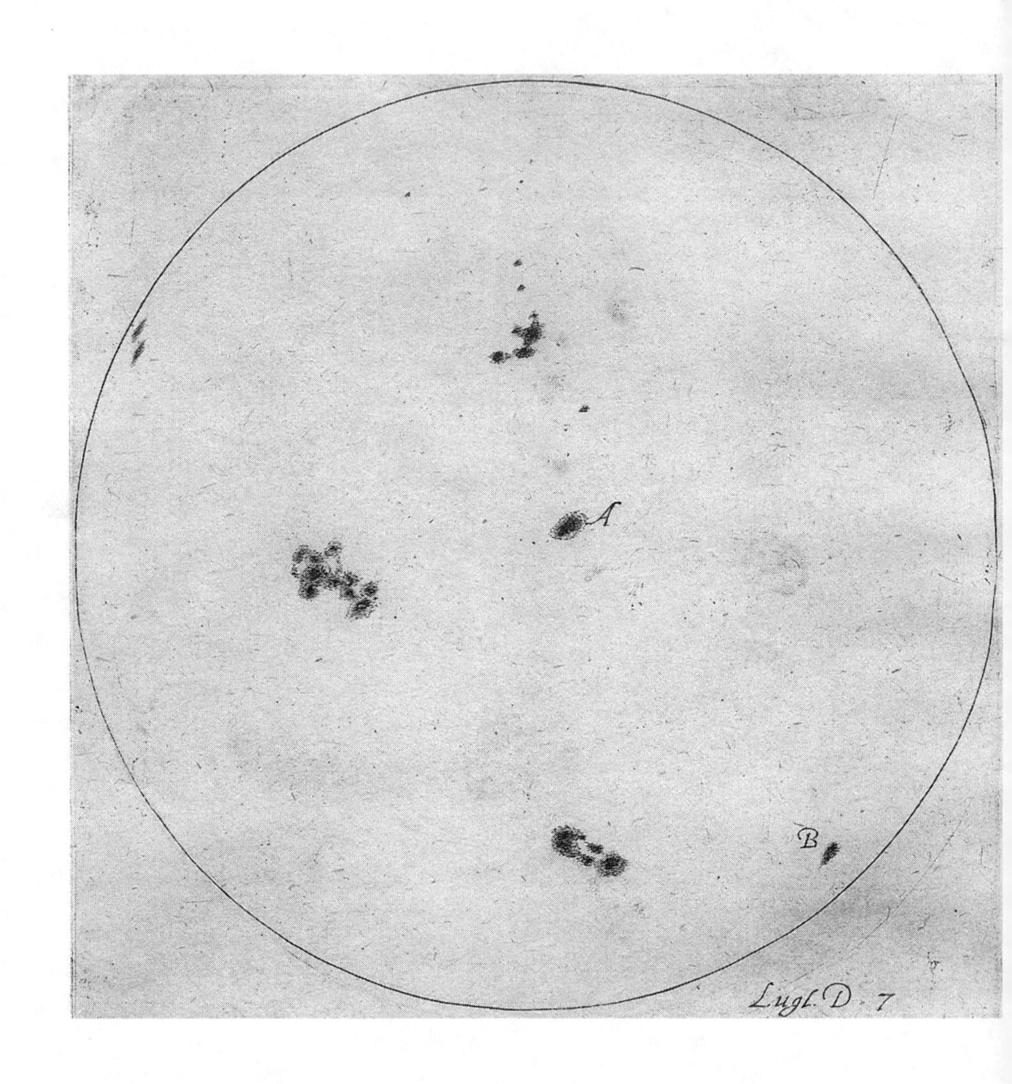
A
B
Lugl. D 7

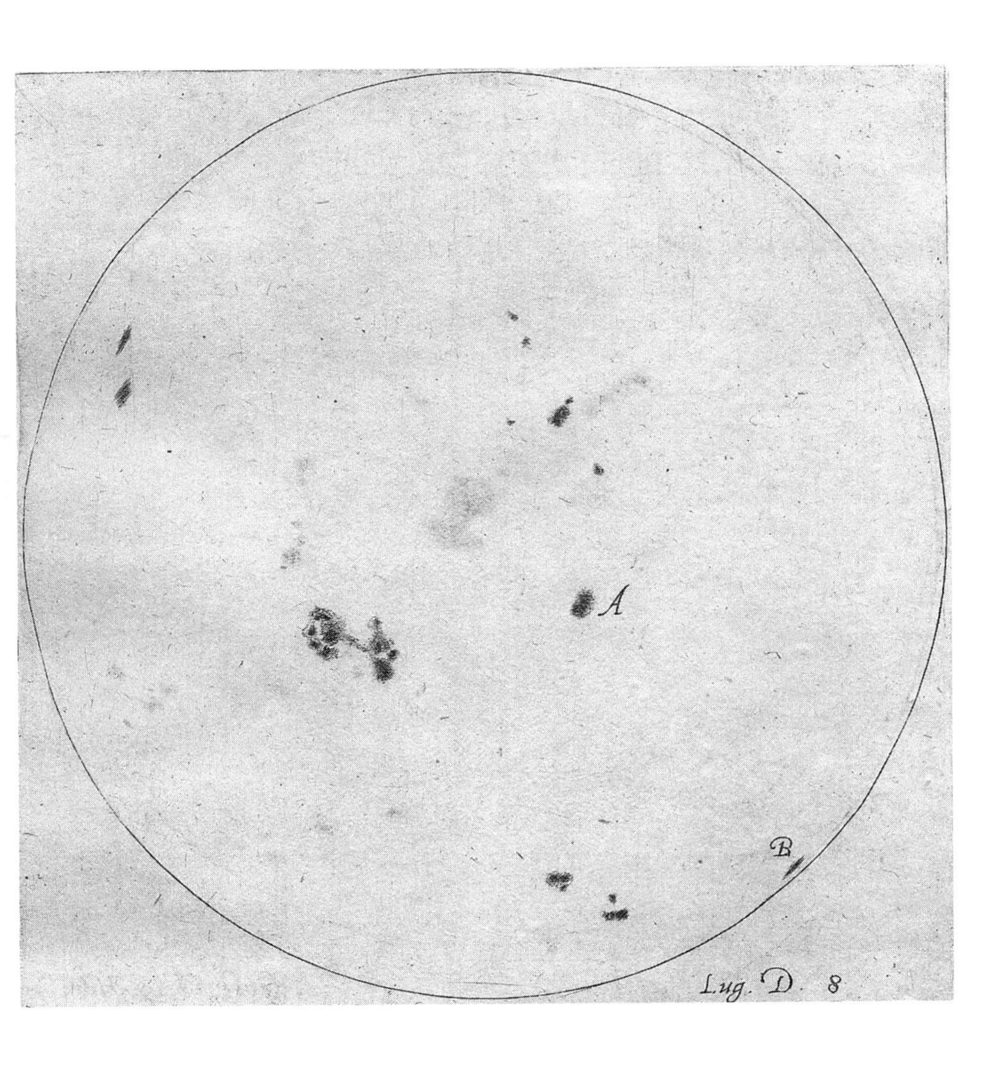
A
B
Lug. D. 8

94

Diſegni della Macchia grande Solare, veduta con la ſemplice viſta dal Sig. Galilei, e ſimilmente moſtrata a molti; nelli giorni 19.20.21. d'Agoſto 1612.

Agost. D. 19. Hor. 14

Drawings of the immense sunspot, observed with the unaided eye by Signor Galilei, and likewise shown to many, on 19, 20, and 21 August 1612.

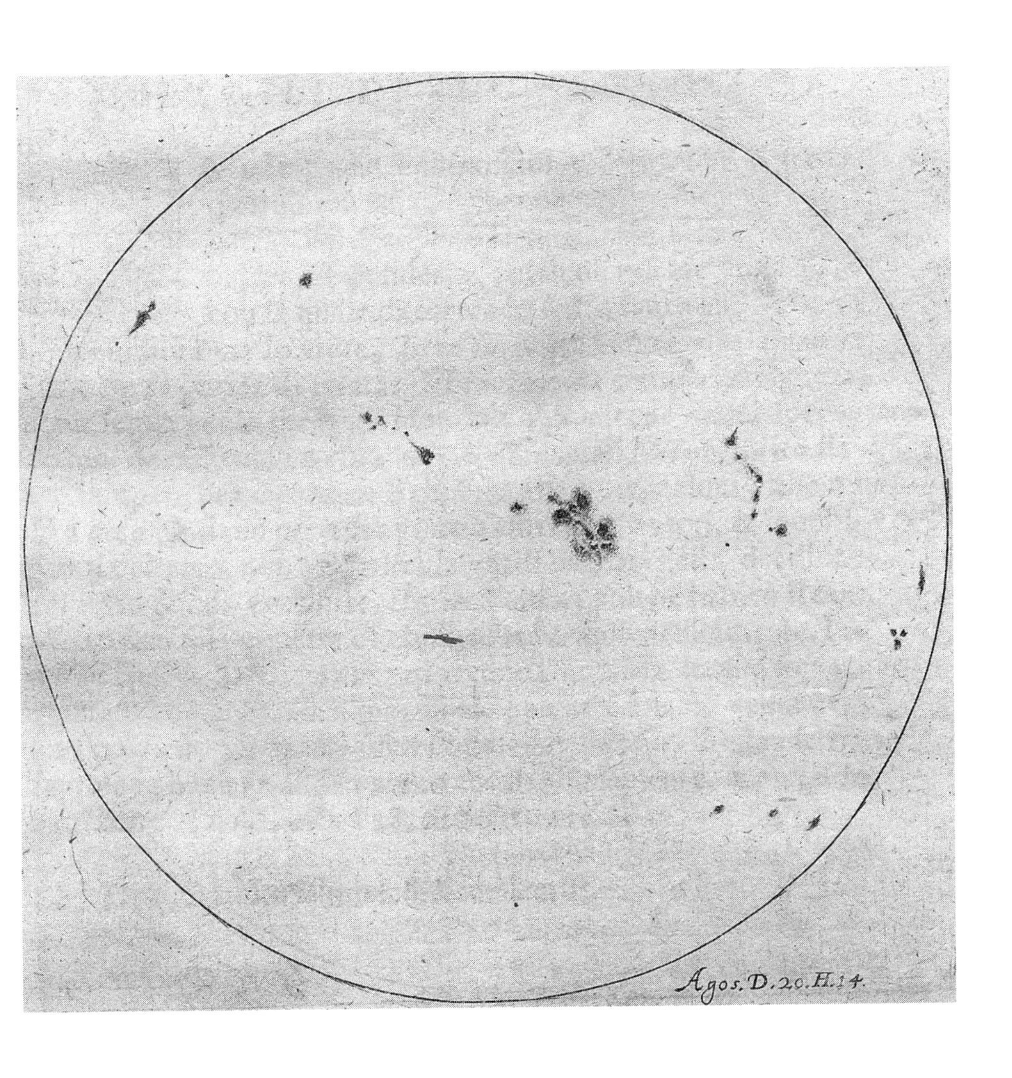
Agos. D. 20. H. 14.

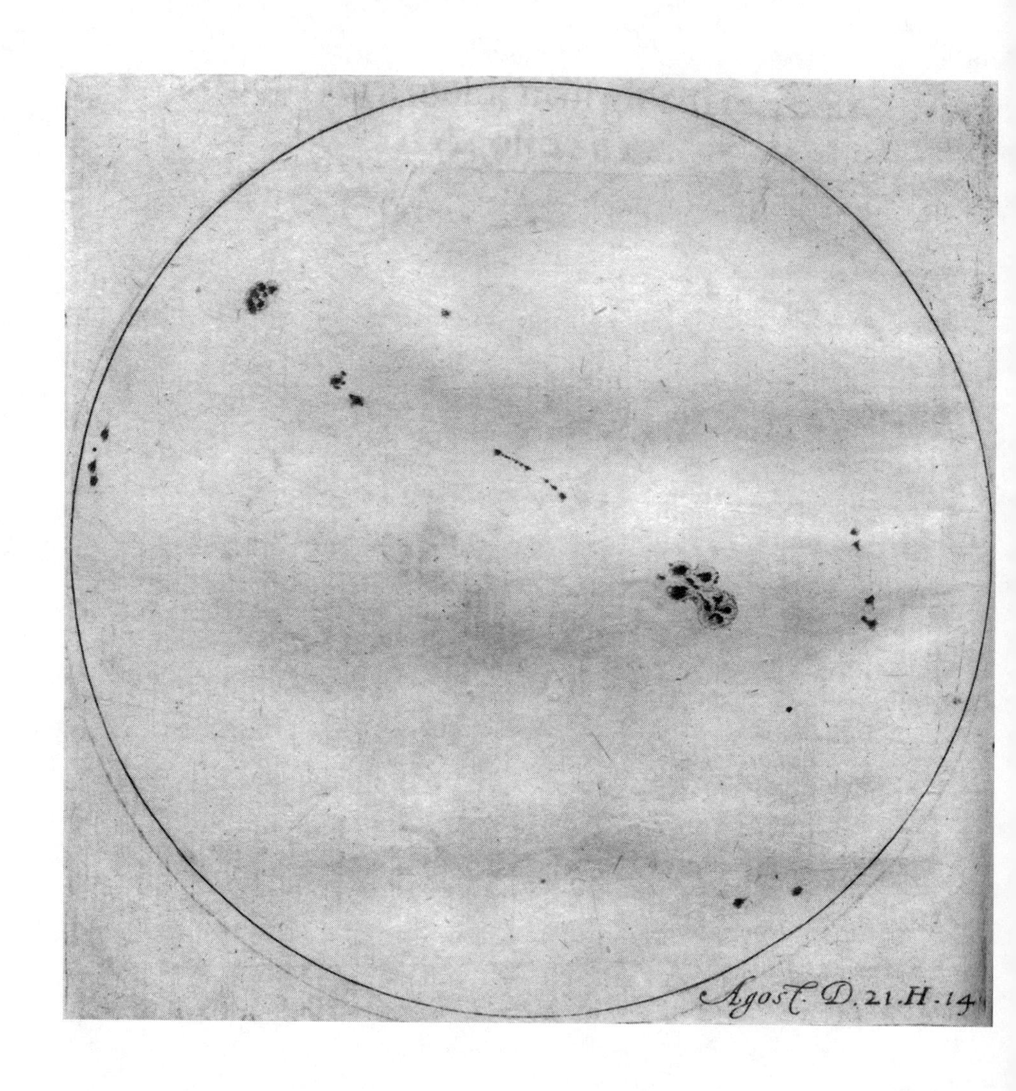
Agost. D. 21. H. 14

9

A more accurate inquiry from Ingolstadt

On 17 January 1612 Scheiner sent a copy of *Tres Epistolae* and the one-page correction to the second letter to Paul Guldin, asking him to share the work with Christoph Grienberger, for he had very few printed copies at his disposal. Father Clavius had been sent his own copy. Scheiner asked Guldin and Grienberger for their reactions, and assured them that his provincial would likewise be glad to hear their opinions. He also mentioned that a Paduan professor named "Vincentius"[1] had observed sunspots, and added that these observations agreed well with his own.[2] In this private letter, an emboldened Scheiner expressed his own views on the nature of sunspots more freely than he had in his printed tract:

> I think that there is an entire sphere of stars around the Sun that the Sun carries with it, and which is revolved about the Sun from east to west along the obliquity of the ecliptic. As far as I have been able to observe, one [complete] revolution has not yet been finished, unless we were to consider it probable that these stars are wandering, for then we would wait in vain for them to show any return to previous positions.[3]

The letter also provides some details of the circumstances under which *Tres Epistolae* was published. Scheiner revealed that Marc

1. Vincenzo Dotti. See p. 76, note 11, above.
2. Scheiner to Guldin, 17 January 1612, Graz, 159, 1, 1
3. Ibid.

Welser had prevailed upon the father provincial to allow the work to be published without mention of either the Society or any individual author, and he added: "You in Rome may not reveal Apelles hiding under the painting, for it would not please the superiors, and Apelles himself does not desire it."[4] It seems unlikely that Guldin concealed Scheiner's name from everybody in Rome, including Grienberger, and we may assume that throughout the controversy, when Grienberger was talking to Cigoli, he knew Apelles' true identity.

By 16 January 1612 Scheiner had finished his next letter to Welser, which focused first, and at great length, upon the embarrassing error about the conjunction of Venus, and subsequently offered more detailed observations of the sunspots. By 14 April, he had finished his fifth letter, a short one elaborating the earlier conjecture that the satellites of Jupiter numbered more than four. These letters, which were not published until the fall of 1612 under the title *Accuratior Disquisitio* (A more accurate inquiry) are discussed below on pages 174–181.

On 19 March Scheiner had written privately to Welser in less assured terms,

> Solar spots were never seen in ancient times. I am gradually beginning to doubt their return; I hesitate about what I will say. For one and only one argument still vexes me, namely that since they are not far away from the Sun (which can be most clearly demonstrated and is in part demonstrated in what I have published) it can scarcely be that they would be obliged to disappear [from sight] for so long. Given this, and also assuming that nevertheless others constantly appear, which occurs regularly, it is most difficult to free oneself from the conclusion that some are compelled to perish, and others to arise. About this more in its proper place, according to custom.[5]

His theory that sunspots were solid planetary bodies orbiting the Sun in regular paths necessitated their eventual return. In January, just after sending off his *Tres Epistolae*, Scheiner could say that he had not yet witnessed a complete revolution of the supposed sphere that carried these bodies; two months later, in March, he was clearly despairing of the spots' reemergence. As he confided to Welser, his entire theory now seemed doubtful, and he would have been quite displeased to learn that his correspondent had promptly sent this confession to Galileo.

4. Scheiner to Guldin, 17 January 1612, Graz, 159, 1, 1. Note that Scheiner had already said the same to Stengel; see p. 51, above.

5. Scheiner to Welser, 19 March 1612; cited in Welser to Galileo, 23 March 1612, *OGG*, 11: 289.

It appears that the reception in Rome of Scheiner's *Tres Epistolae* was lukewarm. Although no correspondence from Guldin or Grienberger to Scheiner survives from this period, Grienberger's letter of 5 February 1612 to Galileo conveys a certain lack of enthusiasm. Grienberger wrote that *Tres Epistolae* had arrived in Rome about two weeks earlier—and must therefore have reached Florence around the same time—and that because Clavius had just died,[6] the response to Scheiner had fallen to him.

> I replied that the author of the letters and observations had not produced improbable things, and that he had ingeniously freed the Sun of spots, and rightly rid the air of impurities and the lenses of flaws; but that at present I had nothing that I would affirm as certain; that I had observed rather notable spots, and as many as seven at once, but my observations were not sufficiently diligent and careful, and that I had sometimes seen similar spots in the air through some fault in the glass; but that I had not ascribed them to the air but rather to the glass by the same indication on the basis of which he removed the spots from the glass; so that I have nothing at hand that I would oppose to the reasons brought forth by the author. I did, however, point out this matter: that on 11 December of last year he searched in vain for Venus below the Sun. That is, during the conjunction it was necessarily above the Sun, as thus far the observations of Venus with the aid of optical tubes appear to require. For its annual changes, just like the monthly changes of the Moon, very clearly demonstrate that it goes around the Sun. But it was then at the apogee of its epicycle, and Magini's calculation and the observations themselves are so persuasive that this cannot be doubted by any reasoning. For always when it approaches this conjunction, it is diminished in apparent magnitude, and now, receding from this conjunction, it appears gradually larger.[7]

Though Galileo still remained ignorant of Apelles' identity and of his affiliation with the Society of Jesus, we may expect that Grienberger sought to give as favorable an assessment as possible of the *Tres Epistolae* without discrediting himself. If the tenor of later letters is any guide, Grienberger and therefore Guldin were considerably more critical when writing to Scheiner himself.

Although we may assume that Welser or a colleague in Augsburg or Ingolstadt had acquainted Scheiner with the general argument of Galileo's first

6. Clavius died on 6 February (Lattis, *Between Copernicus and Galileo*, 26), and Grienberger wrote to Galileo "But the valiant Clavius is now thinking of other matters and hastening elsewhere" (*OGG*, 11: 273).

7. Grienberger to Galileo, 5 February 1612, *OGG*, 11: 273.

sunspot letter, there had been a very significant delay before Scheiner had an opportunity to study these matters in detail in a Latin translation. He had already sent off his fourth and fifth letters to Welser, and now he quickly added a sixth, dated 25 July. Galileo was, at that moment, working on his second letter, which was finished on 14 August. Although Scheiner no doubt learned some of Galileo's points from one or more of his correspondents, it is important nevertheless to bear in mind that he had no direct and detailed acquaintance with Galileo's arguments until he began composing his sixth letter, which responded to the Pisan's first letter. Here Galileo had the advantage; he had seen Scheiner's first three letters (*Tres Epistolae*, January 1612) before he wrote his first and second letters (4 May and 14 August), and he saw Scheiner's last three letters before he wrote his third.

Welser had Scheiner's fourth, fifth, and sixth letters printed in Augsburg *ad insigne pinus*,[8] once again with neither a license nor permission from the censors, although presumably with the unofficial permission from Scheiner's provincial. There was a considerable delay between the dating of Apelles' last letter, 25 July, and the actual appearance of his *Accuratior Disquisitio*, which was sent out in the second half of September.[9]

De Maculis Solaribus et de Stellis circum Jovem errantibus Accuratior Disquisitio (On solar spots and the stars wandering around Jupiter: a more accurate inquiry) began with a long and quite superfluous geometrical demonstration about internal and external angles of triangles. Scheiner then proceeded to use this demonstration in an exhaustive geometrical explanation of the conjunction of Venus of 11 December 1611, the timing of which he had misjudged in his use of Magini's *Ephemerides*. Scheiner showed that it did not really matter whether one took the time in the tables to be the beginning, middle, or end of the conjunction; he should still have seen Venus as a spot on the Sun if it had passed below the Sun, and he stated that his argument in *Tres Epistolae* was unaffected by the error.

Having thus shown himself to be a perfectly capable mathematician, and having reduced his error to an incidental oversight, Scheiner returned to the sunspots. As before, he presented his second series of observations, from 10 December 1611 to 12 January 1612, in a group of small diagrams in which the sizes of the spots were exaggerated, and with the customary apology about their lack of accuracy and "the failing of [his] hand." He made a number of important points, alleging that the spots were rarely spherical, that they were

8. On Welser's press, see p. 51, note 45, above.

9. Scheiner to Guldin, 17 September 1612, Graz, Universitätsbibliothek, MS 159, part 1, no. 2; Welser to Galileo, 28 September 1612, *OGG*, 11: 402.

almost constantly changing their shapes, that they appeared largest in the middle and narrowest near the limb, and that they could not usually be seen at the limb, but appeared and disappeared a little distance from it. He also noted that the spots split up, coalesced, and were often temporarily surrounded by groups of other very small spots, that such groupings were more compact near the limb, and rather loose near the center, and that they had rough edges, and were darker at their center and lighter at their borders. Some spots were darker near the limb than towards the center of the Sun's disc, they moved more slowly near the limb, and their motion appeared to be parallel to the ecliptic.

Scheiner illustrated these conclusions by a discussion of his observations of individual spots, in the course of which he offered a general assumption about the nature of the Sun. Focusing on the speed of those spots moving parallel to and at some distance from the ecliptic, he inferred that they could not be attached to the solar surface, which he assumed was hard and unchanging. If the Sun were a solid body and the spots adhered to its surface, one would expect that a spot at a more northerly solar latitude would take exactly as much time to travel its shorter route across the face of the Sun as a spot on the ecliptic. Because this did not happen, and because Scheiner believed the Sun to be a solid body, he concluded that the spots could not be on its surface.

The spots' constantly changing shapes reinforced this conclusion. Because the Sun was conventionally held to be a hard and unchanging globe, the ceaseless metamorphosis in the spots could be accommodated only beyond the solar body. The fact that the spots were darker near the limb than at the center appeared likewise to support this conclusion, for in the center of the Sun its most powerful perpendicular rays penetrated the material of the spots to some extent, and reached the eye. Such was not the case at the limb, as Scheiner demonstrated with a diagram similar to the one he had used in *Tres Epistolae* to discuss the varying illumination of the spots. If the spots were on the surface in the form of chasms, they would necessarily appear darker near the center of the Sun and lighter near the limb. As Scheiner saw it, the spots were "shadow-casting bodies [that] wander outside the Sun."

Scheiner went into great detail about the large spot μ (see fig. 9.1), which he compared with the spot seen in 1607 by Kepler, who had supposed it to be Mercury, and depicted it in his *Phaenomenon Singulare* of 1609. Here his use of projection was clearly acknowledged, and the technique neatly clinched the argument that spots could not be caused by defects in the glass; the pinhole camera employs no lenses.[10]

10. Mario Biagioli, "Picturing Objects in the Making: Scheiner, Galileo and the Discovery of Sunspots," in *Wissensideale und Wissenskulturen in der frühen Neuzeit: Ideals and Cultures of Knowl-*

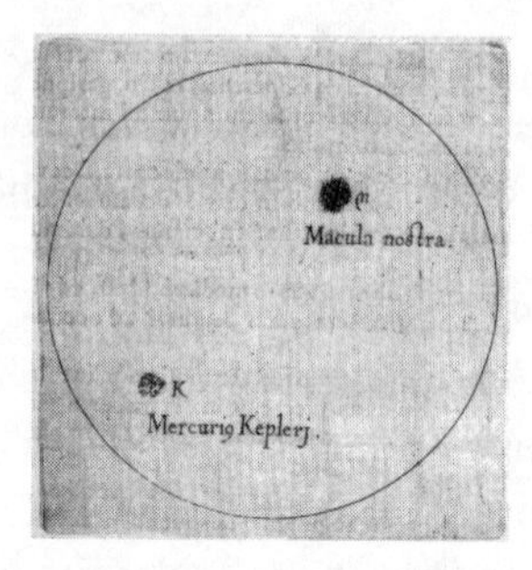

Figure 9.1. Scheiner's comparison of the sunspot seen by Kepler in 1607 and thought to be the planet Mercury, with a sunspot observed by him. From *AD*, 24.

The conclusions of this letter showed Scheiner tentatively making his way toward a new cosmology, one engendered by recent telescopic discoveries. He compared sunspots to the Moon, and concluded that since the solar spots had uneven contours, conjectures about the roughness of the lunar body were not unreasonable. Though he drew the line at speculations about the inhabitants of Jupiter, Venus, Saturn, and the Moon—noting that "it would be absurd to place inhabitants on so many bodies"[11]—it now appeared to him unproblematic to grant that the Earth was not entirely dark but reflected some light, for sunspots were likewise dark bodies reflecting solar light. The fact that the spots allowed the passage of some sunlight also appeared to Scheiner to support a conventional argument for the Moon's translucence, and perhaps to explain both that body's secondary or ashen light, and the fact that it was not entirely dark during lunar eclipses.[12] Scheiner argued that it was likely that heavenly bodies were of all different shapes but appeared round simply because of their remoteness, just as a candle flame seen from a great distance would appear round. Remarkably, Scheiner ended this fourth letter by claiming priority for the discovery of the solar phenomenon and for offering a correct explanation of it.[13]

Although this letter, almost twice as long as *Tres Epistolae*, contained significant new information about sunspots and ended with a plea for prompt

---

edge in Early Modern Europe, ed. Wolfgang Detel and Claus Zittel (Berlin: Akademie Verlag, 2002), 39–96; Biagioli, *Galileo's Instruments of Credit: Telescopes, Images, Secrecy* (Chicago: University of Chicago Press, 2006), chapter 3.

11. For early-modern notions of life on other planets, see Steven J. Dick, *Plurality of Worlds: The Origins of the Extraterresrial Life Debate from Democritus to Kant* (Cambridge: Cambridge University Press, 1982), 44–141.

12. Roger Ariew, "Galileo's Lunar Observations in the Context of Medieval Lunar Theory," *Studies in the History and Philosophy of Science* 15 (1984): 213–226.

13. P. 206, below.

publication, Welser did not oblige Scheiner by seeing it immediately into print. Either his increasing financial difficulties or his ill health or his desire to see Galileo's reaction to the first Apellean work would account for this delay.

Having continued his study of sunspots and other heavenly bodies, toward the end of March 1612 Scheiner observed a star in the same field as the satellites of Jupiter, and his fifth letter, dated 14 April, was devoted to a discussion of these phenomena. The star appeared in a configuration with Jupiter and its satellites, as had the fixed star observed by Galileo and reported at the end of his observations of Jupiter in *Sidereus Nuncius*.[14] As in Galileo's account, Jupiter and the star were moving with respect to each other, but if one assumed that the star was fixed, as Scheiner had initially done, then Jupiter's motion was in the wrong direction. Scheiner concluded at last that the star was actually a satellite of Jupiter. There is no easy explanation for this anomaly.[15] Observing Jupiter and its satellites required what would have been, by the standards of the day, a very fine and powerful instrument, and Scheiner did not see all of the visible satellites during every observation of this series, which stretched from 29 March to 8 April.[16] The field of view of his instrument was small, and its optics not very good. Though one or all of these factors might have caused an error, it is also conceivable that Scheiner observed a passing phenomenon such as a comet. Although judging from its motion it should have remained in the field of his telescope for a few more days, it was not seen after the observation of 8 April.

Whatever the merits of these observations, Scheiner quite naturally assumed that he had seen a fifth satellite of Jupiter, one that, like its companions, disappeared from time to time. Ignoring the fact that the four satellites of Jupiter always reemerged, he then drew a daring parallel between these bodies and his solar satellites. Beginning with the supposed new satellite, he broadened his scope, alleging that some of Jupiter's stars suddenly appeared and disappeared "in almost the same way as the shadows [i.e., spots] on the Sun . . . Just as hitherto one spot always follows another, so the stars of Jupiter also appear." The statement is especially interesting because it is illustrative of Scheiner's entire approach to the subject. By this time, he had been observing the Jovian satellites for about a year, and had of course seen them disappear and reap-

14. *OGG*, 3: 93–94; *SN*, 81–83.

15. Bellino Carrara S.J. here follows the opinion of August Winnecke, "Uber einem von Scheiner im April 1612 in der Nähe von Jupiter beobachteten veränderlichen Stern," *Vierteljahrschrift der astronomischen Gesellschaft* 13 (1878), who states that Scheiner likely observed a variable star. See Carrara, *"L'"Unicuique Suum' nella scoperta delle macchie solari*, 212–213. It appears that the star Scheiner observed was NGC 83225. See Joseph Ashbrook, "Christopher Scheiner's Observations of an Object near Jupiter," *Sky and Telescope* 42 (1971): 344–345.

16. P. 208, below.

pear in that interval. As his letter makes clear, however, he could not see them when they came to within a few minutes of the planet, and their periods still eluded him.

Galileo was, in fact, just then printing the first estimate of their periods in his *Discourse on Bodies in Water*. The pattern of their movement was for Scheiner perhaps somewhat reminiscent of the irregular motion of sunspots. If one were to counter that the number of sunspots appeared to vary constantly, but that the number of Jovian satellites, by contrast, never exceeded four, Scheiner could now maintain that there were at least five bodies about Jupiter, and by implication many more. Jupiter and the Sun were arguably surrounded by swarms of like bodies moving in differently inclined orbits, and the apparent invisibility of the solar satellites beyond the Sun could be attributed to the immense brightness of that globe. As Scheiner presented it, solar satellites were simply part of the larger pattern of the planetary world.

Scheiner's sixth letter—the third of *Accuratior Disquisitio*—also concerned the nature of sunspots. He had now read Galileo's first letter, but he did not address it point by point, choosing instead to focus more generally on all possible objections raised against his own observations. He began with the hypothetical contention that such observations were mere illusions, and in what surely was the most complete discussion of this subject up to that point, he showed the various false appearances that could be caused by defects of the eye, flaws in the lenses of one's instruments, and turbulence and vapors in the air. Scheiner again mentioned pinhole cameras, and this time added that sunspots could also be observed by projecting the reflected image of the Sun onto a white wall or sheet of paper.

He then turned to the actual experiences of others, distinguishing two sorts of testimonies: those borne by eyewitnesses, and those by ear-witnesses, or persons "who prefer to raise their ears rather than their eyes to the secrets of the Sun." In this latter group Scheiner placed a number of prominent persons—among them Federico Cardinal Borromeo, the archbishop of Milan, who, although they might disagree in their judgments about the nature of sunspots, assented to the existence of these phenomena without necessarily having observed them with their own eyes. This is the establishment of a natural fact not by the testimony of eyewitnesses, but rather by the testimony of ear-witnesses. This process is perhaps analogous to character witnessing, but the character in question here was that of a natural phenomenon, not a person.[17]

17. Albert Van Helden, "Telescopes and Authority from Galileo to Cassini," in Albert Van Helden and Thomas L. Hankins, ed., *Instruments: Osiris*, 2nd ser., 9 (1994): 8–29. The problem of witnessing in scientific matters by Robert Boyle and the Royal Society of London, half a century later,

Among the eyewitnesses Scheiner counted Fathers Grienberger and Guldin, who had sent him observations of sunspots made with a pinhole camera—*per foramen inversionis*—between 2 February and 4 April 1612. Scheiner reproduced some of these, stating that he had seen the same spots in the same locations on the same dates. He also introduced Galileo as a witness to his own observations. Reproducing Galileo's illustration in his first sunspot letter of the irregular and changing shape of a spot, Scheiner stated that he, too, had seen this same spot on those dates, and he concluded that Galileo did not disagree with him at all about the rough shapes and configurations of spots, but only in the more precise task of rendering single spots. And because their observations, made in Italy and Germany, agreed so precisely, they ruled out any parallax for the spots. Thus, in about a page, Scheiner neatly disposed of Galileo's first letter on the sunspots.

In *Tres Epistolae* Scheiner had stated that sunspots were as dark as the Moon, a judgment against which Galileo had argued in his first letter. Scheiner now discussed his observation of the lunar eclipse of May 1612, noting that in the center of the Earth's shadow cone "not a speck of the lunar body" could be seen. From this he inferred that the Moon was entirely without light of its own, and because the edge of the shadow showed no irregularities, that the unevenness of the Earth's circumference could not be seen from afar. By implication, then, the mountains on the Moon's periphery could not be discerned from the Earth. Scheiner also reiterated the claim that the sunspots were blacker than any of the nontelescopic spots of the Moon, and that therefore sunspots were as opaque as the Earth.

In May 1612 Scheiner had also observed an eclipse of the Sun, and this observation, too, was discussed in his letter of 25 July 1612. In his particular scrutiny of the sunspots during this event, he found the marks on the solar body to be as dark as the Moon. Scheiner also believed that he saw a weak light coming from the part of the Moon that was then covering the Sun, and from this he argued that the lunar body was somewhat translucent, as were sunspots, as was shown by the fact that they were less dark in the middle of the Sun's disk than at the limb (see pp. 226–228, below). The vexing question of the secondary light of the Moon was now answered; it was nothing but the light of the Sun pervading the Moon. He was skeptical, though not wholly dismissive, of Galileo's hypothesis about the earthly origins of the Moon's ashen glow: "But

has been treated in Steven Shapin, "Pump and Circumstance: Robert Boyle's Literary Technology," *Social Studies of Science* 14 (1984): 481–520; Shapin, *A Social History of Truth* (Chicago: University of Chicago Press, 1995); and Shapin and Simon Schaffer, *Leviathan and the Air-Pump* (Princeton: Princeton University Press, 1985).

to me the reflection, if that is what it is, of terrestrial light does not appear to be so great that it produces this phenomenon, although this line of reasoning is entirely consistent with the principles of optics."[18]

Scheiner had now taken issue with Galileo's statements in *Sidereus Nuncius* about both the satellites of Jupiter and the nature of the Moon. He had also attempted to show that Venus goes around the Sun by an argument independent from that of the phases. If any uncertainties remained about the scope of Scheiner's aim, they were dispelled by his conclusion, in which he noted that the "solid constitution of the heavens cannot endure, especially in the heavens of the Sun and Jupiter," and stated that in the final edition of his works, the prominent Christoph Clavius had warned astronomers that "they must unhesitatingly provide themselves with another system of the world."[19]

Here Scheiner stretched the meaning of Clavius's statement, for "the chorus leader of the mathematicians of his age" had expressed himself rather more cautiously. In the last edition of his commentary on the *Sphere* of Sacrobosco, that of 1611, Clavius had briefly reviewed the new observations of Galileo, referring to *Sidereus Nuncius* as a "reliable little book," and subsequently mentioning Galileo's discovery of the phases of Venus. He had then concluded: "Since things are thus, astronomers ought to consider how the celestial orbs may be arranged in order to save these phenomena."[20] It is worth recalling, in this connection, that Johannes Lanz, Scheiner's former teacher, had written to Clavius in 1607 about the problems posed for the traditional cosmology by new stars and comets. The urgency of the issue had been significantly compounded by the telescopic discoveries, and here Scheiner was probably advocating the world system of Tycho Brahe, or at the very least, a major modification of the Ptolemaic system for, as he argued, "If Venus goes around the Sun . . . and Mercury in all probability does the same, then one and the same heaven must be assigned to three planets."[21]

To sum up *Accuratior Disquisitio*, then, Scheiner had established beyond a reasonable doubt the existence of sunspots. Although he had argued that they were solar satellites, he ended on a cautious note, admitting that they changed shape and that it was perhaps possible that they were constantly arising and vanishing away. His arguments and diagrams put the spots very near the Sun, but the critical problem of their failure to return at regular intervals

18. See William R. Shea, *Galileo's Intellectual Revolution*, 52–53.

19. P. 229, below.

20. *OGG*, 2: 75: "Quae cum ita sint, videant Astronomi, quo pacto orbes coelestes constituendi sint, ut haec phaenomena possint salvari." See Lattis, *Between Copernicus and Galileo*, 198.

21. P. 229, below.

remained. Most crucially, Scheiner went well beyond the sunspots in this tract; he also discussed the motion and appearances of Venus (and by implication of Mercury), the spots and rough surface of the Moon, and satellites of Jupiter, and he even mentioned the appearances of Saturn. The only planet he did not discuss was Mars.

A sort of contagion pervaded his hypotheses: sunspots were dark bodies that did not recede from the Sun far enough to be illuminated by it in a dark sky, but in all other respects they were similar to the dark matter that he believed made up the Moon, Venus and Mercury, and the satellites of Jupiter.[22] A new world system was called for, one where Mercury and Venus, and perhaps other bodies, went around the Sun, and where the regions about the Sun and Jupiter, and probably Saturn as well, were filled with numerous planetary bodies. Whether these were arranged in spheres or had their own motions "like the birds of the air and the fish of the sea," Scheiner believed, remained to be seen.

22. It is worth noting that Scheiner's associate Cysat concluded in 1619 that comets, too, were of "the same material, origin, and nature" as the sunspots; see his *De Loco, Motu, Magnitudine, et Causis Cometæ* (Ingolstadt: apud Elisabetham Angermariam, 1619), 76.

10

Accuratior Disquisitio

To Marc Welser, Magistrate and Prefect of Augsburg

Since you have placed above me, like the ivy crown on a wine cup,[1] the authority of your name, the celebrity of your reputation, and the splendor of your noble birth, my work is now so illustrious that it persuades the thirsty to buy it or at least to taste it, and it has such weight that those of delicate digestion will lay aside their scorn for this vintage. Yet because I have produced a new wine, somewhat turbid and muddy, and have poured out a raw and unformed creation, it is right to cultivate it in the fashion of the fine wine makers, and, like a mother bear, to lick the limbs into beautiful proportions.[2] For ungraceful Venus, from whose form some parts stand out as though perfect while others lie hidden or shine only a bit,[3] still lies neglected.[4] Indeed, matters of such weight among astronomers could

1. Scheiner's opening remarks are perhaps undercut by the common Latin expression, *Vino vendibili non opus est hedera*, "decent wine needs no crown."

2. From Antiquity through the early modern period mother bears were believed to lick their shapeless cubs into ursine form; see for example Pliny, *Natural History* 8: 54, following Aristotle, *Historia Naturalium*. The process was also associated with the method of revision Virgil used on his *Georgics*. See Theodore Ziolkowski, "Judge Bridoye's Ursine Litigations," *Modern Philology* 92.3 (1995): 346–350. The title page of Scheiner's *Rosa Ursina* includes an image of this ursine activity. See p. 322, below.

3. The references is both to the changing appearance of the planet and the imperfection of Scheiner's second letter in *Tres Epistolae*.

4. *Venus enim invenusta iacet adhuc*. The pun *Venus/invenusta* echoes the language of the previous sentence, *venustam . . . proportionem*.

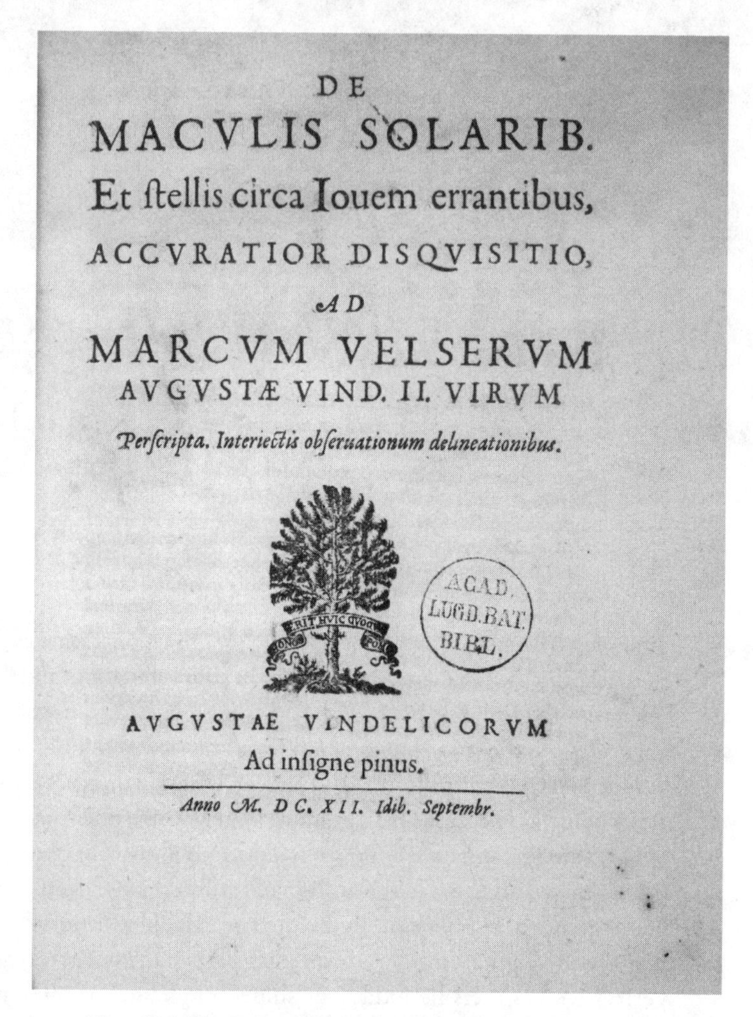

DE

MACVLIS SOLARIB.

Et ſtellis circa Iouem errantibus,

ACCVRATIOR DISQVISITIO,

AD

MARCVM VELSERVM

AVGVSTÆ VIND. II. VIRVM

Perſcripta. Interiectis obſeruationum delineationibus.

AVGVSTAE VINDELICORVM

Ad inſigne pinus.

Anno M. DC. XII. Idib. Septembr.

Figure 10.1. Frontispiece to Scheiner's *Accuratior Disquisitio*. Translation: On solar spots / and the stars wandering around Jupiter. / A more accurate inquiry. / To Marc Welser, / magistrate of Augsburg, / fully discussed and interspersed with drawings of the observations.

not be created in an hour, which was how long it took me to scribble a recently published letter to you.[5] I have therefore returned once more to these concerns, and am just now bringing them to light. I am finishing the entire business of Venus' conjunction with the Sun, drawing solely on the foundations laid by the

5. The letter of 19 December 1611. See above, pp. 65–67.

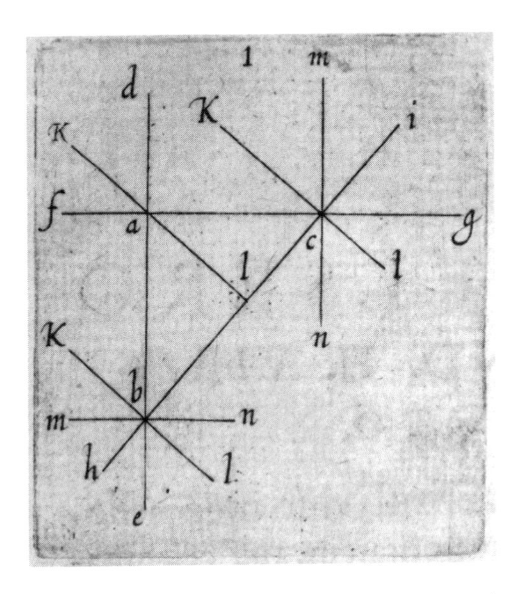

most illustrious astronomer [Giovanni] Antonio Magini in his *Ephemerides*[6] and *Mobilibus Secundis*,[7] after my prefatory remarks.

Lemma

If, when the sides of a right triangle are extended on all sides, a perpendicular is drawn through their intersection to any side of that triangle, it produces in the intersection on the same side of itself or of any side of the intersection three angles equal to the three angles of the given triangle, together and separately.

Let the given triangle be abc and angle bac be the right angle. The sides are extended in both directions: ab to d and e, ac to f and g, and bc to h and i. Now I mean that if through any intersection of sides, a, b, or c, any straight line is drawn that is perpendicular to another side of the triangle, the three angles produced in that intersection through which the perpendicular passes, taken in any way on one side [of a line], will be equal to the three angles of the given triangle, both together and separately.

If we let the perpendicular kl pass first through the intersection a and intersect the line hi perpendicularly at point l, then I assert that the three angles baf, fak, kad, made on one side of the line bd; or fak, kad, dac, made on one side

6. *Ephemeridum Coelestium Motuum, ad Annos* XL (Venice, 1582).

7. *Tabulae Secundorum Mobilium Coelestium* (Venice, 1585).

of the line fc; or the three angles kad, dac, cal, made on one side of the line kl; or dac, cal, lab, made on one side of db; or cal, lab, baf, the three angles made on one side of cf; or, finally, lab, baf, fak, the three angles made on one side of the line lk, are equal to the three angles of the given right-angled triangle abc, taken together as well as taken singly.

Since the three angles baf, fak, and kad, taken together, are equal to two right angles by Euclid 1.13,[8] and since the three internal angles of the given triangle are also equal to two right angles by Euclid 1.32,[9] then the three angles taken together along the same side of the straight line bd will be equal to the three internal angles of the given triangle, according to common notion 1.[10] And thus any three angles taken on one side of a straight line are shown to be equal to three angles of a given triangle. This was the first [part of the lemma].

Again, since the two angles fab and bac are made at point a on the straight line fc by the intersecting straight line ba, by Euclid 1.13, these two will be equal to two right angles, though angle bac is a right angle by hypothesis; baf will accordingly also be a right angle and equal to the other by common notions 7 and 12.[11] Now with this established, the two angles, fak and kad, will be equal to the two angles abc and acb, by common notion 3.[12] Angle fak is equal to angle abc because both are equal to angle lac; fak is the opposite angle [to lac], by Euclid 1.15.[13] Angle abc is equal to lac because in triangle alc the

8. "If a straight line set up on a straight line make angles, it will make either two right angles or angles equal to two right angles." See Euclid, *Elements*, tr. with commentary by T. L. Heath (Cambridge: Cambridge University Press, 1908; reprint, New York: Dover, 1975, 1995), 1: 275.

9. "In any triangle, if one of the sides be produced, the exterior angle is equal to the two interior and opposite angles, and the three interior angles of the triangle are equal to two right angles." Euclid, *Elements*, 1: 316.

10. "Things which are equal to the same thing are also equal to one another." Euclid, *Elements*, 1: 155. Scheiner's term for *common notion* is *pronunciatum*. He takes this term from Clavius, whose edition of Euclid's *Elements* he presumably used: "Communes notiones sive Axiomata, quae Pronunciata dici solent, vel Dignitates." See *Opera Mathematica*, 1: 20. Clavius's version of the first common notion is: "Quae eidem aequalia, & inter se sunt aequalia. Et quod uno aequalium maius est, aut minus, maius quoque est, aut minus altero aequalium. Et si unum aequalium maius est, aut minus magnitudine quapiam, alterum quoque aequalium eadem magnitudine maius est, aut minus" (*Opera Mathematica*, 1: 20).

11. In the Clavius edition, the seventh common notion is: "Et quae eiusdem sunt dimidia, inter se aequalia sunt. Et contra, quae aequalia sunt, eiusdem sunt dimidia" (Clavius, *Opera Mathematica*, 1: 24); the twelfth common notion is "Item, omnes anguli recti sunt inter se aequales" (all right angles are equal to each other) (Clavius, *Opera Mathematica*, 1: 25).

12. If equals be subtracted from equals, the remainders are equal. Euclid, *Elements*, 1: 155; Clavius, *Opera Mathematica*, 1: 23.

13. If two straight lines cut one another, they make the vertical angles equal to one another. Euclid, *Elements*, 1: 277.

angle at l is a right angle (since kl is perpendicular [to bc]), and therefore it is equal to angle bac. But angle lca is common to both triangles alc and abc; therefore the remaining angle lac is also equal to the other remaining angle, abc. Thus the two angles abc and fak are equal to each other by common notion 1. Wherefore the remaining angle kad and acb are equal by common notion 3. The three angles, then, made on one side of the straight line bd are equal to three angles of a right triangle taken singly. This was the second [part of the lemma], and in this way the entire lemma is demonstrated from this example (and there will soon be given a similar demonstration concerning any three other angles formed on one side [of a line] with the help of the two triangles abl and alc).

Now let the straight line kl pass through the intersection c, and be perpendicular to the hypotenuse bc, extended on both sides to h and i. Since therefore kl is perpendicular to hi, the two angles hck and hcl are right angles by definition 10,[14] and as far as their total magnitudes are concerned, the three angles lch, hcf, and fck are equal to them by common notion 19,[15] and the three angles of triangle abc are equal to two right angles by Euclid 1.32.[16] So the three angles lch, hcf, and fck are equal to the three angles of triangle abc by common notion 1.[17] (This is the first part of the lemma.) Furthermore, the angle lch, as it is a right angle, is equal to angle bac as being a right angle, and the angle hcf is common to both, so the remainder, fck, is equal to the remainder of abc by common notion 3. (And this is the second part of the lemma.) And now, if we take the three angles kci, icg, and gcl, on the other side of kl, then the right angle kci will be, as before, equal to the right angle bac by common notion 12,[18] and angle icg will be equal to angle acb, at the opposite vertex by Euclid 1.15,[19] and then the remainder gcl will also be equal to the remainder abc by common notion 3. This same demonstration pertains to all the other three angles made in whatever way on one side of a straight line, in any of the three intersections

14. When a straight line set up on a straight line makes the adjacent angles equal to one another, each of the equal angles is right, and the straight line standing on the other is called a perpendicular to that on which it stands. *Ibid*,153.

15. "Omne totum aequale est omnibus suis partibus simul sumptis" (Clavius, *Opera Mathematica*, 1: 26).

16. "In any triangle, if one of the sides be produced, the exterior angle is equal to the two interior and opposite angles, and the three interior angles of the triangle are equal to two right angles" (Euclid, *Elements*, 1: 316).

17. See note 10.

18. See note 11.

19. "If two straight lines cut one another, they make the vertical angles equal to each other" (Euclid, *Elements*, 1: 277).

Conſtitutio ☉ & ♀ quoad Longitudinem & Latitudinem.

Anno 1611. Menſe Decẽbri.	☉ ♐			Longitudo ♀ ♐		♀ Latitudo. S	D
Die	P	'	"	P	'	P	'
1	8	28	23	5	51	0	26
2	9	29	12	7	7		
11	18	37	18	18	30	0	9
12	19	38	17	19	46		

a, b, c, even if another perpendicular, mn, is drawn to the straight line fg, as one of their three angles is always demonstrated, with the help of a drawn perpendicular, kl or mn, to be a right angle; the second [of the three angles] will either be common to the given right triangle or to one of its angles at the opposite vertex, and so, by necessity the third [of the three angles] will be equal to the third angle of the triangle. By a similar method you will succeed in the case of intersection b, if you draw through it the perpendiculars kl and mn. And in this way the entire lemma is demonstrated, as it was proposed.

The calculation of the conjunction of Venus and the Sun

This happened in the year of our Lord 1611, on 11 December, computed from the *Ephemerides* and *Mobilibus Secundis* of Giovanni Antonio Magini.[20]

At this time the Sun was not far from perigee, so that its visible diameter was greatest, that is, by general agreement, 34′. At this time Venus was at the apogee of its epicycle and therefore closest to the Sun—assuming its path is below the Sun—and farthest from the Earth, and its aspect was smallest, perhaps 1′ or at most 2′ in diameter.

20. Magini, *Tabulae Secundorum Mobilium Coelestium* (1585). Magini gives the time of conjunction for the meridian of Venice as 11h. 12 min. on 11 December. Modern calculations give the solar longitude on 11 December 1611 (Gregorian) as 259°21′, and Venus's longitude as 259°23′. See Tuckerman, *Planetary, Lunar, and Solar Positions*, 823. See also appendix 4.

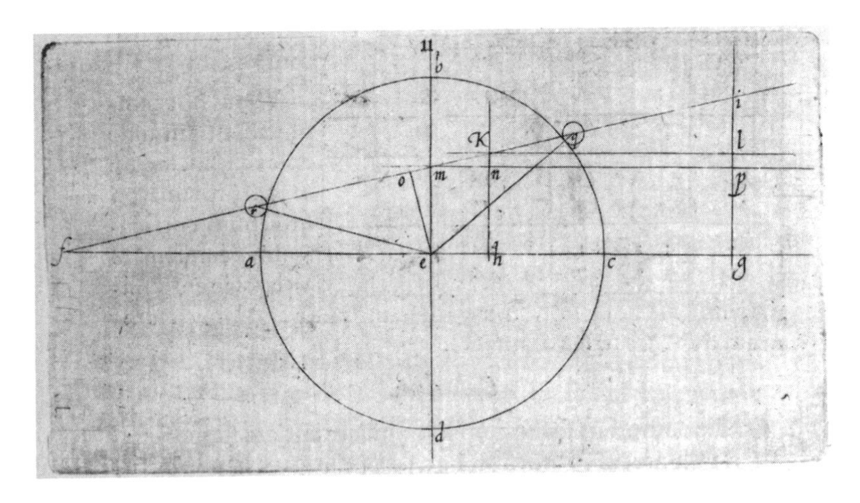

Calculation

All these figures are taken from the determination of Magini:[21]

1. the daily motion of the Sun was 1° [0′] 59″;

2. the daily motion of Venus was exactly 1°16′;

3. the difference by which Venus's motion exceeded the Sun's motion was exactly 15′1″;

4. on 11 December at noon, the center of Venus was 7′18″ distant from the center of the Sun [in longitude];[22]

5. from 1 to 11 December, that is, ten days, from noon of the 1st day to noon of the 11th, Venus decreased 17′ in latitude. Therefore:

6. let the circle abcd in the figure be the Sun, and a the eastern point of the Sun, b the northern, c the western, and d the southern; and let the straight line fg passing through the center e be the ecliptic; and when eh is marked off in it, let it be 7′18″, the distance of the center of Venus from the center of the Sun; and let hg represent ten days, and gi, perpendicular to the ecliptic, 26′, represent the latitude that Venus had on 1 December; but let hk, likewise perpendicular to fg, be the latitude of Venus on 11 December. Then the line ik produced to f will be the path of Venus, and the straight line kl, parallel to the ecliptic, will cut off the line li from the line gi, and this line li will be 17′ because the entire line gi is reckoned 26′ and its segment gl is reckoned 9′ (that is gl is equal to hk, because of the parallelogram hklg): the remainder li will therefore be 17′. For this reason, in the triangle kli two sides, kl and li, are known and

21. See appendix 4.

22. That is, 18°37′18′ − 18°30′.

the angle kli is a right angle, so that angle klg is then also a right angle because the figure klgh is a parallelogram and has a right angle at g since gi is perpendicular to fg by construction; and then by Euclid 1.47,[23] the third side is also known, namely 151′7″. Thus, by the three known sides of the triangle kli (kl being 9010″, li being 1020″, and ik being 9067″) we will easily arrive at the knowledge of the other necessary things. For

7. from the known kl and li and the known eh or mn, by the golden rule, the line nk will turn out to be 49″.[24] Again, since mn is known, from kl and ki by the same rule, the line mk will turn out to be 7′20″. And in this manner the entire triangle mnk has revealed itself to be proportional to triangle kli because of the parallel lines kl and mn and kn and il. Therefore, if

8. kn (which is 49″) is subtracted from the latitude of Venus, hk (which is 9′), the remainder (8′11″) will be the straight line hn or em, the latitude of Venus at the mean or true conjunction.[25] But if a perpendicular line eo is imagined from e at the center of the Sun to the straight line im extended all the way to f, then the triangle eom will be a right triangle because of right angle moe. And since the straight line pm meets the intersection m of the produced lines em and om and makes a right angle pme with the produced line eb, so that it is parallel to the side gi, then, by the above lemma, angle meo is equal to angle pmi. The angle mpi is also a right angle so that the two lines mp and kl are parallel, and then the internal angle mpi is equal to the right angle kli, as is the angle opposite to that same side. Thus the two triangles mpi and eom, as they have two angles equal to each other, also have their remaining angles equal, namely mip and emo, and therefore their sides will be proportional. But the sides mp, pi, and im of triangle imp are known because the side ip is known by its parts: il is 1020″ and lp (which is nk) is 49″, so that the whole of pi is 1069″. But the side im is known by its parts: ik is 9067″ and km is 440″ so that the whole of im is 9507″. And finally the side mp is known by its parts: mn is 438″ and np (which is kl) is 9010″ so that the whole of mp is 9448″. Accordingly, by these sides, using the rule of proportion, together with the known side em (which is 8′11″) we obtain 55″ for the side mo and 8′7″ for side eo. Since by this method I have found triangle emo,

9. I will easily find the sole object of our search, the path qr of Venus under

23. "In right-angled triangles the square on the side subtending the right angle is equal to the squares on the sides containing the right angle" (Euclid, *Elements*, 1: 349).

24. Because kli and mnk are similar triangles, kl:li=mn:nk.

25. Here and elsewhere in this letter, Scheiner consistently uses the term *media conjunctio*, or mean conjunction, when Venus has exactly the same longitude as the Sun's center. This would normally be called *true conjunction*. Scheiner's choice of words is confusing because Venus and the Sun are always in mean conjunction; Venus's mean longitude and that of the Sun are always the same.

the Sun by means of the previously calculated triangle emo and the line eq or er consisting of the sum of the apparent semidiameters at this time of the greatest [semidiameter] of the Sun at perigee of 17′ and the least [semidiameter] of Venus at apogee,[26] namely of 1′, so that the entire eq is determined to be 18′. Because these calculations have been established, and because the angle eoq or eor is a right angle, and the line eo is known (namely 487″), as is eq or er (which is 1080″), it follows also from Euclid 1.47[27] that the side oq as well as or is 16′3″, and the entire path of Venus under the Sun qr, or the duration of the conjunction, is 32′6″, which is two days, three hours, eighteen minutes, and ten seconds, which together make about fifty-one and one-third hours.

10. Now when side mo is subtracted from line oq, it leaves a side of incidence mq of 15′8″, which is twenty-four hours, eleven minutes, eleven seconds. But when the same side mo is added to or, it makes the line mr (16′58″) for the egress of Venus, which is one day, three hours, six minutes, fifty-nine seconds.

11. Furthermore, since the 7′18″ by which the Sun precedes Venus correspond to eleven hours, forty minutes, three seconds, the mean conjunction fell on 11 December at eleven hours, forty minutes, three seconds p.m. When the time of entering is subtracted from this, it leaves for the beginning of the conjunction 10 December at eleven hours, twenty-eight minutes, fifty-two seconds p.m., which is about the middle of the twelve night hours. When the time of egress is added to the eleven hours, forty minutes, three seconds of 11 December, it gives us fourteen hours, forty-seven minutes, two seconds on 12 December for the end of the conjunction, and Venus did exit from the Sun on 13 December at about 5 in the morning, on the civil calendar.[28]

Now that the calculation has been demonstrated, it will hardly be incongruous to add some remarks on the true and genuine form of this conjunction (if indeed it took place below the Sun).

Accordingly, in the adjoining diagram, ABA is the disk of the Sun, whose center is C and whose diameter, coinciding with the ecliptic, is divided into thirty-four equal parts.[29] The little orb D, E, or F is the circle of Venus whose path across the Sun is the straight line DF. The start of the conjunction is at D, the middle at E, and the end at F. With the line GH, 15′ in length and di-

26. In December, the Sun is near its perigee of its eccentric, so it has the largest apparent (semi) diameter and, in this case, Venus was near the apogee of its epicycle, and so had its smallest apparent (semi)diameter.

27. See note 25.

28. The astronomical day began at noon, the civil day at midnight. Therefore 12 December, 14h 47min 2 sec in astronomical time corresponds to 13 December 2 h 47 min 2 sec, or about 3 a.m., not 5 a.m.

29. Scheiner took the solar diameter to be 34′ on 11 December 1611.

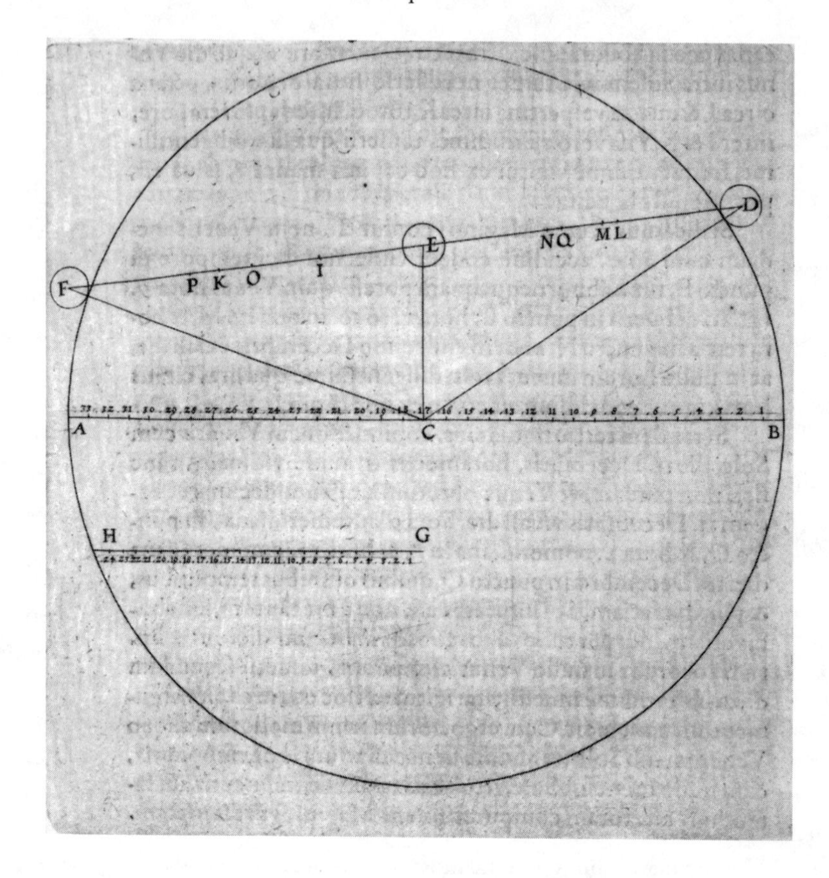

vided into twenty-four equal parts according to the number of hours of the natural day, you will be able to measure geometrically by hours the path DF of Venus, the total duration of its conjunction, as well as the ingress DE and egress EF.

If, then, we assume that the conjunction of Venus with the Sun began at D on 11 December at eleven hours, forty minutes, three seconds of the night, it must be shown that the conjunction necessarily lasted past 13 December, when Venus would necessarily have appeared below the Sun on the eighth morning hour at I, and on the fourth hour of the evening, near K, and at all intermediate times between I and K. But though it was sought constantly and with the greatest diligence, Venus was not seen at all. So from this evidence that portion of my published letter still stands, and it is sound.

If next we were to say with Magini that the mean conjunction of Venus with the Sun occurred at the same time of the eleventh day, at point E,[30] then

30. Magini gives the time as 11hrs 12 min. See note 22.

it can hardly be denied that at nine o'clock Venus must have been at point L, at ten o'clock at point M, and at three o'clock at point N, on that very day, 11 December on the civil calendar.[31] But although it was most diligently sought, it was not found at any of these spots at the above-mentioned hours: and so from this evidence the conclusion is the same [as in my published letter].

If, finally, we should suppose that the conjunction of Venus with the Sun ended on 11 December, at eleven o'clock at night, it could not be that Venus was avoiding our gaze on that [civil] day, 11 December, at nine o'clock before noon at point O, and two o'clock after noon at point P, and ten o'clock before noon on 10 December at point Q, but at all of these times, and at many more, the Sun was observed not by me alone, but also by others through many different tubes. Yet Venus appeared on none of these days and in none of these places, even though, as I said, it was sought most expertly; and so again from this evidence the argument is demonstrated.

Now since Venus should necessarily have crossed below the Sun in one of these three ways, and yet was below the Sun in none of them, as the observations clearly prove, we must either grant that Magini's entire computation, however employed, is worthless—which I do not believe—or else grant, since our observations retain their force and Magini's calculation its value, that Venus proceeded not *below* but *above* the Sun. A threefold cord is not quickly broken,[32] and it was to be tripled so that it would not be broken. If someone should break the first strand, he would break the second with the first, the third with the second, and the first with the third, but he will never break all three together.

Anticipate Venus by one day, and you only delay it the same amount from the Sun or from the course equal to it [Venus]. The conjunction with the Sun, if it is real, will always occur either during my observations or those of some friend. Besides, the calculation of Magini had to be divided in this way, as much to make the force of the argument unavoidable as to compensate, with this triple-membered extenstion, for an error, if any had been committed in it. For just as Mercury observed on the Sun by Kepler in May 1607[33] differed not a little from [Giovanni] Antonio Magini's prediction in longitude as well as in latitude, so we might fear it possible that Venus had dared something similar. Wherefore, most honorable Sir, I beg you again and again—as if I were asking for your patronage in a literary matter, and by the grace

31. See note 30.

32. Here Scheiner draws on Ecclesiastes 4: 12, *Funiculus triplex difficulter rumpitur*.

33. Kepler thought that he was observing a transit of Mercury across the Sun; see above, pp. 20–21.

that makes you outstanding among the most famous men—to find it worthwhile to prevail on Antonio Magini to take up this conjunction of Venus with the Sun once more, as if it were to be computed anew, and to have him communicate with me through you. I ask also that Kepler do the same with regard to Tycho's groundwork, which I wish that we would likewise be able to do someday.[34] And I have asked another colleague to do the same recalculation concerning the hypotheses of others, and I too will attempt it in my moments of leisure. For if all calculations should agree in those four, five, or more days and not wrest Venus away from us at the latitude of the Sun,[35] we will sing hymns of praise. But if—and I am hardly convinced of this—they were to deny that a real conjunction happened, perhaps because of a latitude greater than the one Magini gave, you should know that my entire reasoning is hypothetical and based upon his calculation. If the hypothesis is known and confirmed, let the argument stand, but if the hypothesis is overturned and destroyed, let what was built on it also fall, but let what is true rise and remain standing. This alone is what we are searching and striving for.

The one thing, beyond what has already been said, that can bring harm to this argument is the fact that Venus—being, to be sure, below the Sun—casts either no shadow at all, or one so weak that because of the vigor of the solar light, it cannot be perceived by the sharpest of eyes.[36]

To this last argument I reply that the shadow of Venus, which doubtless revolves below the Sun, will not appear smaller than the full orb of Venus when she is outside the Sun but close to entering it. So, since Venus appears equal to middling sunspots (as will be explained more fully in its proper place), by consequence its shadow will absolutely not be smaller than these spots, especially if what Christoph Clavius (certainly the reigning mathematician of our time) and Tycho Brahe assert is true, that the visible diameter of Venus is as one-tenth in proportion to the Sun.[37] For it is certain that innumerable spots have been seen, and are being seen, that have a very much smaller proportion to the diameter of

34. Kepler's *Tabulae Rudolphinae* were not published until 1627.

35. If Venus's ecliptic latitude is more than about 16′ at the time of conjunction, no transit will occur. Since this was a superior conjunction, however, all this was irrelevant.

36. In fact, following older authorities, Scheiner made the apparent diameter of Venus "at least 3′" (see above, p. 66). As experience was to show, during a transit Venus's apparent diameter is only about 1′. For Galileo's comments on Scheiner's figure, see below, pp. 259–260. Scheiner also raises the possibility here that Venus is transparent and thus would not show as a black spot on the Sun.

37. Clavius, *In Sphaeram Ioannis de Sacro Bosco Commentarium*, in *Opera Mathematica*, 3: 100–102, 117. Clavius's sizes and distances of the planets were about the same as those of Ptolemy and Al-Farghani. See Albert Van Helden, *Measuring the Universe: Cosmic Dimensions from Aristarchus to Halley* (Chicago: University of Chicago Press, 1985), 27, 30, 53.

the Sun—indeed, one-sixtieth and sometimes even one-hundredth—and this will be manifest to any observer.

To the first argument I say that Venus, moving under the Sun, makes a shadow so that the Sun is eclipsed by Venus to an extent proportional to Venus moving below it, and this I demonstrate in the following way:

1. by the consensus of all philosophers and mathematicians, ancient as well as modern. Plato and his imitators, for instance, put Venus above the Sun because he did not turn his attention to this shadow;[38] therefore Ptolemy and his followers denied that a direct conjunction of Venus with the Sun ever occurred;[39] and therefore Clavius, in his *Sphere*, denied that this shadow is so large that it can be discerned by the naked eye,[40] and this agrees with book 2, chapter 7, question 4, article 2 in *De Caelo* of the Coimbra [mathematician][41] and others *passim*.
2. by similarity. Everyone everywhere agrees that the Moon during its course under the Sun causes a visible shadow according to its proportion; from this, it will not appear incongruous if the same is also produced by Venus tarrying below the Sun; indeed, the same phenomenon is reported to have happened when Mercury revolved below the Sun. For 804 years ago, a certain monk saw Mercury under the Sun in the form of a black spot,[42]

38. Plato put Venus and Mercury above the Sun but gave no reason for this; see *Timaeus* 38d. The idea that Venus and Mercury are placed above the Sun because they are never seen to pass below it goes back to Ptolemy, who wrote in IX, 1 of his *Almagest*, "But concerning the spheres of Venus and Mercury, we see that they are placed below the sun's by the more ancient astronomers, but by some of their successors these too are placed above [the sun's], for the reason that the sun has never been obscured by them [Venus and Mercury] either." See *Ptolemy's Almagest*, trans. G. J. Toomer (London: Duckworth; New York: Springer Verlag, 1984), 419.

39. About this argument, Ptolemy wrote (*Almagest*, IX, 1), "To us, however, such a criterion seems to have an element of uncertainty, since it is possible that some planet might indeed be below the sun, but nevertheless not always be in one of the planes through the sun and our viewpoint, but in another [plane], and hence might not be seen passing in front of it, just as in the case of the moon, when it passes below [the sun] at conjunction, no obscuration results in most cases." See Toomer's note in *Ptolemy's Almagest*, 421 n 34. Ptolemy repeated this argument in his *Planetary Hypotheses*; see Bernard R. Goldstein, *The Arabic Version of Ptolemy's "Planetary Hypotheses,"* American Philosophical Society, *Transactions* 57, part 4 (1967), 6–7.

40. Clavius, *In Sphaeram Ioannis de Sacro Bosco commentarius* (Rome, 1570), 95.

41. [Emmanuel de Goes], *Commentarii Collegii Conimbricensis Societatis Iesu in quatuor libros De coelo, Meteorologicos et Parva Naturalia* (Coimbra, 1592). This work was reprinted at least eleven times by 1631. There were printings in Cologne in 1596, 1599, 1600, 1603, 1618, and 1631. See Charles H. Lohr, "Latin Aristotle Commentaries: Authors C," *Renaissance Quarterly* 28 (1975):89–741, at 717–719.

42. "Nam et stella Mercurii XVI. Kal. Aprilis [807] visa est in sole quasi parva macula, nigra tamen, paululum superius medio centro eiusdem sideris, quae a nobis octo dies conspicitur. Sed

as Kepler tells us in his *Phaenomenon Singulare*.[43] And it is demonstrated in the same book that Kepler himself saw Mercury below the Sun on 28 May` 1607.[44] Scaliger also declares the same himself in exercise 72 against Cardanus,[45] according to book 2, chapter 7, question 4, article 2 in *De Caelo* of the Coimbra [mathematician].[46] If Mercury then brings about an eclipse of the Sun, why should Venus not do likewise?

3. by experience. At about the same time when Galileo observed horned Venus above several Italian cities,[47] other mathematicians in Rome admired it and indeed discovered in it the same horned, bisected, and gibbous shape.[48] From this incredible phenomenon we have two ineluctable arguments. The first is that Venus, just like the Moon, is without light of its own and consequently produces a black shadow when below the Sun. The second is that, according to the same phenomenon, Venus goes around the Sun. Since all phenomena thus agree and all reasons are in concert, the prudent man will scarcely dare to doubt it in the future.

Therefore, since we have brought forth this demonstration and, I believe, a fully fashioned Lucifer,[49] let us now return to the father of that light, that is, the Sun. And as in a triumphal procession, let us lead forth his numerous offspring from 10 December (there not being an account since I recently displayed some of my observations) to 12 January, so that this great family might be taken in

---

quando primum intravit vel exivit, nubibus impedientibus minime adnotare potuimus." See *Annales Laurissensis et Einhardi*, in *Monumenta Germaniae Historica Scriptorum*, vol. 1 (Hanover, 1826), 124–218, at 194; issued separately: *Scriptores Rerum Germanicarum in usum scholarum ex Monumentis Germaniae Historicis separatim editi*, vol. 6 (Hanover, 1895, 1950), 123. See also pp. 11–13, above.

43. Kepler, *Phaenomenon Singulare* (note 35), in *JKGW* 4: 83–87. Since he could not justify the date with his astronomical calculations, Kepler changed the year from 807 to 808. Further, since a transit of Mercury lasts at most only a few hours, Kepler argued that *octo dies* (eight days) was a misreading of *octoties*, a corrupted version of *octies*, eight times.

44. Kepler, *Phaenomenon Singulare*, *JKGW*, 4: 92–95.

45. Julius Caesar Scaliger, *Exotericarum Exercitationum Liber XV. De Subtilitate ad Hieronymum Cardanum* (Hanover: Wechel, 1620), 250; Mercury's appearance on the Sun's face is described as "like a spot on the solar body."

46. See note 43.

47. Galileo to Christoph Clavius, and Galileo to Benedettto Castelli, 30 December 1610, *OGG*, 10: 499–505. Scheiner's source of this knowledge, beside what was available through channels of Jesuit correspondence, was probably the preface of Kepler's *Dioptrice*, which was published in 1611 (*JKGW*, 4: 347–348).

48. Christoph Clavius, Christoph Grienberger, Odo Maelcote, and Giovanni Paolo Lembo to Robert Bellarmine, 24 April 1611, *OGG*, 11: 92–93. See also Guldin to Lantz, 13 February 1611, Dillingen Studienbibliothek, MSS 2o, vol. 247, pp. 220–222.

49. Venus, the morning star.

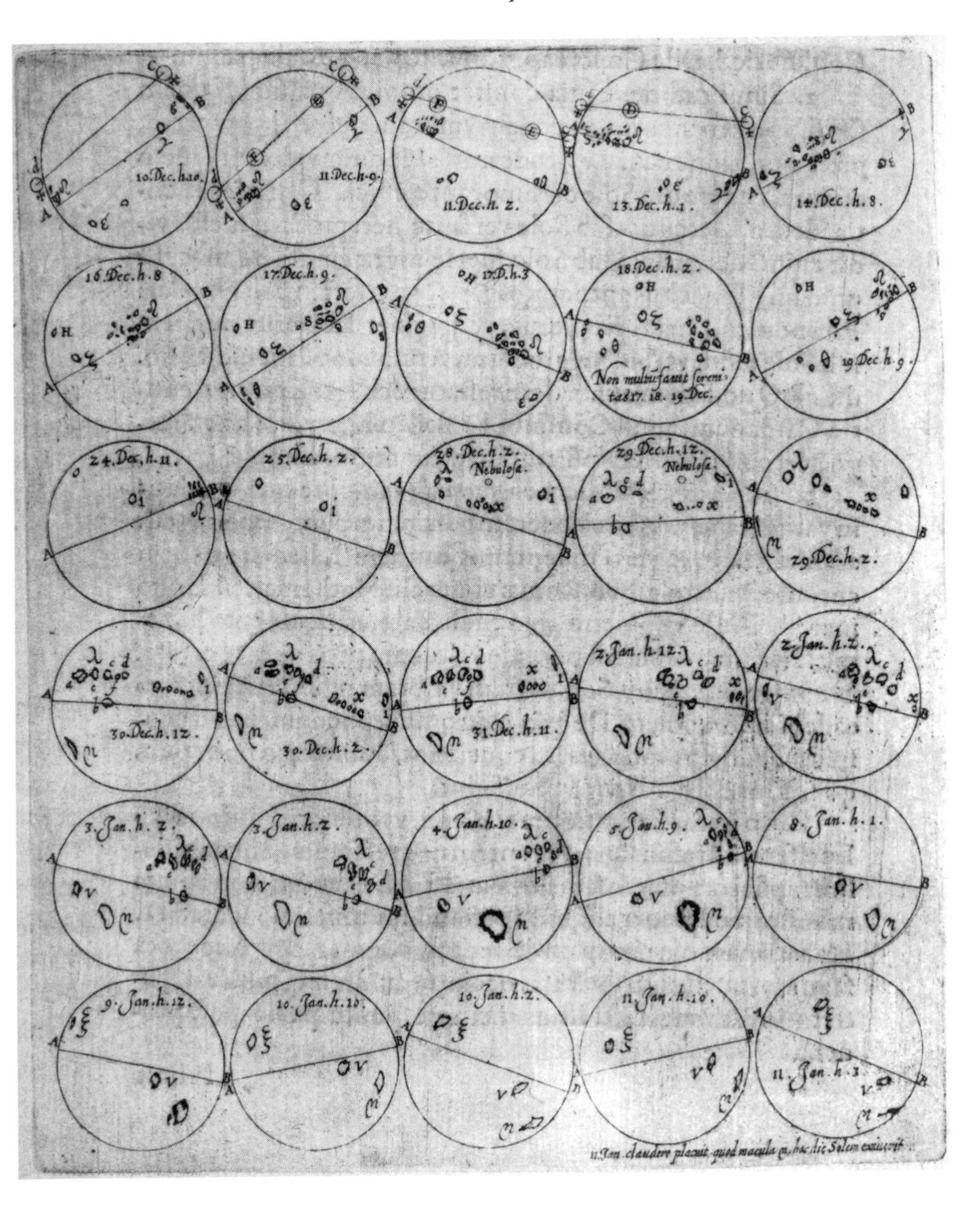

at a glance, and might delight the eyes and mind of the spectator. The reasons for what I am doing here will become clear a bit further on.

During the first four days, the star of Venus should have been observed in conjunction with the Sun at the appointed times on the line cd, which is the path of Venus across the Sun, slightly inclined to the ecliptic AB, and as large as mathematicians generally assume at present and following one of three hypotheses about the facts.[50] According to the first, Venus is carried to D; according to the second, it is carried to E; and according to the last, to F, as it is here depicted in appearance and location. It was also observed that the ecliptic AB closely approaches the course of the spots, for a reason to be given below.

All these observations, [made] as often as the weather allowed—and that was almost always when I observed—are the most accurate possible, although they are perhaps not so precisely drawn on the paper because of the failing of my hand. They have taught me many very remarkable things. For instance:

1. Spherical spots appear rarely, while most spots are miscellaneous in shape, oblong, and polygonal.
2. Very rare is the spot (if it exists at all) that retains the shape that it shows at ingress of the Sun all the way to egress; indeed, there are none that I know of that keep entirely the same size.
3. Most appear largest in the middle of their path below the Sun but smallest at egress and ingress.
4. Most either present or withdraw themselves at a rather large interval from the circumference of the Sun; very few allow themselves to be seen at the very limb of the Sun. But some almost equally large spots suddenly spring up around the middle of the Sun. Others, on the contrary, while just as large, suddenly decay in some way in the middle of their path and vanish from sight (that is, in the space of a night or day).
5. Many of the larger spots display from time to time small ones—here and there, in front, behind, and all around them—when suddenly the latter steal away from view again. And what is even more amazing, one large spot very often becomes a conjoined pair, while two or more frequently combine into one and continue in that form to egress.
6. At ingress, almost all of those that are carried in the same course are encompassed in a very narrow space; near the middle they become separated by a rather large interval; and at the end, when they approach egress, they usually appear to wait for and to band together with each other again, as at ingress.
7. The perimeter of almost all spots is roughened with, as it were, whitish and blackish fibers; and most spots, wherever they appear, are diluted by a greater pallor around the edges than in the middle of their bodies. Indeed, the shape of very many of the spots reminds the observer, as it were, now

50. That is, the three possibilities considered above about the time of the conjunction.

of some blackish snowflake, now of some small, ragged scrap of black cloth, now of a balled-up mass of hair trapped in a large torch; that is, each of these varies according to the thickness or density or opacity of its body, while others are more like a small dark cloud.

8. Some spots are blacker in the middle of the Sun and whiter at the edge.[51]
9. All appear to be carried faster in the middle than in the outer parts of the Sun.
10. The motion of all appears to be parallel to the ecliptic, but I am not yet so certain of this opinion. It is certain that spots that travel across the middle of the Sun have greater delays than those that tend more toward the edge of the Sun. From this follows a new and evident argument: these spots are not on the Sun.

The spots marked Ω were first observed on 10 December at ten o'clock; they were last seen on 24 December at eleven o'clock. In both observations, but especially the first, the bright interval AΩ, seen between the spots marked Ω and the edge of the Sun, A, was sufficiently wide [to account for the motion of] at least one day (if in fact we may argue from the experience of other spots). So the spots marked Ω will have spent at least sixteen days under the Sun, and their passage was as if along the ecliptic, AB. The first appearance of spot μ occurred on 29 December at two o'clock, when it still almost touched the circumference of the Sun, and it was observed touching it again and, as it were, separating from its upper part during its exit on 11 January at three o'clock in the afternoon. Its entire path under the Sun, while parallel to the ecliptic (as is evident from the observations), was usually fourteen days.

It is clear therefore that those spots that run along the diameter of the Sun, the ecliptic, remain longer below the Sun than those whose path moves to the south or north of the ecliptic. It is also indubitable (assuming that the Sun is unchangeable and hard, whether we know at present if it rotates or not) that these spots in no way adhere to the Sun.

After contracting upon entering the Sun, those spots marked Ω expanded during their course, and at the end contracted once again.

As the drawing shows, the spots took on various shapes. Near the ecliptic, however, they proceeded unchanged, from which you have most notably point 6 and some others, especially the next one. So, again, I argue strongly that the position of the spots is outside the Sun: for since the Sun is a hard and unchang-

51. *Quaedam maculae nigriores sunt ad oras Solis, albiores ad extremum.* Scheiner's meaning is not clear here, but from his discussion below it appears that he means darker at the edges and lighter in the center of the Sun.

ing body (according to the general opinion of all philosophers and mathematicians, and about this tradition I will speak elsewhere), it is impossible that such a great variation of dark shapes occurs anywhere except outside the Sun, even when some sort of [solar] rotation is granted. This change in shape is observed much more clearly in the spots marked μ so that they can be easily found, and I tried to transfer the shapes to paper faithfully. When they were first observed on 28 December at two o'clock of the afternoon, only two spots, a and b, had appeared, together with a rather long and thin little peak, c.[52] But on the following day the little peak had separated into two substantial spots, c and d. And although on 28 and 29 December a and b had appeared rather round, spot a had gradually changed, not into an oblong, but into a double spot, as it were; and on the 30th, another spot, e, also came between a and c, and in between c and d another smaller one, f; and for many days some of them showed small lateral spots joined to them.

Because at the same times and under the same aspect, some spots have nothing whatsoever attached to them, while most minute ones may be observed variously joined to [others], I do not therefore ascribe this appearance to a defect in the eye, or in the tube, or in the medium. Indeed, a defect in the glass, the medium, or the eye manifests itself in the same way in all the spots and operates in the same way toward the same place at the same time, as I have very often experienced. These spots grew incredibly up to the middle of their paths, except that spot b was distinguished by the fact that it remained blacker than the rest, of the same size, and always of a spherical shape, except on 2 January. Indeed, all the spots, even on 5 January, when they were bunched together and elongated and much diminished in breadth (except spot b) always appeared rather like black ink. But in the middle of the Sun's pallor they stood out more, which is shown by spot μ, twice as big in diameter as spot a. For although at times it was very black and hung down like a dead mole, in the middle of the Sun it looked thinner and as if sprinkled here and there with light; that happened throughout its entire body when its perimeter appeared more ragged and wooly. From this phenomenon I again draw a powerful argument that these spots are not on the Sun. For otherwise what reason could be offered to explain why some spots, such as μ, would appear black at the edges of the Sun, but rather whitish in the middle? For my part, I assign as the reason the irradiation of the Sun on the side of the spots that face away from us, and since its rays come more directly to us when a spot revolves around the middle of the Sun, it happens that the rays strike the spots more strongly and penetrate them somewhat, but this happens otherwise if the spots are nearer to the Sun's border.

52. These letters do not appear on the drawing for 28 December. See p. 197.

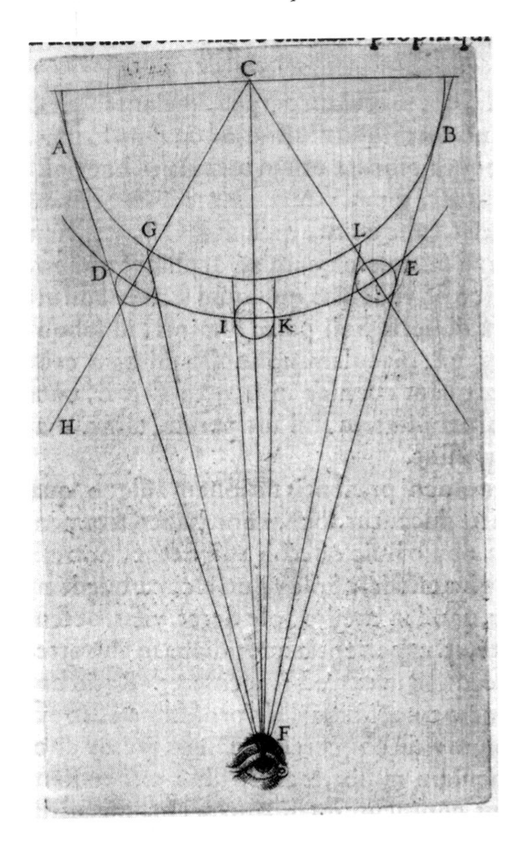

In the figure shown, let AB be the Sun, and about its center C let there be described an arc DE, the path of some spot around the Sun. Now if that spot is at D, between the Sun, AB, and the eye F located on Earth, the rays that descend, or could descend, from the Sun through the spot to the eye are only AF, GF, those that exit from the Sun between A and G, and perhaps a few other points refracted from the vicinity of the spot to the eye. And all these aforementioned rays reaching the eye are very weak, because of the Sun's spherical curvature AG, even to the naked eye. Accordingly, the rays transmitted through the spot will be much weaker and so will make a spot in this location appear barely illuminated to the eye and, following from this, will allow it to remain very dark. This blackness will be greatly emphasized by the contraction of the spot's width into a rather narrow space on account of the motion with which it goes around the Sun, as is shown in the illustration.[53] But ray CH,

53. That is, the spot's width is contracted near the limb because of foreshortening.

which very strongly illuminates the spot and its surroundings by shining on it perpendicularly, is at no time refracted to the eye, and therefore in this location there is no notable whitening of the spot.

Matters are otherwise when the spot steals into point I in the middle of the Sun; since the axis, CF, together with rays IF and KF, comes to the spot as well as to the eye at right angles,[54] it happens that the eye notes whatever the rays bring with them toward the spot from the direction of the Sun's incoming light, and the eye notes this spot somewhat less darkly, and therefore endows the somewhat dilute spot with dazzling whiteness. But this is not what happens at points D and E, because rays BF and LF, on account of their weakness, can have little or no effect on either the spot or the eye.

And I give this explanation for the present phenomenon, an explanation that does not apply if the spots are attributed to the Sun, and one that is not to be assigned as a suitable cause for such a clear effect. Indeed, if these spots were on [the surface of] the Sun, like some sort of chasm, they would have to appear much darker when seen directly (which would happen at the middle of the Sun, as daily experience in other matters attests) than when seen obliquely (which would happen at the edges of the Sun). The reason for this is that in the middle [of the Sun] the entire depth of that cavity would be exposed to sight, while at the border only the outer edge would be. You will say that the direct rays bordering that cavern and sent from the middle of the Sun to the eye would cause the eye to seem to observe some confused light surrounding that chasm. My first response is to ask why this does not happen, and happen much more, when the spot is at egress or ingress, especially because the edge of this large cavity would be seen. My second response is that spot b, with a diameter one-fourth of spot μ, would be blacker in the middle of the Sun than away from the middle, and also blacker than spot μ would be in the middle, although because of its smallness it would have to be entirely absorbed by the surrounding rays. And so, according to this not infrequent phenomenon (for which see also point 8, above) these shadow-casting bodies wander outside the Sun.

About spot μ

This spot has many remarkable peculiarities that I think I should briefly enumerate.[55]

54. IF and KF do not, of course, hit the spot perpendicularly, but they are perpendicular to the eye.

55. Note that while in the text Scheiner speaks of spot μ, in the illustrations it is consistently marked η.

1. It underwent ingress and egress almost on the circumference of the Sun itself, in the shape of a very thin and very black line, not making more white space between itself and the Sun than the breadth it shows to the eye, scarcely equal to the slenderness of the italic letter *l*.[56] Indeed, as the Sun was beginning to set on 11 January, at three o'clock in the afternoon, the upper part of the spot touched the periphery of the Sun, while its lower part extended slightly into the Sun. From this observation of rising and setting,
2. We have the very nearly exact duration of this spot below the Sun, namely thirteen days; for something is to be allotted to that very thin space left at sunrise and sunset, and if we allot much to it, we will have fourteen days.
3. The spot grew sensibly from ingress to the middle of the Sun which was on 4 January, and from 5 January it shrank in the same way until egress.
4. In the beginning its shape was a straight and very thin line, on which a hump on the right side gradually grew toward the middle of the Sun, increasing little by little from the least segment of a circle to a robust semicircle and more. But halfway through its course across the Sun, its right side gradually dissolved into segments smaller than a semicircle, adding to the diameter on the left, as it were, a kind of straight-lined angle until near egress the angle turned back into a line and thickened a bit on top, like a head, [so that the entire spot looked] like the shape of a club. From this you have a new piece of evidence that these spots are phenomena carried about the Sun—otherwise where did this angular bulge on the left come from?
5. The blackness of spot μ greatly surpassed the shadows of other spots and all spots observed so far (with the sole exception of spot b̲), from which we conjecture that it was very thick and dense.
6. In the middle of its course it was of a more dilute whiteness than it was at either end, and it is demonstrated that this happens because the more direct rays sent by the Sun to that same place are able to find some passage to our eyes. From this you will have gathered that these bodies are not altogether diaphanous, but rather that their thickness strongly hinders the transmission of rays.
7. Its perimeter, especially in the middle [of its course], whitened, roughened with innumerable tufts all over.
8. In the extremities it was equally distant from spot μ, and in the middle of its course was further away from it.

56. This is shown especially clearly in the figure for 29 December.

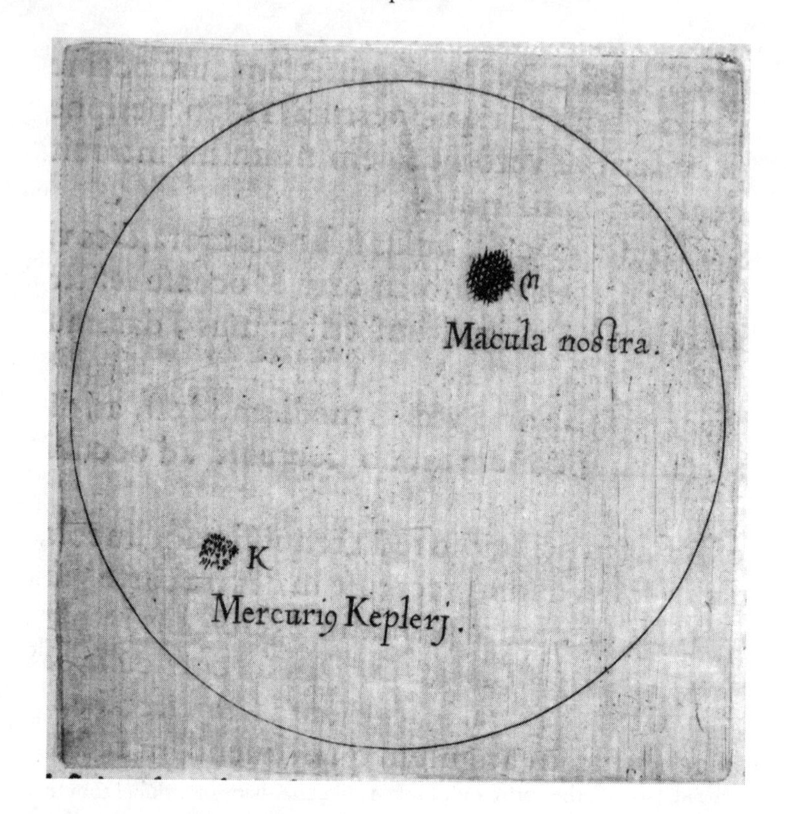

9. Up to this point, this body appears the largest of all those observed. The proportion of its visual diameter is at most [*ut plurimum*] one to eighteen to the visual diameter of the Sun, from which, if what Kepler writes in his *Mercury under the Sun* is true,[57] this spot must be much larger than Mercury, because on a paper illuminated through an aperture held up to the Sun it also showed a larger proportion to its disk. This happened because, being close to the Sun, much more than half of its body was illuminated, for which reason I do not shrink from comparing [the size of] this spot to that of Venus. And as you can see with your own eyes, Kepler's Mercury, K, manifested a smaller proportion [to the Sun's diameter] on the inverted image of the Sun,[58] and spot μ a greater one, which N.[59] showed very clearly to me and to others. And with dividers we carefully

57. *Phaenomenon Singular*, in *JKGW*, 4: 92–95.

58. *Phaenomenon Singulare*, in *JKGW*, 92–93. Kepler projected the image of the Sun on a piece of paper through a little hole in the roof and through a little hole in a window. In such a camera obscura arrangement the projected image is inverted.

59. Presumably one of Scheiner's fellow observers.

measured its diameter and found it smaller than what it should be. For if we had measured the shadow as it projected itself, it would have gone into the Sun's diameter fourteen times. Appreciate, then, this new argument: these spots are neither delusions of the eye nor a mockery of the tube or its lenses, since without a tube they are seen on paper.[60]

10. Spot μ always remained single, contrary to the behavior of other large spots, which thus far have generally unfurled into several shadows, as the depictions of the observations demonstrate. In the middle of its course, however, it sent downward a little tail, and near egress, on 9 January, it showed I know not what sort of appendage on its lower left side. Its motion was equidistant to the ecliptic. And indeed, in due time I will reveal the essence of the motions of these appearances, if God, the Muses, and the lesser gods are willing. But if the drawing of their shadows on paper is not a perfect correspondence, [the flaw] should be attributed to my eyes and hand.

Consequences

From the discussion so far, one may judge that Galileo's [notion] of a rugged Moon is not improbable, since the majority of the spots reveal themselves with this feature. But one may easily reject that hypothesis, whether humorous or serious, about the inhabitants of Jupiter, Venus, Saturn, and the Moon, since it would be absurd to place inhabitants on so many bodies.[61] Yet one will grant without difficulty that some bright light is reflected off the Earth, since all these solar stars persuade us of these points,[62] and they also suggest that the splendor of the Moon visible during an eclipse is caused by the rays of the Sun penetrating it in somewhat obscure fashion.[63] Whether this [explanation] could perhaps be asserted about the secondary light of the new Moon must

60. They cannot be seen with the naked eye directly except in exceptional cases such as the large spots mentioned in Einhard's chronicle (see above, pp. 11–14), and observed by Galileo in August 1612 (pp. 166–168, above). Moderately large spots can, however, be seen when the image of the Sun is projected through a pinhole onto a sheet of paper.

61. Galileo did not speculate in print on inhabitants of the Moon and planets. Kepler, however, was free with such hypotheses. See, for example, his *Dissertatio cum Nuncio Sidereo* (1610). See *JKGW*, 4: 299, 305. See also *Kepler's Conversation with Galileo's Sidereal Messenger* (New York: Johnson Reprint Corp., 1965), 28, 39.

62. In other words, the dark solar satellites showed that there were dark bodies in the heavens that shine with borrowed light. By analogy, light can be reflected by the Earth, which is a dark body.

63. During a lunar eclipse the Moon is not entirely dark. Following his argument about the partial transmission of sunlight through the solar satellites, Scheiner suggests here that sunlight is partially transmitted through the Moon and is apparent during a lunar eclipse.

justly remain undecided [*dubium merito fuerit*];[64] indeed, it is probable that stars[65] are of many shapes but appear round due to the light and distance, as in our experience of a lighted candle whose flame seen from a distance appears spherical but seen from nearby appears pyramidal or conical.

For the moment, I willingly omit many things and judge that my endeavors will satisfy the discerning reader. Now there are two kinds of rivals: those who, because they cannot excel in anything, pluck whatever is brilliant and however they can, and those who, while able to excel, do not, and who no sooner than they have realized that others are undertaking something noteworthy, fly over and soar in so as to carry off another's property. I am keeping both sorts at bay from our work with this letter. For the former will not assert priority, when they perceive things for the most part finished in this supplement; the latter will not arrogate these matters to themselves, when they observe that most assertions have been made and most objections have been resolved.[66] From this, since this phenomenon will be much more easily understood by you than by anyone else, you will soon understand and, unless I am mistaken, you already perceive with your most incisive mind what the progression of the spots will mean, since to you (in my judgment and by the leave of others) a nonpareil has been revealed (though perhaps it should not have been). I have dispatched this letter, so long in its preparation, to you above all as a matter of priority so that like our forefathers you might anoint it with cedar oil,[67] and preserve undiminished the glory of our Germany and your Augsburg, which I trust can be done if publication is no longer delayed. Soon you will receive from me as much or even more about these matters. What these are and where they lead, you will investigate together with me, but I fear that if you do not lead the way, they will almost be wrenched from our hands: for when the outcome of such an important matter is seen by so many, the mathematicians will not be able to restrain themselves. But they will contain themselves when they consider how far we have left them behind, and thus they will either express their own views or will certainly forgo claiming the work of others as their own. And this is entirely in your power to prohibit. God grant that, as we have begun, so in the

64. Presumably Scheiner is referring here to Galileo's argument in *Sidereus Nuncius* that the ashen light of the new and last Moon are due to earth-shine (*OGG*, 3: 72–75; *SN*, 53–57), an argument in which he had been anticipated by Michael Mästlin and which was supported by Kepler (*Dissertatio*, *JKGW*, 4: 301; *Kepler's Conversation*, 32–33).

65. It is not clear whether Scheiner intends planets, fixed stars, or both here.

66. Scheiner is thus staking two claims: that he had been the first to observe sunspots and that he had been the first to give their correct explanation.

67. Cedar oil was used to preserve valuable documents in Antiquity, and was a byword for immortal utterances. See Horace, *Ars Poetica*, 331, and Persius, *Satire* 1: 42.

glory of His name we may happily proceed and finish. Farewell, most glorious man and most munificent Maecenas of the learned.

16 January 1612

At the banquets of lords, it is the custom to set out not only fine dishes but also remarkable things, items that do not feed the stomach but rather delight the eyes and exhilarate the mind. As I myself was permitted, as you know, not very long ago, to attend banquets with the heavenly gods, I saw many astonishing dishes thus far invisible from Earth, and I sampled many things not yet granted to mortals, and though I enjoyed their taste and appearance marvelously, I wanted you to participate in them—you, and others. Then in the past few days, while I was delighting in these habitual feasts, behold, right in front of your unsuspecting [eyes], a certain great inhabitant of that celestial court—I mean Jupiter— offered to us something new and strange, which I myself am sending for you to examine along with me. The god so carried me away that I thought this interlude should be added to the regular, though interrupted, description of observations. Whether it has been done well, I leave to your judgment.

A is the star of Jupiter; the line BC is parallel to the ecliptic; the remaining letters mark the other stars seen near Jupiter at the distances they appear to be from it, and in proportion with its magnitude to them and also to the eye observing them from Earth through a perspective glass at the specified hour. B is the eastern point, C the western. One, two, three, and the rest of the numbers written at the top are situated in the north, while the lower part, opposite them, looks to the south.[68]

All observations were made with the greatest assiduity, always when the sky was very clear and generally very dark, that is, in the absence of the Moon. At these times, Jupiter was observed with various tubes of the best quality, and I have not yet seen any better tubes for observing the Jovian stars.[69] Others also observed the same stars. I mention this so that the trustworthiness of their observations is consistent with these appearances. Single observations are contained in a circle so that the stars to which they pertain can be examined without confusion. Now the following points have been established.

Since it is certain that the little stars on line BC are Jovian and not fixed stars, the sole dispute is whether the little lower star E is a wandering star near

68. The observer is looking south. See figure on p. 208. For a comparison with Galileo's observations of Jupiter's satellites during these days, see *OGG*, 3 (2): 529–530, although Scheiner made his observations generally an hour or so after Galileo made his.

69. See above, p. 45.

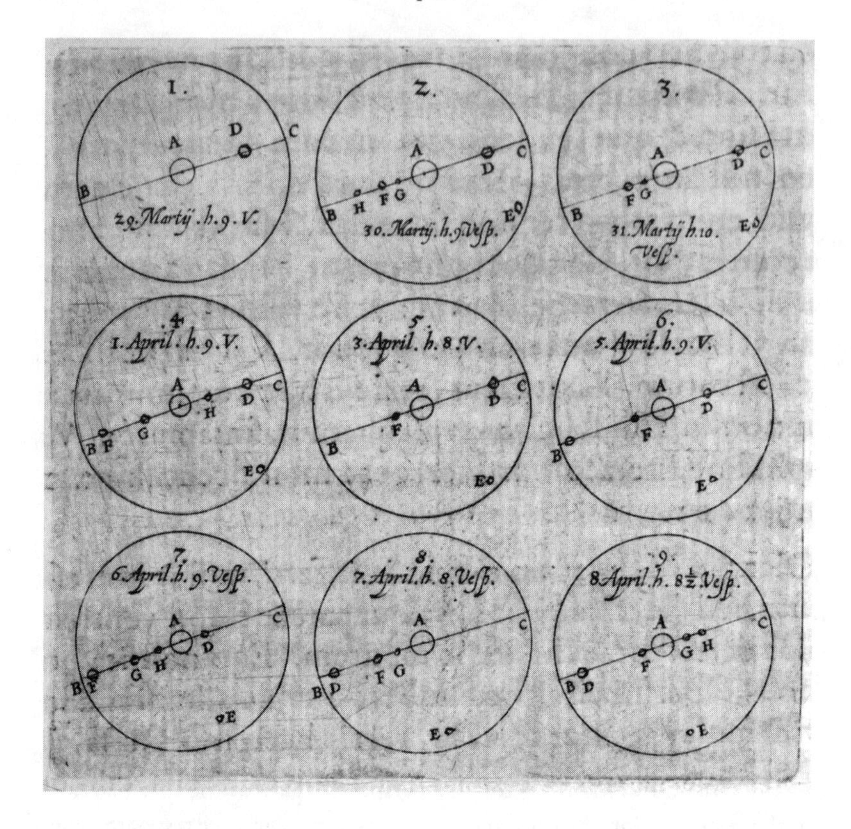

Jupiter, or one fixed in the firmament. For some days afterwards I pondered why, during the observations, I had noted it as "fixed." And, indeed, if I compare the first observation with the middle ones, and the middle one with the last, I am forced to recognize by this evidence an attendant of Jupiter.

My first view of this star was on 30 March, at which time the longitude of star D from Jupiter was, for instance, six minutes, as was the southern latitude of star E, whose longitude from Jupiter was almost eight minutes. The last view of it happened on 8 April, for on the following days it no longer appeared, although it was sought with great diligence, when the smallest Jovian stars were visible, and the sky and all other conditions favorable. At this time, the latitude of the little star E was southerly, the same as on 30 March, but its longitude from Jupiter was just about zero. The center of Jupiter, A, and the star E appeared on 8 April to fall on the same line, AE, perpendicular to the line BC. Therefore, eight minutes were consumed from 30 March to 8 April inclusively, to the conjunction [in longitude] of Jupiter and the star E. During these same ten days, that is, from 30 March to 8 April, Jupiter moved against the order of the signs, from east to west, at least fourteen minutes.[70] It is then impossible that star E was fixed, otherwise on 8 April it would not have been in longitudinal conjunction with Jupiter but would have been far removed from it toward the east, at point I.[71] This, however, did not happen, so it is not a fixed but rather a wandering star near Jupiter. And since from 30 March to 5 April the angle ADE, made by Jupiter, star D, and star E, was greater than a right angle, and after that time was always less than a right angle, it follows that the apparent motion of star E was faster than the motion of star D.[72] And this is one reason that drove me to this conclusion. Please hear now another no less powerful reason.

When the sky is clear and the night dark, fixed stars always appear the same in the brightness of their light and the size of their mass. But after that little star offered itself to our view through the tube on 30 March as very bright and very large—indeed, as large as any star of the first magnitude is to the sharpness of the naked eye, and as large as any Jovian star has been seen thus far—it slowly decreased in both respects during the ensuing days, so that [in brightness and in size] it fell behind the little stars of Jupiter to which it had previously been equal, until finally on 8 April, smaller even than the smallest Jovian stars, it was observed for the last time with great difficulty under a very clear sky and through a most excellent tube, even though in the first days of its appearance it showed itself sufficiently bright and solid even through weaker tubes. From 8 April to the day on which I am writing this, it was not seen at all, although the other little Jovian stars in their different places thrust themselves on us with much less light and apparent size than those possessed by star E. This star is, therefore, not a star of the firmament, otherwise why does it no longer appear?[73] Indeed, if it is a star of the firmament, it will appear on 21 April in the same position with respect to Jupiter in which it appeared on 30 March, as Jupiter's motion is at this time direct.[74] And so this star is not in the firmament, from which it follows that it is a lateral companion of Jupiter.

70. From 30 March to 9 April 1612 Jupiter's ecliptic longitude decreased by 14′. See Tuckerman, *Planetary, Lunar, and Solar Positions*, 824.

71. Point I is not indicated on the diagram of 8 April.

72. There are several things wrong with this statement. First, on 30 March angle ADE was slightly larger than a right angle, from 31 March to 6 April it was acute to the west of Jupiter; and from 6 to 8 April it was acute to the east of Jupiter. Second, Scheiner does not take into account the changes of direction of the motion of satellite D. Clearly, according to his own observations, from 7 to 8 April satellite D moved much faster than star E. See Galileo's comments on this at pp. 288–289, below.

73. It could have been a variable star. See Ashbrook, "Christopher Scheiner's Observations of an Object near Jupiter."

74. At the end of March 1612, Jupiter was nearing the end of its retrograde loop, moving against the order of the zodiacal signs. About 9 April it reached station and resumed its motion with the

We therefore now have a new and fifth lateral star of Jupiter, which I would like to dedicate and offer to you and your family, and since on 30 and 31 March, and also on 1, 6, and 8 April, four other planets of Jupiter shone brightly, it cannot be denied that this one, a brilliant *quinarius*,[75] likewise completes the number of his court.

We also consider that these attendants surround their lord on the flank, just as the satellites of the Sun go around him, for if this star completes its course around Jupiter uniformly, it will necessarily reappear in due time. Although Jupiter now always moves away from us and is much diminished,[76] I do not know whether after ten or eighteen days the appearance of this star will be repeated, since it would then have to be revolving in the lower part of its semicircle.[77] But if it never returns (which I rather fear, and which the other attendants of Jupiter somehow insinuate, since some suddenly appear and others suddenly disappear in almost the same way as do the shadows [i.e., spots] on the Sun),[78] I hardly know what we should say of these little stars. To express the order of their motion from observations of their appearances—of which I have many very accurate ones, my own and those of others—I consider extraordinarily difficult. And therefore I said not in vain in the published *Solar Spots*[79] that the theory of the sunspots and of the stars of Jupiter appeared to be the same.[80] Just as hitherto one spot always follows another, so the stars of Jupiter also appear. Where, then, you ask, do they go, and where do they come from? Here is our work, our struggle,[81] and it is here that Plato bids us to leave

order of the signs. Scheiner's figures are very close to recent values for Jupiter's motion in 1612. See Tuckerman, *Planetary, Lunar, and Solar Positions*, 824.

75. The *quinarius* was a small ancient Roman coin of silver, hence "a brilliant fifth addition."

76. Jupiter was at opposition in February (Tuckerman, *Planetary, Lunar, and Solar Positions*, 824), and therefore in April its distance from the Earth was steadily increasing, which caused it to appear smaller.

77. I.e., it would be in the half of its orbit on the Earth's side of Jupiter.

78. The satellites of Jupiter disappear when they are in conjunction with the planet or are eclipsed in its shadow cone. Their disappearances and reappearances are rather sudden events, on which at this time Galileo was beginning to pin his hopes for determining longitude at sea.

79. *Tres Epistolae*.

80. By 1612 Galileo, Simon Marius, Nicholas Claude Fabri de Peiresc, and his associate Joseph Gaultier de la Valette were making great strides in preparing tables of the motions of Jupiter's moons. See Suzanne Débarbat and Curtis Wilson, "The Galilean Satellites of Jupiter from Galileo to Cassini, Römer and Bradley," *The General History of Astronomy*, 4 vols. (Cambridge: Cambridge University Press, 1984-), 2A:144–157, at 144–147. Only those of Marius were published. See *Mundus Iovialis* (Nuremberg, 1614).

81. *Hoc opus, hic labor est* is the Sibyl's famous phrase describing the easy descent into and difficult escape from the underworld in Book VI of the *Aeneid*. Scheiner may have been more interested in the implication of the subsequent verses, where the Sibyl relates that the task is accomplished only by

off.[82] I am rightly afraid of rushing into judgment in a matter of such importance, but I do not despair of eliciting the truth soon. In the meantime, shine like this star of yours and, if possible, lift yourself from illness,[83] so that in good health you may for a long time shed light on your republic and on us, while your Apelles, familiar to you alone, and unbeknownst to all, might enlighten others.

14 April 1612

What is variously thought by different people about the solar spots, I have depicted on an Apellean canvas. There are some who still doubt the reality of the phenomenon, afraid that they are deceived by their eyes, or by the lenses, or by the interposed air. Most of them, when they cast down this fear, cut off the limbs and leave the head; one denies that [the spots'] parallax can or cannot be observed, another contends that the spots are on the Sun, another that they are always below it, another divests them of their splendor, and yet another of their blackness and density. There are many who deprive those spots that have just ingressed or will soon egress of their slenderness, and we may even find some who will deny the slower motion at ingress and egress and faster motion in the middle. In sum, there is hardly anything said that is not attacked by someone else. To satisfy myself, you, and everyone of the truth of this matter, if that is possible, I will in this letter respond to every objection, though in the briefest possible fashion.

Let me begin with illusions. Every error in the use of the optical tube (which, as it is directed to the Sun, one may aptly call a *helioscope*) must arise either from the eye, from the lenses, or from that transparent body that is between the tube and the Sun. The image that the eye seems to introduce on the Sun appears at times like a spider hanging in the middle of its web, at times like a fly, at times like a blackish sash of varying width, very ragged on its bottom edge, flowing across the entire Sun, at times like a dark cloud, and at times like some drops verging on blackness—all of which can be seen in the adjoined figures. In A you have spiders and flies, in B undulating sashes, in C clouds, and in D drops. And even through the best tube these all appear frequently and very clearly

those favored by Jupiter or begotten by the gods and raised to the skies by ardent virtue. See *Aeneid* VI: 129–131.

82. In *Timaeus* 38, after a brief and ambiguous discussion of the order and motions of the planets, Plato notes that a detailed analysis of this issue would need to be taken up later, and in moments of greater leisure.

83. Welser was ill and in great pain, suffering from gout. He was to die from that condition on 23 June 1614.

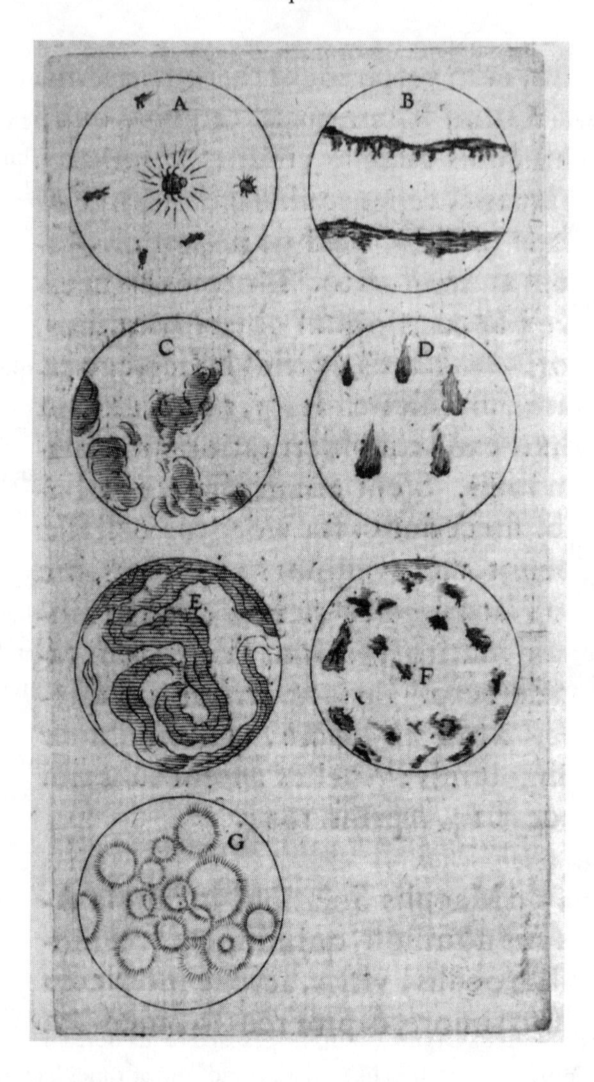

on the Sun. And so it is obvious that these arise from the agitated aqueous humor of the eye alone, for such are frequently presented to those who have moister eyes or those who have the benefit of drier sight, usually after a meal. They arise next because the right eye may render the same thing one way and the left another, at the same time and through the same helioscope; because often one eye may present nothing but the unadorned Sun and the things seen below it, while the other proffers these monsters; because at the same time and with the same tube, one man may see these [specters] and another may not; because in the space of one or two minutes, more or less, the same man sees this same thing either disappear or change its place on the Sun while everything else remains the same; because all these visions are generally driven away by tightly shutting the eyelids or rubbing the deluded eye; and last of all because if they appeared on the Sun, they would also appear when the tube is moved to some other bright or illuminated object that is near and well known to us (provided we have not earlier disabused our vision of them in the ways mentioned).[84] Indeed, I am not the only one who very frequently experiences these trifling spectacles, but all others besides me who have obtained and used this instrument, even one of modest quality. Therefore, those who are ignorant of this deception may easily attribute to the Sun what is in their eyes. And because these mockeries of the eyes are liable to change within days or even hours or minutes, they may ascribe what appears to be firmly on the Sun to the inconstancy of vision. I believe that what you once indicated in a letter—that somewhere in Italy the Sun was observed as if cut by black and nearly perpendicular lines—had this as its origin. And so that no one may doubt that these appearances are usually derived only from the eye and not at the same time caused by the lenses or the air, consider this: you will see all these things on a dark night at a sufficiently great distance from a candle or a burning lamp; and through the same tube, or even without any tube, you will see one thing with the right eye, another with the left, since it rarely happens that both eyes agree on the same image and you will see another thing with both eyes open, and yet another one with either; and you will see one thing and another person something else.[85] All eyes of each and every viewer, however, will see blackish spider webs or sashes of vapor undulating crosswise in the bright center, or clouds or mists weakening vision, or abundant little drops interrupting the light in various ways, just as the eye observes these same things on the Sun through a tube (though, of course, they have to do with the eye itself, as has been sufficiently demonstrated).

Another error is caused by the lenses of the optical tube. For they are either not of spherical roundness, and thus deform the shape of the object, or they have not been brought to a sufficient polish, and thus introduce mists or evenly scattered clouds, because the rays[86] of the optical pyramid, falling upon rough glass, may not find a passage, or may disturb the arrangement, and thus produce confusion in the eye. Or the lenses are defective due to swirls and bubbles; the former so overwhelm the object that the eye may thoroughly persuade itself

84. I.e., by tightly shutting the eyelids or rubbing the deluded eye.

85. The possibilities here are: (1) both eyes open, what is seen with the right; (2) both eyes open, what is seen with the left; (3) both eyes open, what is seen by both simultaneously; (4) left eye closed, what is seen with the right: and (5) right eye closed, what is seen with the left.

86. *Species* usually has the meaning of likeness or image, but in this case is best translated as "rays."

that what is in the lens is in the object, while the latter induces the opposite error.[87] The bubbles are either entirely transparent or they are not: if the first is the case, every single bubble radiates like an individual Sun; if not transparent, each one assails the eyes with what looks like single pieces of coal. And this happens by the inversion of the image, so that the bubbles that are in the right part of the lens will appear to the eye to be in the left side of the same lens.[88] But this is better understood from the figures in which E shows the swirling tracks of the glass, which corrupt the entire image of the object. This is evident when you transmit the Sun through a similar lens onto a smooth wall or reflect it onto a wall from a similar lens, for the entire image of the Sun will undulate with these tracks. It happens in the same way to the eye when the likeness of a visible thing is imparted through such a lens. From this we may also understand the reason things are not reflected as clearly from moving as from still water. Figure F shows the effects of opaque bubbles, which are not very different from the drops falling in the eye and the likenesses of spider webs shown in circles A and D above, except that those sorts of illusions are easily removed, while these, inherent in durable bubbles, never. In lens G transparent bubbles appear; each one of them pours forth rays like a small Sun and quite hinders clear vision.[89] So the errors due to the glass follow logically. Now since both eyes of one man, or even of various men, are employed in turn, they meet with entirely the same errors, and one or both eyes of anyone applied to the tube at any time will again meet the same error, and if at the same or a different time those lenses are removed from the tube and in their place others are inserted, those errors seen earlier are no longer perceived. Moreover, if those defective lenses are rotated in the tube, the apparitions wheel around with them, the order, number, place, and size being preserved all the while. Further, a tube directed everywhere else on the Sun, even in the clearest air, carries with it the same appearances. More remarkably, if you look through a tube at a window of your room facing you, or through a tube that is outside and directed toward a nearby white wall, or you hold a sheet of very white paper up to it, you will still observe all these appearances as before. These arguments are more than sufficient to prove that some appearances do not arise from the thing observed, or from the air, or from the eye, but from the lenses. And to be certain that the

87. I.e., what is in the object is in the lens.

88. Since the Galilean telescope presents an erect image to the eye, Scheiner is referring here to a single convex lens.

89. The objectives of Galileo's telescopes contain many small bubbles and a few precipitates. The bubbles scatter light but do not seriously interfere with the quality of the image, while the precipitates do not affect image quality at all. See Vincenzo Greco, Giuseppe Molesini, and Franco Quercioli, "Optical Tests of Galileo's Lenses," *Nature* 358.9 (July, 1992): 101.

bubbles in the glass lenses were causing this apparition, I smeared some small pieces of wax around certain ones and over others. And I found that some bubbles were entirely filled by the coat of wax, while others showed the usual appearances. In this it seemed to me most extraordinary that bubbles otherwise so small that they nearly escaped vision seemed to resemble very large pieces of coal. And this happened because of the nearness of the bubble to the eye, which draws it in under an angle made larger by the refraction of the aqueous humor as well as by that of the vitreous. For in its anterior convex surface, before a sensation is elicited, the refraction of the incident rays [*species*] dilates the narrow beam of radiation due to the convexity of the humors, and thus a greater angle of vision offers something otherwise small as exceedingly large to observation. So I conclude in passing two things: first, a thing can be represented in the eye much larger than it actually is; second, it is possible for the eye to perceive an object contiguous to its cornea, since those bubbles are very close to it. Indeed, to ascertain the truth of these things, I applied the telescope to my eye in the usual way and kept an eye open (which can be done) while inserting a smooth reed, and I drew it gently back and forth across the corneal membrane, and I saw it most consistently. Due to this most certain experience, the repeated assertion of Aristotle, that a sensible thing placed over the sense does not produce a sensation,[90] must be explained in the case of the eye if it were wholly covered, for in this instance the eye excludes all light necessary for seeing, as is evident in the case of the eyelids. Or [the assertion] is said to have surely been made in regard to that sensation that it makes and is ordinarily accustomed to make with the attention of the mind, for we perceive many things that we do not or cannot notice since a larger sensible object prevents a smaller one of its kind from being seen. Since the appearance of those bubbles, by which I was earlier more astonished, takes place in accordance with the inversion of the rays [*species*], so that the blisters in the upper part of the concave glass are seen in the lower, and those that are on the left will occupy the right part of vision, it happens that these rays [*species*] in themselves are exceedingly weak. This is both because they are inverted and because they are thin, since they exceed the width of the object from which they arise and since they make use of very weak light. So I give the following reason why things that are farther from the eye keep those most near from being noticed: the former emit rays more direct, more focused, and brighter, while the latter are all weaker. But it seems best to submit to your eyes, too, this ocular experiment: in the adjoining figure, let the concave glass be A, opposite to which the eye, B,

90. *De anima* 2. 7. 419a12–22. See also David C. Lindberg, *Theories of Vision from Al-Kindi to Kepler* (Chicago: University of Chicago Press, 1976), 6–9.

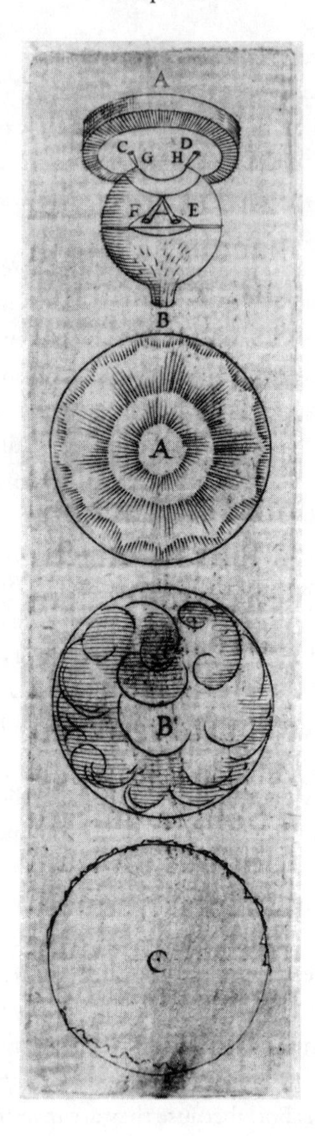

sees two bubbles in the concave lens, C in the left part of the glass and D in the right part, so that the left bubble, C, falls at E in the right part of the crystalline humor and D falls at F in the left part of the same humor close to the points of inversion G and H. And since the distance GC is smaller than GE, it is then necessary that the base of the optical cone GE is greater than the base of the cone GC and therefore that bubble C appears much larger at E than it is at C. But there will be more details about these things at another time.

Further, I recall the misrepresentation arising from lenses, and that spectacle that comes about when lenses are at an improper distance from each other. For either they are extended too far and separate the Sun into its rays of various colors, or they are contracted too much and condense it into clouds; you may consider both cases in the adjoining figures, in one of which, A, the Sun is represented as amplified too much, while in the other, B, it is too compressed and compacted into whitening clouds of uneven boundaries. From this it has arisen, I think, that many have observed a not negligible roughness on the Sun, and there will be more about this a bit later. For the same reasons, a certain author in his *Nodus Gordius*, because of a poor, hasty, and even ignorant observation—one in which the star of Jupiter was kindled into a little three-pronged torch—denied there were Jovian stars.[91]

Varying atmospheric mixtures between us and the Sun can lead to a third kind of error concerning the spots, but I have nothing in particular to deplore here. It makes its influence felt in two ways: one in coloring the Sun and the spots, and the other in making the Sun rough or making the spots tremble. For thin clouds increase the blackness of the spots, and thick vapors color the light of the Sun, and these dense and viscous vapors also make the Sun at the limb like a very white cloud with somewhat ragged borders, bright yet agitated vapors roughening it in many places around the periphery. This was the principal reason that the edge of the Sun also seemed to many observers to be rather full of indentations.[92] But it is certain that this illusion is imposed on the Sun by interposed solar vapors, for whenever the periphery of the solar disk appeared broken, it was soon restored, and whenever whole, it was soon broken. This happens with constant change until these vapors quiet down or the Sun emerges from them toward its meridian altitude and then shines steadily and with a perfect roundness. The figure of the Sun with its wavering circumference is offered at the letter C. The remaining errors are very similar to those in the previous [figures]. But the restlessness of these vapors is frequently reflected in the spots themselves, for the spots often somehow bubble up in their place, shake, and vibrate with a kind of nodding. But all these things relate to the roguery of the interposed vapors.

91. Jacob Christmann, *Nodus gordius ex doctrina sinuum explicata. Accedit appendix observationum, quae per radium artificiosum habitae sunt circa Saturnum, Iovem et lucidiores stellas affixas* (Heidelberg, 1612), 42: "At the beginning of the month of December, through both cylindrical radii [i.e., telescopes], the body of the star of Jupiter appeared in three distinct scintillations and represented two diameters of glitter. Indeed, the body of Jupiter was observed altogether on fire, as it was separated into three or four fiery balls from which thinner hairs spread downward like the beard of a comet." On Christmann, see H. Ludendorff, "Über die erste Verbindung des Fernrohres mit astronomischen Meßinstrumenten," *Astronomische Nachrichten* 213, no. 5112 (1921): 385–390.

92. David and Johannes Fabricius saw the solar limb in this way. See p. 30, above.

And these, indeed, are the factors that can obscure the clarity, undermine the truth, and infect the health of this most celebrated phenomenon, but I make light from these shadows, knowledge from errors, and medicine from poison. That infamous scorpion, even if its sting may seem to hurt a little, exudes, when forcefully compressed, an oil by which the wound is gently healed.[93] Come then, let us first rip the masks from these monsters, let us use lenses that are without the defects we have discussed, let us diligently clean our eyes and apply them properly to the tube, a tube, I repeat, free from these defects, and under a clear sky let us admit the Sun into our eyes. And in this case, should any shadows appear, I affirm that they will not be shadows but true perihelial bodies,[94] because they satisfy none of the conditions that I have established concerning deceptions, but move perceptibly every day under the Sun from east to west, either in the plane of the ecliptic or parallel to it, against the order of the signs. And I do mean under the Sun, since they move in a higher semicircle above the Sun from west to east following the sequence of the signs. And this argument is impossible to refute. Once more, having seen these things, let us bind up more tightly these specters that cast I know not what cataracts on [the eye of] the astronomer, and presently apply soothing oil. Since faults of the air either do not affect the solar spots at all, or remove them entirely from sight, the spots cannot be said to be defects in the air. But the aperture of the tube, if too large or too small, equally removes spots from vision so that from this source no danger threatens either. Single bubbles, single grains of sand in the lenses, and single drop-like threads floating in the eye counterfeit spots, for he who perceives these flies together with the spots will by no means distinguish them except by previously applied remedies of rotation, translation, and compression.[95] And that oil of the compressed scorpion cures the injured eye, and separates falsehood from truth. Now the solar spots will always be observed stable below the Sun while the other [false spots] will revolve on every side and wander to every location. And this argument is impossible to refute. From this, many observers will know how to judge any black bodies they have seen in very pure air, even if according to an eyewitness the spots are not in the air, but rather either in the eye or in the lenses.

If I now show that the solar spots can also be seen without a tube, by the eye of any man, what will any opponent offer other than a sham objection?

93. This belief apparently originated in Europe in the Middle Ages. See Gary A. Polis, ed., *The Biology of Scorpions* (Stanford: Stanford University Press, 1990), 472–474.

94. I.e., bodies close to or around the Sun.

95. Apparently Scheiner means compression of the image that has been magnified too much by shortening the separation between the objective and ocular. See above, pp. 216–217.

Certainly, neither the eye, nor the glasses, nor the air can be blamed. Accept it, therefore. If through a round hole of about this size—O—or a bit larger,[96] the Sun is admitted perpendicularly onto a clean sheet of paper or some other white plane, it shows itself and all the bodies below it in proportion to the distance, position, and number that they retain among themselves and to the Sun. And I have made many observations in this manner, and to anyone who so desired I have shown, whenever visible, spots so large, dense, and black that they were quite apparent even through thin clouds. And this argument is devoid of all suspicion of deceit. It is not necessary, as many erroneously believe, that the place of observation be all that dark; for I observe these things in places where I can write and read. A large distance from the aperture of inversion is helpful. On the other hand, if you hold a clean mirror up to the Sun and reflect the solar image from the mirror onto a clean wall or sheet of paper at the required distance, you will see solar spots, in number, arrangement, and size with respect to each other and the Sun. And this long-sought method of observing, which I learned from a certain very good friend, is also absolutely without any blemish or error.

Finally, in addition to the experience and the weight of the reasons proffered in this and in my previous letter, there is also the approval of the most learned men of our age, some of whom are ear-witnesses,[97] and others eyewitnesses to this phenomenon. Of the ear-witnesses, that is, those who prefer to raise their ears rather than their eyes to the secrets of the Sun, there are very many who by their authority must justly persuade any obstinate person and lead him away from his error. Through you I came upon the judgment and names of these most distinguished men and will recall them to mind with what I judge to be a not unpleasant memory. The luminaries of our age accepted the essence of this phenomenon very willingly: in Italy, the most reverend and illustrious Cardinal Borromeo, archbishop of Milan;[98] Andrea Chiocco, a physician of Verona;[99] Giovanni Antonio Magini, most celebrated,

96. In the original the circle is about 0.25 cm in diameter.

97. *Auriti testes.*

98. Federico Borromeo (1564–1631), nephew of the saint and cardinal Carlo Borromeo (1538–1584), founder of the Ambrosian library in Milan. He was made a cardinal in 1587 and became archbishop of Milan in 1595. In 1613 Galileo sent him a copy of *Istoria*, for which the cardinal thanked him without expressing an opinion about the nature of sunspots. See *OGG*, 11: 498, 511. On Borromeo's interest in observation, see Giliola Barbero, Massimo Bucciantini, and Michele Camerota, "Uno scritto inedito di Federico Borromeo: *L'Occhiale Celeste*," *Galilaeana* 4 (2007): 309–341; on sunspots in particular, see 311, 324, 325, 326.

99. Andrea Chiocco (1562–1624) studied medicine and philosophy at the University of Padua and practiced medicine in his native city of Verona. Among his lost works is a *Discorso sulle apparenze solari*, a defense of more traditional views of the sunspots. In 1617 he was named censor

and shining with his own light;[100] the Very Reverend Angelo Grillo;[101] Ottavio Brenzoni;[102] Leonardus Canonicus;[103] and some others unknown to me by name; in Mainz, Reinhard Ziegler, rector of the Society of Jesus;[104] the very learned Simon Stevin, in the Netherlands;[105] in Bohemia, Johannes Kepler, the Imperial Mathematician;[106] in our Germany, Johannes Praetorius, now professor at Altdorf, formerly the astrologer of Emperor Maximillian, according to a trustworthy account;[107] Johann Georg Brengger, doctor of medicine in Kaufbeuren,[108] and many others who do not at present come to mind. And

---

of books by the Holy Office. He contributed a poem to Giovanni Alfonso Magini's *Tabula Tetragonica* (Venice, 1612), and perhaps it was Magini who brought Chiocco's opinion to the attention of Scheiner. Chiocco was, however, also a correspondent of Welser; see Welser, *Opera*, 1: 43–44. More generally, see C. Colomero, "Andrea Chiocco," *Dizionario biografico degli italiani*, 25: 11–12.

100. See above, p. 65, note 26.

101. Angelo Grillo (1557–1629), a Benedictine monk and abbot, as well as a prolific poet and madrigalist who wrote under the pseudonym of Celio Liviano. He was a favorite of Maffeo Barberini and Torquato Tasso, an enthusiast of the new telescope, and a sometime correspondent of Chiocco, Galileo, and Welser.

102. Ottavio Brenzoni (ca. 1576–1630), a physician of Verona, wrote occasionally about astronomical and astrological matters to Galileo from 1605 to 1610; see *OGG*, 10: 137–141, 152–153, 216, 224, 269–270, 272, 309.

103. It seems likely that "Leonardus Canonicus" is a misprint for "Laurentius Canonicus," or Lorenzo Pignoria, *canonico* of San Lorenzo in Padua since 1609, and the first man in Galileo's circle to have notice from Welser about the observations made in Germany.

104. Johannes Reinhard Ziegler (1569–1636) entered the Jesuit order in 1588. He received his doctorate in theology at the University of Mainz in 1606 and served as the university's rector from 1609 to 1612. He taught the mathematical sciences, philosophy, and ethics at the university. He edited Clavius's *Opera Mathematicae* of 1612. From 1612 until his death he was the confessor to three successive archbishops of Mainz. His scholarly reputation was in the mathematical sciences. See Albert Krayer, *Mathematik im Studienplan der Jesuiten: Die Vorlesung von Otto Cattenius an der Universität Mainz (1610/11)* (Stuttgart: Franz Steiner Verlag, 1991), 48–50 and passim.

105. Simon Stevin (1548–1620). His works do not contain any mention of sunspots. Perhaps Scheiner had in mind the little tract *De Maculis in Sole Animadversis, &, tanquam ab Apelle, in Tabula Spectandum in Publica Luce Expositus, Batavi Dissertatiuncula* (Raphelengen: Plantin, 1612), published anonymously in reply to Scheiner's *Tres Epistolae*, but written by Willebrord Snellius. See Rienk Vermij, *The Calvinist Copernicans*, 44–45, and pp. 316–318, below.

106. Kepler's reactions to Scheiner's *Tres Epistolae* are contained in a long letter to Matthäus Wacker von Wackenfels written early in 1612. See *JKGW*, 17: 7–15. A copy of this letter is contained in Dillingen Studienbibliothek MSS 2°, 247, the volume that contains the correspondence of Johannes Lanz (449). In his *Prodromus* Scheiner mentioned owning a manuscript of Kepler's concerning the sunspots; see Luigi Ingaliso, *Filosofia e cosmologia in Christoph Scheiner* (Soveria Mannelli: Rubbettino, 2005), 147 n. 35.

107. Johannes Praetorius (1537–1616) was professor of mathematics at the University of Altdorf from 1576 until his death. His sunspot observations were published by Johann Caspar Odontius, *Maculae Solares ex selectis Observationibus Petri Saxonii Holsati* (Nuremberg, 1616).

108. Johann Georg Brengger took his degree in medicine from the University of Basel in 1588 and became a physician in Augsburg and Kaufbeuren. He was a student of mathematics, and he

although varying in judgment and disagreeing among themselves and with me, they all nevertheless agree with pleasure on the main point that this experience is truly manifest in the object itself and is not some trick of the glass or the eye, even if they have never observed it with their eyes. They see to it that what is asserted with reason—the true work of the philosopher—is not disproved by rash persuasion but is cautiously weighed with maturity of judgment.

Now I turn to those who will confirm the same things not only by approbation but also by perception, and Italy has furnished numerous such individuals. Indeed, Christoph Grienberger of the Society of Jesus, a famous mathematician,[109] began seeing them on 2 February, on the feast of the Purification of the Holy Virgin. And Paul Guldin, also in Rome, and a famous mathematician of the same society,[110] observed spots on the Sun from 18 until 22 March, and since the spots he observed include worthy considerations, they are to be reconsidered in more profound detail. Now all were made with a pinhole camera,[111] and therefore the shape, place, and magnitude of the spots must be judged sufficient for drawing many conclusions. So, from 16 March to 4 April, the appearances of the Sun were thus:

It was necessary to lay out these observations, both so that you too may see how very little I fear the critic, since I scarcely doubt the observations made on these days by others, and so that Paul Guldin may see that he agrees with me as much as I agree with him, which I believe, is down to the last detail.[112] And because almost everything in these phenomena borders on the miraculous, they are shown in the course of these days. Spot a[113] was seen on 16 March by me and by a certain very learned man, a Roman professor of mathematics,[114] with and without a tube, and at that time it was equal in diameter to Jupiter at its largest. But gradually it decayed in size and shape; it appeared cleft in two on 18 and 19 March, and trifold on the 20th, and then returned to a simple shape, until after the 23rd it ceased to be seen. But from this appearance I do not immediately dare to infer that these bodies—which are exceptionally large—are themselves augmented and diminished or that they rise up from the inner depths and perish, since the same spot a, on 22 March, stole away from the helioscope and

corresponded on this subject with Christoph Clavius, and (through Marc Welser) with Galileo about the heights of lunar mountains. See *OGG*, 10: 460–462, 466–473; 11: 121–125.

109. See above, pp. 171–173.

110. See pp. 47–49, 171–173, above.

111. *[P]er foramen inversionis.*

112. Scheiner sent Guldin a copy of *Accuratior Disquisitio* on 17 September 1612. See Scheiner to Guldin, 17 September 1612, Graz, Universitätsbibliothek, MS 159, no. 21.

113. This is the spot at the top left of the diagram for 16 March.

114. This may have been Odo van Maelcote; see p. 47, above.

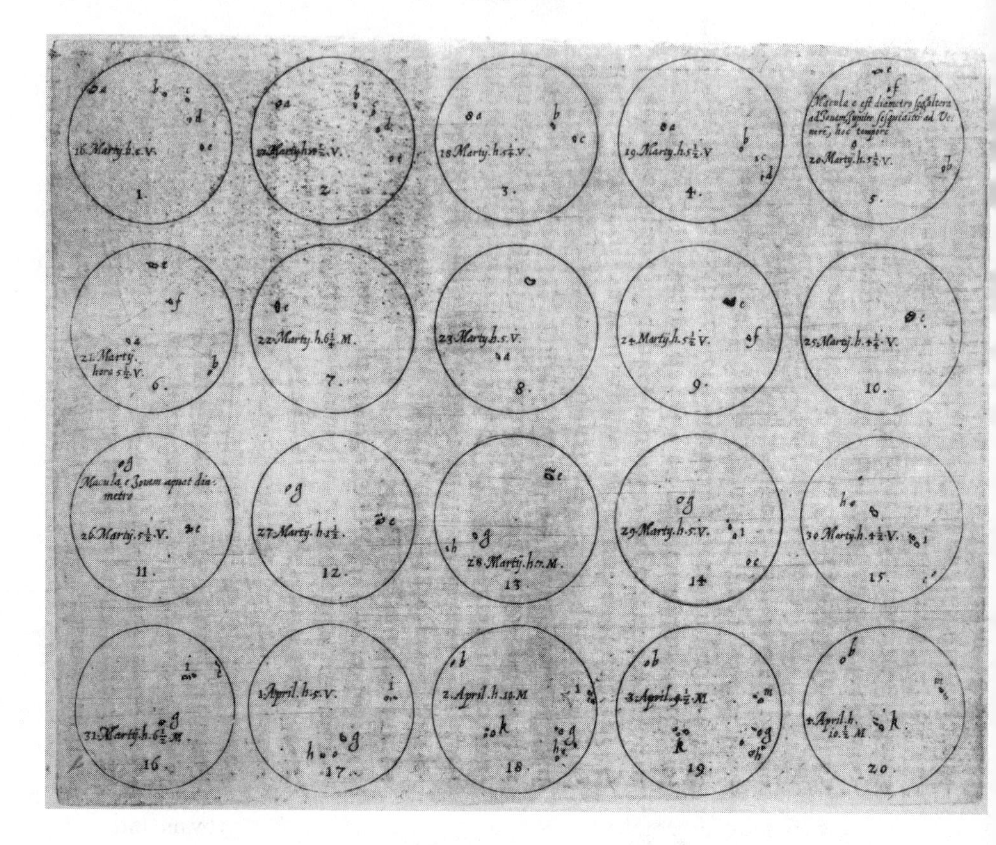

on the 23rd appeared again. But f̲, after a concealment of two days, returned on 24 March in another place as a small and obscure shadow. Since this often happens on other occasions even in the smallest and thinnest of these bodies—in such fashion, that if it were appropriate, I would report the hour, day, and month—I am forced to conclude, against what many believe, that these bodies can hardly be rising and perishing, but rather that appearances, disappearances, reappearances, and aspectual transformations of this kind result from other causes, which can be ascribed to motion, to rarity and density, to position with respect to the Sun, to reciprocal illumination, to change of the accidental medium, and, finally, to their particular shape. I have said all this not because I would not desire or would not be able to abandon one opinion for another, if the truth of the matter has led us there, but because we are following the opinion that has hitherto been more customary and more accepted by philosophers. Now as for spot a̲, it was seen again, first by the above-mentioned professor on 17 March and then also by a certain other very learned man whose great

chronology we will, I hope, soon see.[115] Further, it imparted to us an image of such dense blackness that when the Sun's circle was projected on paper through clouds obscuring the Sun almost entirely (as you see in the adjoined figure), the image, though very black, made its way to the eye. It was, therefore, darker than the clouds, for a less dark thing is not visible at all through a darker one, just as a fine web is not visible through a coarsely woven sack, although the sack may be visible to the eye through the web. Spot e showed this same characteristic to an ever greater extent, and most of the larger spots do the same right up to today. The matter needs only observation, and I have many eyewitnesses to it.

From the spots g and h above, you may gather the unevenness of motion, for spot g entered the Sun on 26 March—indeed, even before this but without being seen—and the entry of spot h occurred on 28 March. But the egress of both seemed to occur at the same time, that is, 4 April. You easily see what follows from this, namely, that these shadows are not on the Sun, unless you want to conclude that the Sun is more changing than the sea. For since spot e moved forward under the Sun for a minimum of twelve whole days, g at the very most eleven, and h at most nine, it is impossible that they are indeed on a rotating Sun, unless it is very much corrupt in some of its parts. Just as spots a and f disappeared before egress, so the three spots l and the two spots m and some others did not appear in the beginning. Their slowness of motion at ingress and egress, their swiftness in the middle, and how they change, you may learn from numerous spots but especially from spot e, which from its ingress grew for several days but then gradually lost size, exhibiting slenderness on both sides, as is shown in the drawing. Nearly all these observations were made not only with a tube but also with the Sun projected through an aperture onto a sheet of paper held perpendicularly, and thus the disk of the Sun, cast on the paper, supplied the true location and motion of the spots, and the tube directed to

115. Matthäus Rader. See pp. 40–41, 49, above.

the Sun supplied the shape.[116] From this, I suppose these observations could be exactly the sort required, especially by you, so exalted in all things. In just this way, long ago I compared my observations of the spots with those of the learned Vincent of Padua,[117] and sent them for you to see. But the renowned modesty of a certain very noble and learned Venetian is not to be passed over; when his own name was suppressed, he assumed that of Protogenes, and his own name is as worthy of celebration as is this pseudonym.[118] It is he who relates, in his judgment concerning the spots, among other things, that *the consequences of these observations are the following*:

1. That these appearances are not merely in the eye.
2. That they are not a flaw in the glass.
3. That they are not a sport of the air; they are neither in the air nor revolving in any heaven that is much lower than the Sun.
4. That they are moved around the Sun.
5. That they are not far from the Sun, because otherwise they would appear illuminated at a great distance from it, as are the Moon, Venus, and Mercury.
6. That they are very flat or thin bodies because along the longitude of the [Sun's] sphere their diameter is diminished but along the latitude it is conserved (this is because they become thinner near the swelling of the solar perimeter).
7. They are not to be admitted among the number of the stars,
   1, because they are of an irregular shape,
   2, because they change their shape,
   3, because they all undergo an equal motion and, since they are not very far from the Sun, should already have returned several times, contrary to what has happened,
   4, because spots frequently arise in the middle of the Sun that at ingress escaped sharp eyes,
   5, because sometimes some disappear before having finished their course [across the Sun].

And that excellent Protogenes learnedly observed and wrote down these things, for the most part agreeing with my findings; those who are trying to

116. In other words, the location of the spots on the Sun was mapped on the projected image in the camera obscura, while their shapes were ascertained by direct telescopic observation.

117. Vincenzo Dotti (1576–1629), known for his architectural drawings. He began observing sunspots toward the end of 1611. See Paolo Gualdo to Galileo, 11 November 1611, and Fortunio Liceti to Galileo, 16 December 1611, *OGG*, 11: 231, 244.

118. The ancient Greek painter Protogenes was Apelles' legendary imitator and rival.

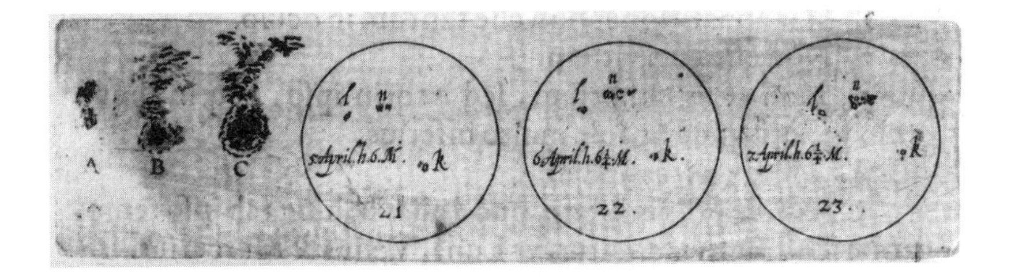

overthrow the better part of these arguments may take instruction from him, if they do not wish to learn from me.

The astronomers, however, are struggling over two of these issues—whether these bodies are thin, and whether they are impermanent or stars—and the dispute is still before the judges. Likewise the debate over whether they move unequally could hardly be greater, even though I have so often observed them moving in that fashion. But if this is true, as I think it is, the question of why the same configuration of these spots does not return has achieved some closure. And we must not forget another witness, greater than all. For on 5 April, Galileo Galilei observed the spots as in drawing A, but on 6 April as in B, and, finally, on 7 April as in C.[119] And I, during these three days, found that the Sun was the same, and the proportion of sizes and shapes is correct, both with respect to each other and to the Sun.[120] From this it is clear that Galileo does not disagree with me at all about the main shapes and the conformation of all the spots with respect to each other, but only departs from me somewhat in the precision suitable for single spots. This could result either from the strength of the light or from the shortcomings of the tube, or from the intervention of the medium, or, finally, from weakness of the eyes.[121] For I have very often had the experience that at one and the same time I observe spots separated from each other and in the next instant observe them confused and joined to one another by, as it were, one movement. Why and in what manner this may happen I would now demonstrate except that the great length of the letter forbids it. Thus the observations that Galileo produced from 26 April to 3 May[122] likewise agree completely with mine, by which it may be confirmed that these phenomena are entirely without parallax with respect to the Sun, since in parts of the world as distant from each other as our Germany and his Italy they appear in the same place on the Sun.

119. Galileo's first letter on sunspots. See above, pp. 99–100.

120. Galileo does not specify the location of this spot except to say that during those three days it was always far from the solar circumference. See p. 99.

121. In other words, Scheiner assumes that the Sun's light in Italy was stronger, the atmosphere more transparent, and Galileo's telescopes and eyes better.

122. See pp. 99–100, above.

I here omit innumerable other eyewitnesses of this phenomenon living here with me, men skilled in mathematics as well as theology and jurisprudence.

The recent lunar eclipse that happened in May has instructed me in the matter I treat here. It began before nine o'clock in the evening by about half a quarter [*dimidio veluti quadrante*]; it ended at twelve o'clock at night and so in this way exceeded the calculation in duration and likewise in magnitude, since it was at least eight digits. But I am not merely stirring up this matter here, for these other issues offer something: the terrestrial shadow was most tenuous farthest away from its center and so conferred a certain mixing of sunlight with it upon the Moon, as was manifest to those observing. But closer to the center it was so dense that it offered not a speck of the lunar body to be observed, either with the naked eye, or with common eyeglasses, or with eyes equipped with a tube. The perimeter of the Earth's shadow was circular, and it did not exceed the blackness of the ancient lunar spots in darkness.[123] Because of this, the confluence of the terrestrial shadow with these spots offered an uneven perimeter to the eyes such that we suspected that it arose from prominences of the Earth. But when the eclipse waned, we saw that these shadowy protuberances remained on the Moon and were the ancient spots. Finally, before the end of the eclipse, using a tube, we saw a small segment of the Moon through the attenuated shadow of the Earth, although we had frequently tried in vain to do this through the middle portion of the shadow. From these observations, I conclude that the Moon does not possess any light of its own, that the uneven contours of the Earth are not perceptible by observation from afar, and that most sunspots are bodies no less opaque than the Earth,[124] since their shadow appears blacker than any of the ancient lunar spots or, indeed, any of the new spots, as my innumerable fellow observers, men very experienced in these matters, freely and willingly acknowledge. However, the inconstant wavering of the terrestrial shadow on the Moon that I so frequently observed cannot come about except through the shifting agitation of the vapors between the Earth and Sun, [a movement] that cut off the solar rays in changing fashion and thus made them tremble and vibrate.

In the same month an eclipse of the Sun was observed to have begun at about 10 a.m., and it ended at 12:45. Altogether it lasted approximately two and three-quarter hours, and it scarcely attained seven digits (about which, however, more at another time). The things worthy of notice and relevant to the matter at hand are these: The tube made no distinction whatsoever as to darkness between that

123. I.e., the large spots visible to the naked eye.

124. Galileo's annotation: Sunspots [are] as opaque as the Earth.

part of the Moon which covered the Sun and that part which extended beyond it, but neither did it distinguish the entire Moon in any way from the rest of the sky surrounding the Sun or, finally, from a body of any kind whatsoever. But around the middle of the eclipse, during the space of half an hour, the tube showed us that the Moon's perimeter, which covered the Sun, was somehow enveloped by a golden circumference extending beyond the Sun on both sides to the length of about one digit in a golden circular arc, and this appearance was no illusion. Next, the tube also showed us sunspots as black, indeed, blacker than the Moon itself appeared[125]—as we all assert, moreover, according to tradition—for the Moon was approaching a more dusky color. This is confirmed by the fact that the Sun, projected through an aperture onto a sheet of paper, also distinctly represented the shadows of the spots. And the tube accomplished this when the sky was very clear. Eyes without a tube, however, either by themselves or aided by common spectacles, perceived something else and more wonderful; I mean to say that my eyes and those of many others, by their own testimony, perceived this more wonderful thing at every moment of this eclipse. Indeed, we saw—all who strove to see—that the part of the Moon that covered the Sun was entirely like crystal or some transparent glass but was so uneven that in one place it had turned all white and in another it was just beginning to whiten; and I steadily observed the entire Sun like this but with one great difference:[126] the part covered by the Moon shone through with a very gentle and exceedingly feeble radiance. And, indeed, it was not at all possible to confirm this observation with a tube until near the egress of the Moon from the Sun, when one person most firmly asserted that he saw through a tube the entire circumference of the Sun, even though the Moon still occupied some portion of it.

If these appearances are not illusions—and we do not consider them to be—you will understand, I believe, that the sunspots are bodies no less dense and opaque than is the Moon,[127] and so they cannot yet be identified as either vapors or clouds. And you understand that the Moon itself is transparent throughout, more or less according to its greater or lesser density (which is also the case with many spots, and this explains the disintegration of many of them).[128] This

125. Galileo's annotation: The part of the Moon that covered the Sun and that part that went beyond it were equally obscure, and the entire Moon was as obscure as the ambient heaven, and the spots were then equally black as the Moon, indeed, even blacker.

126. Galileo's annotation: The part of the Moon that was covering the Sun was pellucid like crystal, and the Sun's body could be seen through it.

127. Galileo's annotation: Spots are no less dense and opaque than the Moon.

128. Scheiner argues that, like the Moon, sunspots are translucent with "rarer and denser" parts. See Roger Ariew, "Galileo's Lunar Observations in the Context of Medieval Lunar Theory,"

being so, that vexing question about the secondary light of the new Moon is easily answered; indeed, it is nothing other than the light of the Sun pervading the Moon[129] and refracted by it into our eyes. This light is weak, because it is refracted and because it penetrates the Moon, but the other light, because it is reflected to us by the Moon's surface,[130] is stronger and brighter. However, the more the Moon recedes from the Sun, the gentler this refraction becomes, and, on the contrary, the stronger the reflection becomes. From both of these points, the cause of the diminished light of the former and the increased light of the latter is evident. But to me the reflection, if that is what it is, of terrestrial light does not appear to be so great that it produces this phenomenon,[131] although this line of reasoning is entirely consistent with the principles of optics.[132] It will be worth the effort during future eclipses to pay attention to this point. From the same experience you may understand how the Moon and thus the spots are incomparably blacker than any non-stellar celestial body moving around the Sun, since what is between us and the Sun and what is positioned near the Sun are of the same nature. But the Moon exceeds in blackness what is located directly between us and the Sun, as is evident from experience. It is plain that the Moon is also blacker than what is near the Sun, even if the blackness of both may appear equal.

Finally, so that I may end this letter, it is still in doubt whether the spots are on the Sun or outside it, whether we may consider them capable of being generated or not, and whether we may call them clouds or not, but it certainly

---

Studies in the History and Philosophy of Science 15 (1984): 213–226. When the Sun's light passes through the rarer parts, only the scattered denser parts are seen, and thus the sunspot appears to have disintegrated.

129. Galileo's annotation: The entire Moon is transparent more or less according to greater and lesser density. The secondary light of the Moon is light from the Sun penetrating it. But if this is the case, how do the lunar mountains, much more insubstantial, emit such black shadows? Indeed, those valleys nearest the terminator would have to shine with the brightest light, but the interposition of so many mountains impedes the light. And at the quadratures that secondary light would have to resemble the light of the new Moon and be brighter near the terminator.

130. Galileo's annotation: If the Moon is transparent and reflects the light more strongly, it would have to be not as transparent as glass, but glass that thick is not transparent.

131. Galileo's annotation: He doubts that there is any reflection of terrestrial light. At any rate, he supposes the terrestrial light to be extremely weak and incapable of illuminating the Moon. But the contrary is the case, both because the Earth is larger than the Moon and because when it is illuminated, the Moon is illuminated by the Sun from a nearer place.

132. In *Sidereus Nuncius* Galileo had explained the secondary or "ashen" light of the Moon by means of reflected light from the Earth, or earth-shine. See *OGG*, 3(1): 72–75; *SN*, 53–57. Scheiner is here clearly arguing against Galileo. For a recent discussion of the importance of earth-shine in the *Sidereus Nuncius*, see David Wootton, "New Light on the Composition and Publication of the *Sidereus Nuncius*," *Galilaeana* 6 (2009): 61–78, on 73–78.

appears to follow, according to the general opinion of the astronomers, that the solidity and [current] constitution of the heavens cannot endure, especially in the heavens of the Sun and Jupiter. And therefore Christoph Clavius, the *choragus*[133] of mathematicians in this age, should rightly and deservedly be heeded. In the final edition of his works, he warns astronomers that, while it is a very old problem, on account of such new and hitherto invisible phenomena, they must unhesitatingly provide themselves with another system of the world.[134] For if Venus goes around the Sun, as was suggested in the first painting of Apelles and gradually established from its daily transformations, as Tycho Brahe already taught some time ago, and as the Roman mathematicians and Galileo observed at about the same time, although in different places, and as we have already observed, and if Mercury in all probability does the same, then one and the same heaven must be assigned to those three planets. But all these matters will be investigated more carefully in due course.

In the meantime, I must not pass over in silence the fact that these satellites of the Sun—of whatever nature they finally may be, whether servants born at the master's home or purchased elsewhere—inflict a serious wound upon divinatory astrology and especially genethlialogy.[135] (I pose no objections to predictions about the weather).[136] For since these bodies are of enormous size, they undoubtedly affect the Sun in various ways, dividing, refracting, reflecting, diffusing, and concentrating the light directed at us, and at the same time distributing the Sun's natural affection to these lower regions, and so they are very powerful. But if, in the opinion of the astrologers, a meeting of Mercury with the Sun has so much influence on our region, then what will be the influence of so many constant conjunctions of the Sun with these bodies, the majority of which are as large as, or larger than, most planets? Since up to now practitioners of judicial astrology have known nothing about these bodies, it is evident that their knowledge, boastfully displayed for so long, was a merely fortuitous and rash prediction and, in a word, a ludicrous vanity that struck

133. From the Greek term for "chorus-leader."

134. In the final edition of his *In Sphaeram Ioannis de Sacro Bosco Commentarius* (Mainz, 1611), Clavius briefly mentioned the new phenomena revealed by the telescope. He ended the passage as follows: "Since things are thus, astronomers ought to consider how the celestial orbs may be arranged in order to save these phenomena" (75). See James Lattis, *Between Copernicus and Galileo: Christoph Clavius and the Collapse of Ptolemaic Cosmology* (Chicago: University of Chicago Press, 1994), 180–216.

135. Scheiner's use of the term "satellites" here draws on its primary meaning in Latin, the follower or attendant of a superior person. "Genethlialogy" refers to the casting of nativities.

136. Scheiner's distinction is between divinatory and judicial astrology, which concern human fate, and natural astrology, which pertains to weather.

fear into the hearts of inexperienced boys. But, if I am not mistaken, these things and many others will receive their proper place and venue for discussion. I have only wanted to point it out here so that these revealers of future events might see what they are doing if they overlook important causes that, in their own judgment, certainly inhere in these phenomena.

And I have gladly communicated this completion of my earlier work to Your Excellence so that you would know how wrongly this great phenomenon is called into doubt by some, and how mistakenly reviled by most. For all the other things that I have shown in the first painting are consistent.[137] We are still as a loss about one thing alone: whether these bodies are generated and perish or whether they are eternal.[138] While we investigate this question with that diligence and acuity of sense that belong to man or might be his, enjoy, O most eminent man, these duly aired matters. Farewell to you in God, from your Apelles, from our house, and from the entire community of letters.

In Munich, where this letter is to be read and assessed by every most learned man of your circle, I offered it on 25 July 1612.

Yours,
Apelles waiting behind the picture
or, if you prefer
Ulysses under the shield of Ajax

137. Galileo's annotation: He says that all the rest of the things posited in the first painting are consistent.

138. Galileo's annotation: He has said a thousand times that they are stars, but now he is uncertain whether in fact they are generated and perish or not.

11

Galileo responds and goes into print

Galileo appears to have been far more successful than Scheiner in obtaining the collaboration of others in his presentation of the sunspots. He was especially reliant on Lodovico Cigoli in Rome (fig. 11.3), because observations produced by the projection technique would reliably establish that those in other places were seeing and recording the same thing. Cigoli, having been given instructions by his fellow painter Sigismondo Coccapani, was producing these sorts of observations by June 1612, and they were mentioned in Galileo's sunspot letters.[1] By July, Galileo was receiving sunspot observations from as far away as Brussels; in October, he obtained some made in Sicily at Castelli's request.[2] In the meantime, Cesi was working on expanding the circle of membership in the Lincean Academy, and Marc Welser was among the proposed new members. Cesi was also proposing in this period an "epistolary volume" about the celestial novelties, in which Galileo's sunspot letters would be included.

Galileo's second letter, dated 14 August 1612, was finished only three weeks after Scheiner completed his *Accuratior Disquisitio*, dated 25 July. The timing makes clear that Galileo's first and second letters had as their target Scheiner's *Tres Epistolae* alone, as does a

1. Cigoli to Galileo, 30 June and 14 July 1612, *OGG*, 11: 348–349, 362. On Coccapani see Elisa Acanfora, "Sigismondo Coccapani disegnatore e trattatista," *Paragone* 40 (1989): 71–99.

2. Daniello Antonini to Galileo, 21 July 1612, *OGG*, 11: 363–365; Don Sigismondo di Cologna to Benedetto Castelli, 10 October 1612, *OGG*, 11: 412–413.

Figure 11.1. Engraved portrait of Galileo, by F. Villamena, 1613. From *Istoria e Dimostrazioni*.

remark made by Castelli's correspondent in Sicily in the autumn of that year, who asked to be given, in recompense for his observations, "those answers that the most excellent Signor Galilei wrote a few months ago to that Apelles."[3] The Jesuit astronomer's *Accuratior Disquisitio*, by contrast, was written in response to Galileo's first letter alone. Scheiner thus did not respond to Galileo's second letter, and Galileo's third letter would address the shortcomings of this second tract.

3. Don Sigismondo di Cologna to Benedetto Castelli, 10 October 1612, *OGG*, 11: 413

Figure 11.2. Portrait of Federico Cesi by Pietro Facchetti, 1610–1612.
With permission from the Accademia Nazionale del Lincei.

While Galileo's second letter reached Federico Cesi (fig. 11.2) in Rome on 8 September, it took considerably longer to come to Welser in Augsburg, copies of the text and appended drawings having been made by Gianfrancesco Sagredo in Venice first.[4] Upon receipt of the work in early October, Welser wrote to Faber that he thought the second letter merited publication along with the first, but that he anticipated problems with the many drawings that accompanied

4. Gianfrancesco Sagredo to Galileo, 22 September 1612, *OGG*, 11: 398.

Figure 11.3. Self-portrait of Lodovico Cardi da Cigoli, c. 1606, oil on canvas. Courtesy of the Bridgeman Art Library.

it. Welser proposed reducing them to a smaller format, as Apelles had done, so that they would not take more than half a folio page. He also suggested that the eventual publication would be further enhanced by a prefatory letter from Cesi.[5]

Cesi's original plan to assemble an epistolary volume entitled *Helioscopia* and devoted to "celestial novelties" of various sorts underwent significant modification in this period, and sunspots became the sole focus. Unlike the

5. Welser to Faber, 4 October 1612, *OGG*, 11: 407.

hesitant Welser, Cesi immediately saw Galileo's drawings as an essential part of the argument as well as a display piece for the Lincean Academy; he was convinced that the sensory experience they recorded would vanquish Galileo's rivals.[6] There is no question that at an early stage Cesi understood very well—possibly after some suggestions by Galileo, whose side of the correspondence does not survive—the scientific necessity of producing accurate renditions of the projected spots, and he had no hesitations about commissioning engravings equal in size to the original drawings.[7] Though the volume would be expensive, Cesi urged Galileo not to take such considerations into account. Promising to reproduce Apelles' drawings as well, he and Cigoli met with three copper engravers, and chose the Alsatian Matthäus Greuter, who was also well-known as a draftsman and publisher, and then resident in Rome.[8] The drawings themselves were to be printed in quarto, with four figures per folio, and all assembled at the end of the volume.[9]

Over the next five weeks, Cesi and Cigoli supervised Greuter's work in Rome, while Galileo turned to composing a third letter in Florence in response to the *Accuratior Disquisitio*. The core of this third letter was a detailed quantitative demonstration that Scheiner's arguments about the distance at which the supposed satellites orbited the Sun could not be correct. Galileo proved this by supposing the radius of the orbit of one of these bodies was twice the radius of the Sun, and then showing that the observed motions of sunspots could not possibly be derived from this arrangement. Whereas Scheiner's perspective demonstration of the spacing of the spots and the foreshortening at the limb had ultimately been a verbal argument with a diagram, Galileo's argument, using the same sort of diagram, was truly quantitatively mathematical.[10] For instance, where Scheiner had used diagrams to illustrate the foreshortening of sunspots at the limb, Galileo demonstrated by numerical calculations that if the spots were distant from the Sun's surface by any appreciable distance they would not show the foreshortening observed.

Early in October 1612, Cesi sent Galileo two proof prints of observations, and in November he reported that he had returned some of the plates and the

6. Cesi to Galileo, 8 September 1612, *OGG*, 11: 393. See also Pasquale Guaragnella, "Icone e linguaggio verbale in Galilei: Sulla questione delle macchie solari," *Scienze umane* 6 (1980): 267–278.

7. Cesi to Galileo, 14 September 1612, *OGG*, 11: 395.

8. For Greuter's work, see F. W. H. Hollstein, *Hollstein's German Engravings, Etchings and Woodcuts*, vol. 12, ed. Tilman Falk (Amsterdam: Van Gendt & Co, 1983), 107ff.

9. Cesi to Galileo, 29 September 1612, *OGG*, 11: 404.

10. Shea, *Galileo's Intellectual Revolution*, 51–63.

original drawings to Greuter so that certain inaccuracies in them might be corrected.[11] Cesi proved a demanding employer, sending even the corrected engravings back to the artist for still more modifications.[12] The printing of the plates was in any case finished by the beginning of November, and attention was focused on Galileo's third and final response to Apelles. He had received a copy of the *Accuratior Disquisitio* rather promptly, for by 8 October he wrote to Cesi that he planned to answer it.[13] Cesi agreed that the latest missive from "the hidden Apelles" merited a response, but counseled haste: "the Germans are very quick and can easily anticipate you."[14]

At about this time, the middle of October, it became known in Rome that Apelles was a Jesuit.[15] This complicated matters somewhat, particularly in the relationship between Galileo and the German Jesuit Christoph Grienberger of the Collegio Romano, whose reactions had been thus far mostly favorable.[16] In early November, Cesi advised Galileo not to attack Germans in general, for they were in his judgment "most hospitable to letters and scholars, and [they] sustain their glory with an abundance of books and editions. The Linceans, above all, must maintain their friendship, for [the Germans] are liberal in philosophizing, and I see that they honor the Italians greatly in situations where they have no particular passion or envy."[17] That same day, Cigoli, doubtless in consultation with Cesi, repeated the warning almost verbatim.[18] Galileo soon acknowledged that the delay in finishing the lengthy third letter involved precisely the question of tone, for he was torn between the desire to expose the weaker points of his Jesuit opponent's approach and a fear of incurring Welser's displeasure. He had no such qualms about Apelles' feelings; as he told Cesi:

> It is amazing to see, in his treatise, the audacity and frankness with which he persists in asserting that his treatment of the material is very different from what I have written of it, even though anyone can see that he copied it from my *Nuncius*. I was certainly completely astonished by the boldness with which

11. Cesi to Galileo, 6, 13, and 28 October and 3 November 1612, *OGG*, 11: 409, 416, 420, 422.

12. Cesi to Galileo, 20 November 1612, *OGG*, 11: 428.

13. Cesi to Galileo, 13 October 1612, *OGG*, 11: 416.

14. "I Germani sono prestissimi, e facilmente prevengono." Ibid.

15. Cigoli to Galileo, 19 October 1612, and Cesi to Galileo, 28 October 1612, *OGG*, 11: 419, 420.

16. On 23 November, Faber wrote to Galileo that "Father Grienberger . . . told me that he had not yet seen the latest little work of Apelles. But in truth, although he knows that [Apelles] is a Jesuit, he agrees much more with you than with Apelles, for it appears to him that the arguments with which you raze to the ground the foundation that they are stars are very effective. But as a son of Holy Obedience, he does not dare to give that judgment" (*OGG*, 11: 424).

17. Cesi to Galileo, 3 November 1612, *OGG*, 11: 423

18. Cigoli to Galileo, 3 November 1612, *OGG*, 11: 425.

> he handles me, as they say, privately, and I can only imagine what he would say to defend himself publicly.[19]

In reading the *Accuratior Disquisitio*, Galileo was quite clearly displeased by Apelles' claim to have discovered a fifth satellite of Jupiter and by the use to which his rival put these Medicean Stars (see pp. 287–289, below). Scheiner, for his part, appears to have been entirely unaware of Galileo's anger; in a letter of 9 November, Welser asked Johannes Faber to relay the following message from the Jesuit astronomer:

> I received the letter, together with Galileo's observations. I am incredibly delighted when I see that his agree with mine, and mine with his, to a hair's breadth. You will see, compare, be astonished and delighted when you realize that, at such a distance between places, the one agrees so beautifully with the other as to the number, order, place, size, and shape of the spots. For if I agree as much with Galileo, or he with me, about the substance of these bodies, a more beautiful union cannot be imagined. In the meantime, where we differ in opinions, we will unite our spirits through friendship, especially since we both strive for one goal, which is the Truth, and I have great faith that we will bring it out.[20]

As if in anticipation of the coming quarrel, Cesi reminded Galileo on 10 November that "corrections made with gentle words and keen reasons sting more than those made with harsh words,"[21] and he now thought it wiser not to include the *Accuratior Disquisitio* in the book, the printing of which was at last at hand. There had been a delay because the first line of Welser's first letter, a truncated version of Matthew 11:12, *Regnum caelorum vim patitur, et violenti rapiunt illud*,[22] had not passed the censor, presumably because of the threatening tenor the Scriptural verses lent to the astronomical treatise.

It was in this period, November 1612, that Galileo discovered that Saturn had lost its "companions," the lateral appendages that he had taken to be separate bodies upon discovering them in July 1610. Cesi immediately passed on the astonishing news to his fellow Linceans as well as to Christoph Grienberger, the Jesuit mathematician of the Collegio Romano, and even to Antonio

19. Galileo to Cesi, 4 November 1612, *OGG*, 11: 426.

20. Scheiner to Welser, October/November 1612, in Welser to Faber, 9 November 1612, *OGG*, 11: 428.

21. Cesi to Galileo, 10 November 1612, *OGG*, 11: 429.

22. "And from the days of John the Baptist until now, the Kingdom of Heaven suffereth violence, and the violent take it by force." Matthew 11: 12.

Bucci, the scholar who wrote the report for the censor (see pp. 371–372, below).[23] There was some danger that the Aristotelians could use this apparent disappearance to argue that the sunspots were satellites, just as Apelles, in the interest of maintaining the perfection of the heavens, had relied on the emergences and disappearances of Jupiter's satellites in order to explain the behavior of sunspots (pp. 286–287, below). Galileo added a section to his third letter to address this new phenomenon, making a complicated and successful prediction only recently explained by Deiss and Nebel.[24]

A much more worrisome issue was Galileo's commentary on Scripture and the perfection of the heavens in his second letter. Cardinal Conti had written him that Scripture did not support the perfection of the heavens and could, indeed, be interpreted to support its corruption (see pp. 349–352, below). Galileo had, therefore, written in the original draft of his second letter:

> Who is it that, after having seen, observed, and considered these matters, would want to persist in a belief that is not only false, but erroneous and repugnant to the indubitable truths of the Sacred Scripture as well? For Scripture tells us that the heavens and the entire world are not only generable and corruptible, but also generated and dissoluble and transitory. Notice how Divine Goodness, in order to retrieve us from such an immense error, inspires some people with the necessary approaches.[25]

Galileo subsequently expanded the passage and softened its tone:

> Who is it that, after having seen, observed, and considered these matters, would not be willing (once every doubt occasioned by apparent physical reasons has been dismissed) to embrace a belief that so conforms to the indubitable truths of the Sacred Scripture? For Scripture in so many passages quite openly and clearly shows us the unstable and fallen nature of the celestial material, without depriving of their deserved praise, however, those sublime intellects who with subtle speculations managed to harmonize sacred dogma with the apparent discordances of the physical discourses. This supreme theological authority having been removed, there is now good reason for [these minds] to yield to the natural [i.e., scientific] reasons of other grave authors,

23. Cesi to Galileo, 24 November 1612, *OGG*, 11: 438. On Galileo's observations and interpretations of the appearance of Saturn, see A. Van Helden, "Saturn and His Anses," *Journal for the History of Astronomy* 5 (1974): 105–121.

24. Bruno M. Deiss and Volker Nebel, "On a pretended observation of Saturn by Galileo," *Journal for the History of Astronomy* 29: 3 (1998): 215–220.

25. *OGG*, 5: 138, note, MS A.

and even more to sensory experience, to which I don't doubt Aristotle himself would have given way. For we see that he not only recognized experience as among the methods capable of reaching conclusions about natural phenomena, but he even accorded it the most important place. Therefore, if he argued for the immutability of the heavens because no perceptible alteration had been detected there in ages past, it is quite credible that had his senses demonstrated such changes to him as are now manifest to us, he would have held the contrary opinion, the one to which we ourselves are summoned by such marvelous discoveries. Notice how Divine Goodness, in order to remove all doubts from our minds, inspires some people with the necessary approaches.[26]

This pious elaboration still did not satisfy the censors, and Cesi reported to Galileo on 30 November,

in sum, one can only dislodge the Peripatetics little by little. I wrote according to the law [*in jure*] (so to speak) on the basis of the evidence, adducing ten passages in Scripture and as many from the Patristic writers that confirm what you say, that the corruptibility of the heavens conforms to Scripture and is indicated by it. It was not enough, and they replied that those same passages had been very thoroughly interpreted by others in a Peripatetic way; so we must have patience. In a word, they don't want you to say anything about Scripture in that passage. But let me know what you want done.[27]

Galileo sent off his third sunspot letter to Marc Welser on 1 December 1612 (p. 253, below) and the printing of the entire book was then well under way, though hampered by the censors' demand for further changes. When Galileo's revised text finally arrived, the Lincean Francesco Stelluti, closely involved in the publication, anticipated no further difficulties with the censors;[28] the only remaining difficulty seemed to be the sullenness of the printers and their ignorance of Tuscan.[29]

Galileo had in the meantime been working on an appendix to the third letter, an ephemeris of the satellites of Jupiter. While it is not known when he conceived this project, the context—the familiar one of priority—is clear. As soon as the existence of Jupiter's moons had been accepted, some observers

26. *OGG*, 5: 138–139, note, MS B.
27. Cesi to Galileo, 30 November 1612, *OGG*, 11: 439.
28. Stelluti (?) to Cesi, December 1612, *OGG*, 11: 453.
29. Cesi to Galileo, 28 December 1612, *OGG*, 11: 450.

began to draw up tables of their motions so that their future positions could be predicted. Galileo was under some pressure to maintain his lead, for he had made the first approximations of the periods of the satellites by the spring of 1611[30] and continued to refine these. He published figures in his *Discourse on Bodies in Water* in the spring of 1612 that closely approximate modern values.[31] While Kepler thought it miraculous that Galileo had been able to do this in such a short period of time, because it was so difficult to distinguish one satellite from another,[32] Galileo wished to construct tables as well, on the basis of which ephemerides of the satellites could be drawn up.[33]

By the summer of 1611, Kepler had estimated the period of the third satellite (Ganymede) to be eight days,[34] and Galileo also heard from Cigoli—ever prompt in relaying challenges and slights—that Giovanni Antonio Magini, who had many ephemerides to his name, had said that "having or not having first discovered these things is of little importance, but what is important now is to find the course of these four stars of Jupiter, and that in this will lie all the glory." Cigoli also alleged that Magini worked constantly and with due diligence on these observations, and hoped to conclude the work, and the implicit contest, very shortly.[35] Moreover, the periods themselves could rather roughly be determined from the series of observations of the satellites in Galileo's own *Sidereus Nuncius*, as Giovanni Battista Agucchi had demonstrated to Galileo in the autumn of 1611. Even though he had not observed them himself, Agucchi had considered publishing this information, and asked Galileo for further data on the satellites.[36]

But defending a claim to priority was not the sole reason why Galileo chose to continue his investigations of the satellites even while ailing and preoccupied with his book on floating bodies, the emergent sunspots, and the ongoing arguments about the nature of the Moon. He had begun seriously to consider

30. Galileo to Piero Dini, 21 May 1611, *OGG*, 11: 114. Galileo here estimated the period of the outermost satellite as a little over sixteen days, and that of the innermost as less than two days.

31. *OGG*, 4: 63–64; *Discourse on Bodies in Water*, tr. Thomas Salusbury, ed. Stillman Drake (Urbana: University of Illinois Press, 1960), 1; Drake, *Galileo at Work: His Scientific Biography* (Chicago: University of Chicago Press, 1978), 167–168.

32. Galileo determined the periods to be one natural day, 18 hours, and about a half; 3 days, 13 hours, and about a third; 7 days and very nearly 4 hours; and 16 days and very nearly 18 hours. See *OGG* 4: 63–64.

33. Antonio Santini in Lucca wrote to Galileo on 20 July 1611, "I am pleased to hear that you've set to work on the theory of the periods and tables of the new planets discovered by you" (*OGG*, 11: 155).

34. Kepler to Wickens, July 1611, *OGG*, 11: 167; *JKGW*, 16: 390.

35. Cigoli to Galileo, 23 August 1611, *OGG*, 11: 175.

36. Agucchi to Galileo, 14 and 29 October 1611, *OGG*, 11: 219, 225. See also Drake, *Galileo at Work*, 175–176.

using the positions of Jupiter's moons to determine longitude at sea, a problem so critical for navigators that the Spanish Crown had a few years previously established a prize for its solution. Nicholas-Claude Fabri de Peiresc in Aix en Provence had already set up a "longitude bureau," coordinating observations of the moons sent to him by correspondents in various places to determine longitude differences on land.[37] Errors of hundreds of miles on maps of the Mediterranean were being discovered, and the longitudes of places in the Orient were in error by as much as 25 degrees even on the latest versions that Gianfrancesco Sagredo was then buying for Galileo.[38] In December 1612, Sagredo wrote to Galileo giving prices of maps and atlases by Hondius, Jansonius, and Plancius,[39] and it was also in this year that Galileo had first observed eclipses of Jupiter's satellites.[40]

Although neither Galileo nor other observers had initially considered these eclipses very important, in the long run they proved more useful in the determination of longitude differences than observations of the formations of satellites, simply because in the latter case the motions involved were too slow. The sudden disappearance or reappearance of a satellite, being almost instantaneous, meant that timing the eclipses became the dominant method of determining longitude.[41] Beginning with Galileo, astronomers tried also to use this method for determining longitude at sea, and since here one needed to know one's longitude immediately, the local time of satellite eclipses had to be compared with *predicted* times at the reference location, and this meant that accurate tables of eclipse times had to be at hand. This was the first requirement, and one that occupied Galileo for years. But while astronomers had worked for more than two millennia on the models of the planets, modeling the motions of the Medicean Stars was, in 1612, still in its infancy. It was not until the last quarter of the seventeenth century that eclipse predictions for the first, internal satellite (Io), reached a degree of precision adequate to this purpose. In the meantime, Galileo's continuing observations and work on the satellite tables

37. Pierre Humbert, *Un Amateur: Peiresc, 1580–1637* (Paris: Desclée de Brouwer et Cie, 1933), 81–89. For the calculations of the orbital elements of Jupiter's satellites made by Peiresc's associate, Joseph Gaultier de la Valette, see Bibliothèque Inguimbertine, Carpentras, France, MS 1803, fols. 189r–223r.

38. Albert Van Helden, "Longitude and the Satelllites of Jupiter," in *The Quest for Longitude*, ed. William J. H. Andrewes (Cambridge, MA: Harvard University Collection of Historical Scientific Instruments, 1996), 85–100, at 88.

39. Sagredo to Galileo, 15 December 1612, *OGG*, 11: 449.

40. Galileo noticed that one of the satellites had entered Jupiter's shadow during his observation of 18 December 1612 (*OGG*, 3: 451); he found eclipses in his calculations for 18 March, 4 April, and 4 May 1612. See *OGG*, 3: 527, 530, 534.

41. Van Helden, "Longitude and the Satelllites of Jupiter," 85–100.

had advanced sufficiently for him to be able to draw up a short-term ephemeris that was reasonably accurate for a few months.

But there was yet another reason for these ephemerides. In his *Accuratior Disquisitio*, Apelles had announced the discovery of a fifth satellite of Jupiter, one whose complicated motions, he alleged, made predictions of its visibility and invisibility, as well as of its locations, impossible for the time being. And on the basis of this supposed discovery, as well as his own observations of the other four satellites, Apelles had introduced a dimension of unpredictability in satellite behavior, a feature that would also account for the unpredictability of the satellites swarming around the Sun in the guise of spots. The satellites of Jupiter thus became an argument in support of the satellitic nature of sunspots (pp. 287–289, below). Galileo's ephemerides were therefore also designed to emphasize the fundamental difference between Jupiter's moons and the Sun's supposed satellites; anyone who cared to could observe Jupiter through a telescope and see that the satellites appeared precisely when and where Galileo had predicted they would.

As the printing progressed, Galileo drew up an ephemeris for February and March 1613, hoping to publish an explanation of it in Latin for separate distribution in other countries.[42] The printing of the sunspot letters went more slowly than expected, and though Cesi was most impressed by Galileo's predictions of the satellites' positions for all of March, he feared that the sluggish pace of production would mean that the book would only emerge sometime after that month.[43] Though Galileo was once again in poor health, he set to work at Villa delle Selve on extending the ephemeris through April, and added a number of last-minute changes he wished to make in the text of the third letter.[44] The ephemeris was eventually printed as an appendix to this letter covered the period 1 March to 8 April, but most readers saw them only later (pp. 300–304, below).

The censors in Rome demanded further changes in the third letter; though the astronomer had argued that because he was not treating theological material *ex professo* he could surely invoke God rhetorically, they demanded that Galileo "speak of natural phenomena, without mixing in any supernatural allusions at all."[45] Galileo would have also liked to include letters to him from Christoph Clavius and Christoph Grienberger in general support of his observations,[46] but as Cesi was lukewarm about the prospect, and Grienberger unwilling to have his letter published in this context, in the end neither item

42. Galileo to Cesi, 5 January 1613, *OGG*, 11: 459–460.

43. Cesi to Galileo, 11 January 1613, *OGG*, 11: 462–463.

44. Galileo to Cesi, 25 January 1613, *OGG*, 11: 465–469.

45. Cesi to Galileo, 26 January 1613, *OGG*, 11: 471; Galileo to Cesi, 25 January 1613, *OGG*, 11: 465–466.

46. Cesi to Galileo, 1 December 1612, *OGG*, 11: 444.

appeared.[47] The name of the work also presented difficulties: Cesi, endorsed by Ioannes Demisiani, who had coined the word *telescopium* in 1611, preferred *Helioscopia* as a title,[48] but Scheiner had already used that term in his *Accuratior Disquisitio*.[49] Galileo evidently wanted a title that began with *Lettere*, but Cesi favored a Latin word, or in a kind of capitulation to the preference of the Accademia della Crusca for Tuscan words, something like *Scoprimenti solari* or *Contemplazioni solari*.[50] Galileo continued rather stubbornly to refer to the book as *Lettere Solari*;[51] Cesi proposed *Solar Discoveries, complete in three letters, together with those of the masked Apelles*.[52] Galileo's eventual suggestion, made in early January 1613, is identical to the title page under which the book finally appeared, with one exception: he referred to himself as simply "Chief Mathematician" to Grand Duke Cosimo de' Medici, rather than as the Tuscan ruler's "Philosopher, and Chief Mathematician."[53] In view of Galileo's earlier insistence on the title of "Philosopher and Mathematician" when he was negotiating his appointment at the Tuscan Court in 1610,[54] this is a remarkable omission. There is no further mention of this subject in the correspondence between Galileo and Cesi, but the word *Filosofo* appears squeezed in, as though it had been added to the title page as an afterthought (fig. 11.4). It is, however, not missing in the inscription surrounding Galileo's portrait (fig. 11.1).

The title page of Galileo's work insists, in both textual and iconographic terms, on the importance of the Lincean Academy; the affiliation of both the addressee and the astronomer precedes mention of their social rank (as magistrate [*duumvir*] of Augsburg and as Florentine nobleman), and of their professional positions (as Imperial Advisor and as Grand Ducal Philosopher and Mathematician). A wreath of laurel, a traditional emblem of victory and of intellectual achievement, surrounds the sharp-eyed lynx. The conventional crown of the political sovereign, rather than circumscribing the evergreen laurel wreath, is embellished and to some extent dwarfed by it, as if to emphasize the relative intellectual freedom of the Linceans, and the durable luster these scholars presumably brought to their rulers.

47. Cesi to Galileo, 1 and 14 December 1612 and 1 and 15 February 1613, *OGG*, 11: 444, 446, 475, 481.

48. Cesi to Galileo, 29 September and 6 October 1612, *OGG*, 11: 404, 409; Edward Rosen, *The Naming of the Telescope* (New York: Henry Schuman, 1947), 60–68.

49. Cesi to Galileo, 28 October 1612, *OGG*, 11: 420.

50. Ibid.

51. Galileo to Cesi, 4 November 1612, *OGG*, 11: 425.

52. Cesi to Galileo, 10 November 1612, *OGG*, 11: 429.

53. Galileo to Cesi, 5 January 1613, *OGG*, 11: 460.

54. Galileo to Belisario Vinta, 7 May 1610, *OGG*, 10: 353.

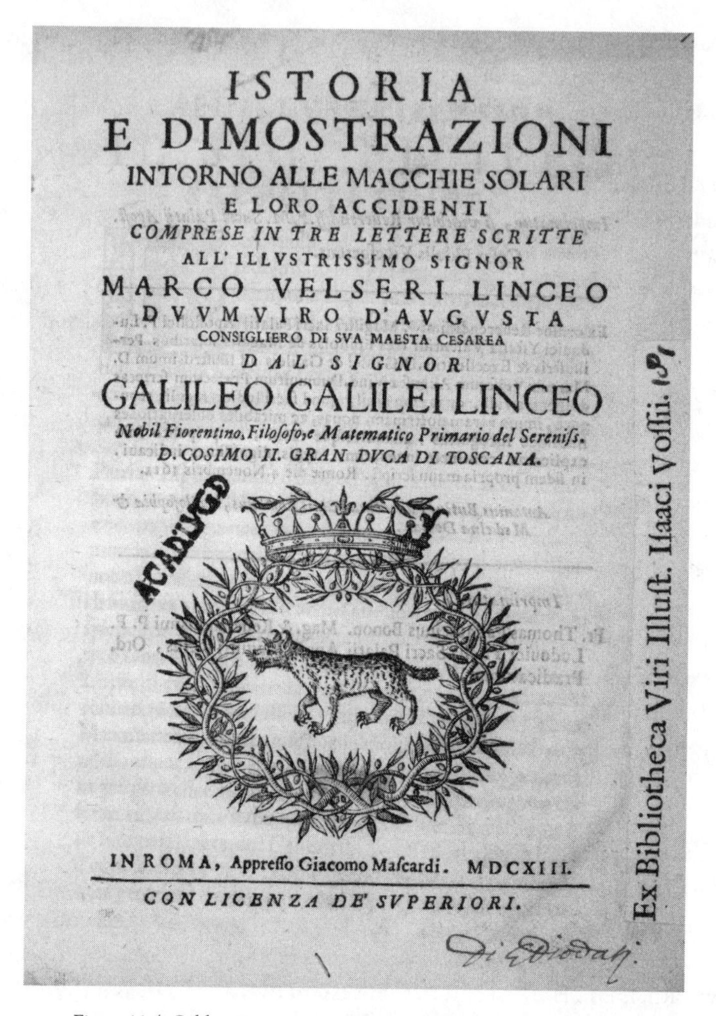

ISTORIA
E DIMOSTRAZIONI
INTORNO ALLE MACCHIE SOLARI
E LORO ACCIDENTI
COMPRESE IN TRE LETTERE SCRITTE
ALL' ILLVSTRISSIMO SIGNOR
MARCO VELSERI LINCEO
DVVMVIRO D'AVGVSTA
CONSIGLIERO DI SVA MAESTA CESAREA
DAL SIGNOR
GALILEO GALILEI LINCEO
Nobil Fiorentino, Filosofo, e Matematico Primario del Serenifs.
D. COSIMO II. GRAN DVCA DI TOSCANA.

IN ROMA, Appreſſo Giacomo Maſcardi. MDCXIII.
CON LICENZA DE' SVPERIORI.

Figure 11.4. Galileo, *Istoria e Dimostrazioni* (Rome, 1613), frontispiece.

The Linceans' preoccupation with Galileo's *virtù* (a word that implies both prowess and piety) and his enemies' envy, reiterated in one of Cigoli's allegories and in countless letters by Cesi, is taken up in Galileo's portrait (fig. 11.1), presented directly after the prefatory letters.[55] There, the allegorical figure of Virtue, crowned with laurel, smiles down upon the astronomer; the toothless harpy Envy, wreathed in snakes, occupies the lowest region of the page. The two

55. See Miles Chappell, "Cigoli, Galileo, and *Invidia*," *Art Bulletin* 57 (1975): 91–98; and Eileen Reeves, *Painting the Heavens* (Princeton: Princeton University Press, 1997), 172–183.

putti above the niche of the portrait are especially crucial to the Linceans' long-term agenda of maintaining Galileo's contested claims to priority. The figure on the right, while holding the traditional chronicle and trumpet of Fame, holds the instrument not to the mouth, but to the eye and toward the apparent source of light, an iconographical touch that does not suggest that Galileo invented the telescope, but rather that he was the first to achieve renown in deploying it. The figure on the left writes in a book and holds a geometrical sector, the latter being the instrument which Galileo (among others) refined and manufactured, only to see it appropriated by Baldessare Capra.[56] Thus while the *putti* embody, in general fashion, the chief activities of Galileo's sunspot study—observation, measurement, and calculation—they also allude to his contested role in developing and perfecting particular instruments. While the uppermost reaches of the work suggest a faint bias against Northern usurpers—a bias really only brought into focus in Galileo's reaction to the claims made by Scheiner's final letter, and tempered somewhat by Cesi—the lower portion, with its Greek fretwork, frames the astronomer's portrait, and suggests the venerable Greco-Roman foundation of his endeavors.

The final problem was the preface. Cesi had asked Galileo for an outline of the points he thought ought to be included,[57] and on the basis of these notes, Angelo de Filiis, who had become the librarian of the Linceans in 1612, wrote a draft.[58] It was doubtless at the suggestion of Cesi that de Filiis made *Invidia* or Envy his theme and stressed Galileo's priority in discovery of the new telescopic phenomena, including sunspots. Though the tone struck Galileo as offensive, Cesi defended de Filiis' composition, replying to the astronomer:

> The letter I recently received from Your Lordship about the preface to the work perplexes me and the other Linceans who are here. We value your advice, but the need we see to disabuse indifferent people—who far outnumber your friends and enemies combined—of the ideas disseminated by the envious and by other adversaries who are now defrauding you of your achievements, does not allow us to agree entirely with you. Fair and sound minds are few in number, and among them, not many in Germany, France, Flanders, and even in nearby Naples have accurate reports of the sequence of celestial discoveries. Your books have not reached everywhere; Your Lordship has not seen everything into print. And I can say for certain that many in those places

56. For discussion of this episode see Mario Biagioli, *Scientists' Names and Science's Claims*, forthcoming.

57. Cesi to Galileo, 3 November 1612, *OGG*, 11: 423

58. Cesi to Galileo, 15 February 1613, *OGG*, 11: 481–482

> have exhibited things discovered by you, and if some of them haven't gone so far as to appropriate them entirely, they nonetheless have made no mention whatsoever of Your Lordship. Therefore it is not a bad thing to take some action, to do something that might clarify and mortify at once. We can make the preface more serious; we can produce the same effect with less emotion and less evidence. Your Lordship should consider all this, and decide in the manner you think best.[59]

Though Cigoli found the style of de Filiis' preface somewhat inflated, he agreed with the necessity of the message. He also suggested that the apparent need to protect Galileo's priority was not confined to his astronomical exploits, but extended to his work in hydrostatics, likewise conducted with Salviati at Le Selve; only the Linceans' lack of expertise in this area had prevented them from addressing the invidious treatment accorded the *Discourse on Bodies in Water* in this text. His concerns, in sum, did not differ from those of Cesi.

> Signor Galileo, when one foresees evil and can escape it, it's rightly called "prudence," but when people have so flamboyantly revealed themselves, the time is up, and you have to face the storm and show that you're alive. . . . Your Lordship should attend to writing about your achievements with all haste, without letting these charlatans interrupt you, but in the meantime do not forbid Prince [Cesi] to print this letter to the reader. The things that have already come out are unknown to an infinite number of people because you have made so little of them. Now, have them all printed in both Latin and Italian, so that these people die of envy, and let it be the talk of the town. Allow the letter to be printed, because we all desire it, and we think it necessary for a variety of reasons. You did not seek it out, and it is outside [Tuscany]; it is in Rome, and it will scald [your enemies] more from here than from anywhere else. Let yourself be persuaded; give us your *placet*, and soon, because time is running out. Don't say no in any way, because we will all be disappointed.[60]

At length Galileo allowed a much shortened and somewhat less inflated version to be printed (pp. 374–378, below); as anticipated by Cesi and Cigoli, it was still strong enough to give offense to many in Rome.

Istoria e dimostrazioni intorno alle macchie solari e loro accidenti came off the press near the end of March 1613. Advance copies without the preface

59. Cesi to Galileo, 22 February 1613, *OGG*, 11: 483–484.
60. Cigoli to Galileo, 26 February 1613, *OGG*, 11: 484–485.

had been sent with a traveler to Germany,[61] and it is possible that Welser never saw the aggressive preface of the work. In November 1612, Cesi had proposed to print 2,000 or 3,000 copies, some of which were to be included, without Scheiner's contribution, in the proposed epistolary work about celestial novelties. This number was reduced in February 1613 to 2,000, of which 1,000 would contain Apelles' tracts; in the end the total print run was 1,400 copies, half of which included the Apellean texts.[62] These copies had a separate title page, and Apelles' tracts were introduced by a Latin note, supposedly written by the printer, Giacomo Mascardi, but in reality composed by Angelo de Filiis,[63] explaining the need for including these tracts; Galileo's texts, of course, referred to them repeatedly. The total cost of the book was 258.70 *scudi*, Galileo's annual salary being 1000 *scudi*.[64] Galileo's section contained 44 engravings, 39 of sunspots and 5 of the tables of the Medicean Stars, at a cost of 44 *scudi*. The frontispiece with Galileo's portrait, perhaps designed by Cigoli, was engraved at a cost of 6 *scudi*, possibly by Francesco Villamena.[65]

61. Cesi to Galileo, 22 March 1613, *OGG*, 11: 489–490.

62. Cesi to Galileo 24 November 1612 and 15 February 1613, *OGG*, 11: 438, 482; and "Spese per la Pubblicazione dell' *Istoria e dimostrazioni intorno alle macchie solari, OGG*, 19: 265–266.

63. Cesi to Galileo, 14 and 28 December 1612, *OGG*, 11: 446, 450. The note gave as the justification for the reprinting that very few copies of *Tres Epistolae* and *Accuratior Disquisitio* were available in Italy or in other regions, and that Galileo's frequent references to the two tracts made it necessary to include it. In the copies that included Scheiner's tracts, the censor added: "I have indeed seen some of Apelles' letters on this subject, as well as the Disquisitions sent to Signor Welser; they contain nothing that might offend [the reader], and I rule that these might be published as well" (*OGG*, 5: 12).

64. The figures are given in decimal fractions. For a discussion of the purchasing power of these sums, see R. E. Spear, "Scrambling for *Scudi*: Notes on Painters' Earnings in Early Baroque Rome," *Art Bulletin* 85.2 (2003): 310–320; on the cost of producing engravings in this period, see Michael Bury, *Print in Italy* (London: British Museum, 2001), 44–51.

65. *OGG*, 19: 265–266. See Antonio Favaro, "Iconografia Talileiana," *Atti del Reale Istituto Veneto di scienze lettere ed arti* 72.2 (1912–1913): 995–1055, on 1001–1002. See also Ada Alessandrini, "Originalità dell'Accademia dei Lincei," *Atti dei Convegni Lincei* 78 (1986): 77–177, at 99–127.

12

Galileo's third letter

Third letter of Signor Marc Welser to Signor Galileo Galilei

Most Illustrious, Excellent, and Worthy Lordship,

My grave indisposition continues to torment me, so that I cannot visit my friends with long and frequent letters, as would be my duty and desire, particularly toward Your Lordship, with whom I so enjoy discussion. But the impossibility prevents me from doing so, and it is to be counted as a gain[1] when God grants me leave to greet you briefly with a few lines, as I do with the present. I send Your Lordship some new speculations of my friend about celestial things, the printing of which I have permitted chiefly out of regard for the observations, which I allow myself to believe will be agreeable to all who love and seek the truth. As for the rest, I will not risk inclining more toward one side than toward the other, since my affliction does not even permit me to apply my mind as I should. I hear that Your Lordship has written a second long letter about this matter, and sent it to me; I have not yet seen it but am expecting it with singular

1. *[E]t in lucro reputandum est*. Welser's Latin expression here is probably based on Terence's *Phormio*, II, 246: *Quiquid praeter spem eveniat, omne id deputare esse in lucro*: "Whatever happens beyond one's hopes, must be regarded as so much profit." Welser suffered from gout, and would die from it in 1614.

eagerness. In the meantime, I end by kissing Your Lordship's hand most cordially and wishing you well.

From Augsburg, 28 September 1612
Your Very Illustrious and Most Excellent Lordship's
Most affectionate servant
Marc Welser L[inceo][2]

Fourth letter from Signor Marc Welser to Signor Galileo Galilei

Most Illustrious, Excellent, and Worthy Lordship,

Your Lordship's second letter of 14 August finally has appeared, sent to me by Signor Sagredo. Please believe that it was received like manna, so great was my eagerness to see it. I have not yet had time to read it carefully, but I can tell you sincerely that I derived great pleasure from the quick glance I have given it. And although I am always aware that I am unqualified to serve as a judge in so important a case, and much less so now that my infirmity does not even permit me to apply my mind at length to any speculation, I will be so bold as to say that Your Lordship's discourses proceed with great verisimilitude and probability. That they arrive precisely at the truth human frailty cannot affirm, until Blessed God permits us to survey from above what now we contemplate from below, in this vale of misery. I offer many thanks to Your Lordship for the favor that you grant me on this occasion, and in having both letters printed as soon as possible, as I understand he has decided to do, Prince Federico Cesi is undertaking something worthy of his rank and calling as a benefactor of virtues and of literature. The illustrations [*figure*] of the observations will cause some difficulty, but if they are reduced to a smaller form they will occupy little space. I would have greatly preferred that Apelles had seen this work before printing his last discourses, and yet I think that in some respects it is perhaps better this way. I will not fail to communicate it to him after I have savored it somewhat myself, but he suffers from the great inconvenience of not understanding Italian, and the translations, besides proceeding slowly, often not only lose the vigor of the original, but also corrupt the meaning, unless the interpreter is very skilled.

Signor Sagredo kept the treatise *Discourse on Bodies in Water* for a few days, having been asked to do so by a friend of his, a senator who begged to be al-

2. Welser had just been elected to membership in the Accademia dei Lincei. See p. 373, note 3, below.

lowed to read it; this may well have been Protogenes.[3] I dispense with it all the more easily, as I have had the good fortune of seeing another copy, the reading of which so converted me—and I am not ashamed to confess it—that what originally seemed a paradox now appears indubitable, and so well armed and fortified with reasons and experiments [*isperienze*] that surely I cannot see how and where your adversaries are able to attack it, although I imagine that they cannot leave the subject in peace. May Your Lordship continue to do honor to yourself and to our era by retrieving one truth after another from the dark pit of ignorance, and may you remain undaunted by the envious and the rivalrous, but keep me always in your favor. May God bless you.

From Augsburg, on 5 October 1612
Your Very Illustrious and Most Excellent Lordship's
Most affectionate servant
Marc Welser, Lincean

Third letter of Signor Galileo Galilei to Signor Marc Welser about the solar spots, in which Venus, the Moon, and the Medicean Planets are also treated, and the new appearances of Saturn are unveiled

Most Illustrious Sir and Most Honorable Patron,

I feel I must respond to two most welcome letters from Your Most Illustrious Lordship, one written on 28 September, and the other on 5 October. With the first, I received the second [set of] discourses of the masked Apelles, and in the second you inform me of the receipt of my second letter about the solar spots, which I sent to you on 23 August. I will respond first briefly to the second and then I will come to the first, considering in somewhat more detail certain particulars contained in this reply of Apelles. Because I have considered his first letters, and he has seen my considerations, I am in some sense under an obligation to add some things concerning my first letter and his second reply.

As for the last letter from Your Lordship, I have learned with delight that after a rapid glance you have accepted as likely and very probable the reasons I have adduced in order to confirm the conclusions that I am trying to demonstrate. But the proof is whether a second and subsequent reading will persuade you, for it is not out of the question that some people, although of very per-

3. On Protogenes, see Eileen Reeves, "Faking It: Apelles and Protogenes among the Astronomers," *Bildwelten des Wissens. Kunsthistorisches Jahrbuch für Bildkritik* 5.2 (2007): 65–72.

spicacious judgment, sometimes at a first glance welcome as a work of passable perfection what later, when examined more carefully, seems to them to be of much more modest value, and especially when personal affection for the author and a preconceived good opinion affects the unbiased and plain judgment. Therefore, I will be awaiting, with a mind still in suspense, your further judgment; it will appease me until, as Your Lordship so very prudently says, we are destined, by the grace of the true, pure, and immaculate Sun, to learn in Him, in addition to all the other truths, the one that we now seek, dazzled almost to the point of blindness, in that other material and impure Sun.

But in my opinion we should not therefore refrain entirely from the contemplation of things, however distant they may be, unless we have already decided it was the best resolution to prefer all our other occupations to every act of speculation. For we either want to try to penetrate the true and intrinsic essence of natural substances by speculating, or we want to satisfy ourselves by finding out a few of their properties. I consider investigating the essence of the nearest elementary substances an undertaking no less impossible and a labor no less vain than that of the most remote and celestial ones. And I seem to be equally ignorant of the substance of the Earth as of that of the Moon, of earthly clouds, and of the spots of the Sun. Nor do I see that, in understanding these nearby substances, we have any advantage other than the abundance of details; all are equally unknown, and we wander through them, passing from one to the other with little or no acquisition [of knowledge]. And if, upon inquiring into the substance of clouds, I am told that it is a moist vapor, I will then wish to know what vapor is. Perhaps I will be informed that it is water, attenuated by virtue of warmth and thus dissolved into vapor, but being equally uncertain of what water is, I will in asking about this finally hear that it is that fluid body flowing in rivers and that we constantly handle and use. But such information about water is merely closer and dependent on more [of our] senses, but not more intrinsic than [the information] I had earlier about clouds. And likewise I do not understand any more of the true essence of Earth or of fire than I do that of the Moon or of the Sun. Such knowledge is available to our comprehension when we enter the state of beatitude, and not before then. But if we want to limit ourselves to learning certain properties, it does not seem to me that we need despair of grasping these in the most remote bodies any more than in the closest ones. At times, indeed, such knowledge will be more exact in the former than in the latter. And who does not know the periods of the motions of the planets better than he does that of the waters of sundry seas? Who does not know that the spherical figure of the Moon was recognized much earlier and more quickly than that of the Earth? And is there not still a controversy over whether the Earth itself remains immobile or wanders, while [at the same time] we are quite certain of the motions of not a few stars?[4] I want to conclude from this that although one would attempt the investigation of the substance of the solar spots in vain, this does not mean that certain of their properties—their location, motion, shape, size, opacity, mutability, appearance, and disappearance—cannot be learned by us and then serve as our means better to speculate upon other more controversial conditions of natural substances. These, then, elevating us at last to that ultimate purpose of our efforts—that is, to love for the divine Craftsman—may preserve in us the hope of being able to learn every other truth in Him, the source of light and truth.

The debt of thanking you is among the many other obligations I have to Your Lordship, for if I have investigated any true propositions, [such labor] has been the fruit of your commands, and those same commands will be my excuse if I do not succeed in bringing to completion this new and most difficult undertaking.

Regarding what you tell me about the Most Excellent Prince Federico Cesi's idea, it is in fact true that I sent His Excellency copies of the two solar letters, but not with the intent that they be published, because in that case I would have applied greater care and diligence to them.[5] For although I desire and esteem Your Lordship's assent and applause alone as much as I do that of the entire world, all the same, because of your goodwill and the kind disposition of your spirit toward me and my affairs I anticipate an indulgence that I must not expect from the scrupulous inquisitions and severe censures of many others. And there are a few things that I have neither entirely digested nor determined on my own. Chief among these is the occurrence of the spots in certain places of the solar surface, and not in others. For if we imagine the progress of all the spots as straight lines—a necessary argument for the axis of their rotations to be perpendicular to the plane that passes through the centers of the Sun and the Earth, a plane that is none other than the ecliptic—it remains, in my opinion, worthy of serious consideration why it is that they fall only within a zone that in width does not extend more than 20 or 30 degrees on either side of the largest circle of these rotations, so that hardly one in a thousand spots strays beyond its confines, and then only by a small amount. In this the spots imitate the laws of the planets, whose movements beyond the great circle of the annual rotation are limited by similar intervals.[6] This and a few other considerations have made

4. *Stelle* referred both to stars and planets.

5. Cesi began making plans to publish Galileo's letters in an "epistolary volume" upon receipt of Galileo's first letter. See p. 233, above.

6. Galileo clearly means the annual path of the planets in the zodiac. However, he states "*cerchio massimo della conversione diurna*," which would mean the *celestial equator*.

me postpone publishing this material in a longer treatise. Nevertheless, Prince Cesi may decide for himself; he is the absolute master of my writings, because my certainty of his flawless judgment and of the zeal he has for my reputation assures me that if he permits them to appear in print he must have judged them worthy of the light.

As for Apelles, it still distresses me that he had not seen my second letter before the publication of his *Accuratior Disquisitio*, and that my ambiguity and sloth in writing have not been able to keep apace with his decisiveness and promptness. It is, however, perfectly true that much of the delay was due to the fact that my letters were detained in Venice for more than a month because of the excessive regard that the Most Illustrious Signor Giovanni Francesco Sagredo had for them; he wanted some copies to remain in that city where I thought I would be sufficiently honored by the mere fact that he had read them, and because of the many figures [in them], this took a long time. I am also sorry for the difficulty my having written in our Florentine tongue has caused Apelles. I did this for several reasons, one of which is that I do not wish somehow to neglect the richness and perfection of a language that is adequate for treating and explaining the concepts of all disciplines, and therefore writing in this idiom, rather than in any other, is most appreciated by our Academies[7] and by the entire city. But besides, there was another private concern of mine, and that is not to deny myself Your Lordship's replies in that language, for these are read by me and by my friends with much greater delight and wonder than if they had been written in the purest Latin. And when reading letters of such elegant locution, it seems to us that Florence extends its borders, or rather its ramparts, all the way to Augsburg.

What Your Lordship writes about your reading my treatise *Discourse on Bodies in Water*—that those things that first appeared to be paradoxes in the end turned out to be true and manifestly demonstrated conclusions[8]—you should know that this happened here as well to many people who, because of their other views, are considered to be persons of perfect taste and sound judgment. Only a few strict defenders of every Peripatetic trifle remain opposed; from what I understand, they were from early infancy educated and nourished in their studies on the belief that to practice philosophy is, and cannot be other than, to become closely acquainted with Aristotle's texts so that passages can quickly be collected in great numbers from various of his works, and thrown

7. The Accademia della Crusca in Florence, which concerned itself with the Tuscan language. Welser would be elected to this academy in 1613.

8. *Delle cose che stanno su l'acqua*, or *On the things that are on the Water*. See Welser's fourth lettter to Galileo, pp. 252–253, above.

together as the proofs to whatever problem might be proposed. Therefore they do not ever want to lift their eyes from those pages, almost as if this great book of the world had not been written by Nature to be read by anyone other than Aristotle, and as if his eyes had to make observations for all future.[9] Those who submit themselves to such strict laws remind me of certain rules to which, for humorous effect, capricious painters sometimes restrict themselves, representing a human face or some other shape by a jumble of agricultural instruments alone, or just with fruits or flowers from this season or that.[10] As long as these oddities are offered as jokes, they are nice and pleasing, and where the artist manages to choose and apply one thing or another more becomingly to the counterfeited part, they show his greater acuity. But if someone, perhaps because he had consumed all his studies in a similar style of painting, then wanted to draw the general conclusion that every other manner of imitating was imperfect and blameworthy, surely Cigoli and other celebrated painters[11] would laugh at him.[12] Of those who oppose my views, some have written and others are now writing, but thus far only two works have seen the light, one by an Unknown Academician, and the other by a lecturer of Greek at the University of Pisa, and I am sending both to Your Lordship with this letter.[13] My friends are of the opinion—and I do not disagree with them—that since more solid objections have not appeared, there is no need to respond further, and they think that any further attempts to calm those who still remain uneasy would be no less futile than such efforts would be superfluous for those who are already persuaded. And I myself must regard my conclusions as true and my reasons as valid, for without losing the assent of any of those who from the beginning were with me, I have won over many who were of the contrary opinion. Therefore, we will wait for the outcome, and then it will be determined which opinion seems more appropriate.

I come now to Your Lordship's other letter, and I regret very much the dis-

9. On Galileo's deployment of the metaphor of the Book of Nature in these years, see Mario Biagioli, *Galileo's Instruments of Credit* (Chicago: University of Chicago Press, 2006), 219–259.

10. Giuseppe Arcimboldo and his followers.

11. Galileo had written "Cigoli and il Passignano [i.e., Cresti]" in his original draft of this letter. On their observations of the sunspots, see pp. 75–76, 233.

12. Erwin Panofsky, *Galileo as a Critic of the Arts* (The Hague: Martinus Nijhoff, 1954), pp. 16–20; and Erwin Panofsky, "Galileo as a Critic of the Arts: Aesthetic Attitude and Scientific Thought," *Isis* 47 (1956): 3–15, at 6–7.

13. *Considerazioni sopra il Discorso del Sig. Galileo Galilei intorno alle cose, che stanno in sù l'Acqua o che in quella si muovono . . . Fatte a Difesa, e Dichiarazione dell'opinione d'Aristotile da Accademico Incognito* (1612), *OGG*, 6: 145–196; Giorgio Coresio, *Operetta intorno al galleggiare de corpi solidi* (Florence, 1612), *OGG*, 4: 199–244. On this latter scholar, see Francesco Paolo DeCeglia, "Giorgio Coresio: Note in merito a un difensore dell'opinione d'Aristotele," *Physis* 37.2 (2000): 393–437.

quiet that the persistence of your illness, along with the affliction itself, brings to your many friends and servants, and most of all to me. For I, too, am likewise afflicted by many of my familiar indispositions; because they almost constantly keep me from all activity, they remind me that with time's rapid pace, anyone who wishes to leave behind some trace of his passage through this world must constantly keep busy.[14] Whatever the course of our life may be, we must accept it as the greatest gift from the hand of God, in Whose power it also lay to do nothing at all for us. Indeed, not only must we receive our particular destiny gladly, but we must be infinitely grateful for His goodness, which in such fashion turns us away from excessive love of earthly things and lifts us to that of the celestial and divine.

Your apologies for being so brief in your letters are superfluous to me, for I am always happy merely to hear that that I remain in your good graces; indeed, I should apologize for my wordiness or, to put it better, beg you to forgive me, and I would do so if I were doubtful of the indulgence that I expect from your courtesy.

With the letter from Your Lordship, I received the second tract of the masked Apelles, and I sat down to read it with great curiosity, excited by the name of the author as well as by the nature of the title, which promises a more accurate inquiry not only on solar spots, but on the Medicean Planets as well. And because the comparative expression "more accurate inquiry" can refer only to other inquiries about the same material [i.e., the Medicean Planets], it cannot be doubted that it has to do in some way with my *Sidereal Message*,[15] which is also pertinent and which Apelles does not neglect. Hence I began hoping that I would find this entire argument resolved, having been able to touch only upon its general contours in my *Message*. Besides what was promised in the title, I found the observation of Venus explained more fully than in the first letter, and also certain details about the Moon. In all these matters I discover many of Apelles' views quite opposed to mine, and various reasons and implicit responses to the arguments I adduced in the first letter I wrote to Your Lordship. Because of the esteem I have for the author I must neither pass by nor ignore these matters; since I do not have a painting in front of me to hide me and to block the view of those who pass before and behind me, I should at least greet him in passing [*per termine*]. And because the development of the entire argument has thus far been discussed before Your Most Illustrious Lordship, I will in declaring myself once again put forth as briefly as I can what occurs to me in this regard. And following the order established

14. On Galileo's illnesses in these years, see Drake, *Galileo at Work*, 165, 175, 180, 229.
15. *Avviso Sidereo*. Galileo often used this title in his Italian writings. See *SN*, pp. ix–xi.

by Apelles, I will consider the ultimate goal of his first part, which is to demonstrate that the revolution of Venus takes place about the Sun and nowhere else. He bases his entire demonstration, just as he did in his first letter, on a morning conjunction[16] around 11 December 1611 of that star with the Sun, adding this time a quantitative investigation of its motion under the solar disc, conveyed by calculations and geometrical demonstrations. And here two doubts arise in me: one about his way of handling such demonstrations, which would not prove entirely satisfactory to a rigorous mathematician, and the other about the usefulness of such a display and procedure to the author's primary intention.

As to the manner of demonstration, I pass over [the fact] that an astronomer more scrupulous than I am could object to seeing arcs of circles treated as though they were straight lines, subjecting them to the same criteria. But I do not wish to take this into account because in our particular case the arcs used are not so great that the error in computation would turn out to be of a noticeable excess. But I would rather have liked Apelles to have been a somewhat more committed geometer in the lemma that he proposes, and also in the rest of his demonstration, and I do not see his reason for fashioning out of what is a simple general proposition, demonstrable in a few words, a lemma in the form of a particular proposition and explained at such length. For in every triangle it is the case that if its sides are extended and through the intersection of two of them a parallel to the opposite side is produced, the three angles made on one side of that parallel (or on one of the extended sides) are each equal to the [corresponding] interior angles of the triangle. (I will not add, as Apelles does, that the said angles not only taken one by one, but that also all three together are equal to all three [of the interior angles] together because I would be saying something too obvious and superfluous.) For let the two sides AC and BC of the triangle ABC be extended to G and I, and through the intersection C let MN be drawn parallel to AB. It is clear that the three angles made on one side of the extended line ACG are in this way equal to the three internal angles of the triangle; that is, angle MCA to angle A because they are alternate angles, the exterior angle MCI to the interior angle B, and the remaining angle ICG to the remaining angle ACB because they are vertical angles. And if instead of angle ACM we were to take NCG, the other part of the conclusion will be manifest, the three angles—MCI, ICG, and GCN—being on the same side of the parallel MCN. It also happens, in the particular case of the right triangle, that these parallel lines are also perpendicular to the sides of the triangle, and

16. Galileo means the conjunction of Venus with the Sun following Venus's morning phase, when it is visible in the early morning sky, preceding the Sun. After this conjunction, Venus will reappear in the evening sky.

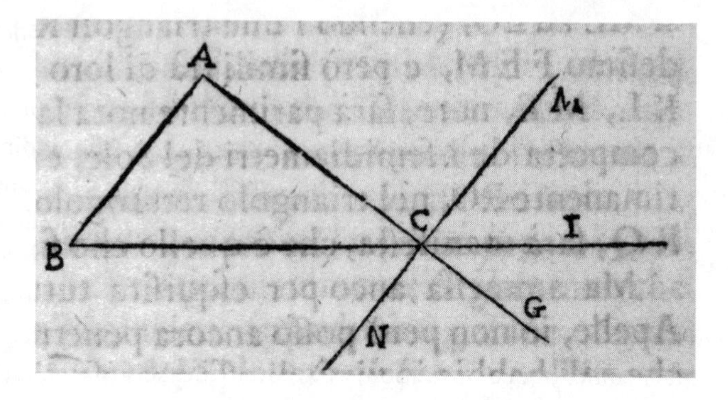

this much was sufficient for the use that Apelles makes of the lemma. And I will even say, by his leave, that the entire lemma was superfluous, seeing that his application of it in his main problem follows immediately from a single proposition of the first book of Euclid.[17] For, referring to his figure[18] and demonstration, the latter and the lemma do nothing more than show that angle OME is equal to angle MIP, which is obvious because they are external and internal angles of the straight line OMI, which cuts the two parallels EB and GI. And permit me also to say that the entire method should have been shortened, not only by the removal of the lemma, but also by a considerable restriction of the rest of the demonstration, the final conclusion of which was to find the length of the line RQ, supposing that GH, HE, KH, and IG are known. Now, because we know the lengths of KH and IG, those of IL and LG also become known; and because IL is to LG as IK is to KF and GH is to HF, and since IL, LG, and GH are known, HF will therefore also be known. But EH is given, and therefore the remainder, EF, is also known. And because FE is to EM as KL is to LI, because of the similar triangles FEM and KLI, and the three [quantities] KL, LI, and FE are known, therefore EM will likewise be known. Moreover, because in the right triangle KLI the sides KL and LI are known, KI will also be known. And since IK is to KL as ME is to EO—the two triangles KLI and MEO being similar to triangle FEM, and thus similar to each other—and the three lines IK, KL, and ME are known, EO will likewise be known. But ER is known, being composed of the semidiameters of the Sun and Venus, therefore the remaining side RO of the right triangle ERO, and its double, RQ, will be known, which is what was sought.

17. Euclid, *Elements*, Book I, Proposition 29, 1: 13: "A straight line falling on parallel straight lines makes the alternate angles equal to one another, the exterior angle equal to the interior and opposite angle, and the sum of the interior angles on the same side equal to two right angles."

18. P. 189, above.

But even if we grant that the entire demonstration of Apelles is excellent, I am still unable entirely to fathom what he has claimed to have gained by virtue of this over anyone who might persist in denying the revolution of Venus around the Sun, for either his adversaries will admit that Magini's calculations are correct, or they will consider them to be doubtful and wrong. If they find them dubious, Apelles' effort is pointless because it does not demonstrate that Venus truly was in corporeal conjunction [with the Sun]; but if they grant them as true, then further computation is unnecessary, because just the difference between the motions of the Sun and the star [i.e., Venus], together with its latitude taken from the same ephemerides,[19] will suffice to explain how this conjunction necessarily lasted so many hours such that the observation could be repeated numerous times. Nor was it necessary to make a triple observation of the beginning, middle, and end of the conjunction, it being very well known that calculations are based upon the midpoint of the conjunction. For if these observations were erroneous, they would not necessarily be corrected by referring them to the beginning or the end of the conjunction, there being no reason whatsoever that one might imagine it not possible to err in the calculation of a conjunction by an interval greater than that of the duration of the conjunction itself. But I do not believe that his opponents would resort to denying the accuracy of the astronomical computations, especially because they have safer refuges, such as those that I proposed in the first letter. And just as for the most expert in the science of astronomy it was enough to have understood what Copernicus writes in his *Revolutions* to assure themselves of the revolution of Venus about the Sun[20] and of the truth of the rest of his system, so for those who reason with less than average ability it was necessary to remove the strongholds mentioned by me above. Of these refuges, I do not see that Apelles has ever grazed more than a couple, and it also seems to me that these two are not wholly demolished.

I said in the first letter that adversaries would be able to take refuge in saying that Venus is not seen below the Sun either because of its smallness, or else because it is so bright itself, or else because it is always above the Sun.

What Apelles produces to deny this first escape route to his opponents is not enough, because to begin with they will not agree that the shadow of Venus under the Sun must appear as large as its body does when illuminated and just beyond the Sun, for this extraneous radiation makes the stars appear much

19. See p. 188 and note 20, above.

20. Copernicus, *De Revolutionibus*, I, 10; *Nicholas Copernicus on the Revolutions*, 18–19. See also Roger Ariew, "The Phases of Venus before 1610," *Studies in the History and Philosophy of Science* 18 (1987): 81–92.

larger than they are in reality. This is obvious in the case of Venus which, when it is a thin sickle and consequently separated from the Sun by just a few degrees, appears nonetheless to the unaided eye as round as the other stars, concealing its shape with its splendid rays; it cannot, therefore, be doubted that Venus then appears much larger to us than it would when deprived of its light. When, by contrast, it is located under the brilliant disc of the Sun, there is no doubt that its dark little body would be not a little diminished—I mean as regards its appearance—by the interference of the Sun's splendor. It is most erroneous, therefore, to conclude that it would appear equal to spots of middling size. And who knows if these spots, in order to be visible in the brilliant field of the Sun, are not much larger than their apparent size? In fact, Apelles himself can serve as the best witness of this, if we call to mind what he wrote in the third [letter] of his *Tres Epistolae*, at the second corollary, to wit: *They are very large; otherwise the Sun, illuminating them, would entirely absorb them by its great size.*[21] And the same thing must be asserted of the body of Venus. Therefore one commits a compound error in comparing the size of bright Venus to that of the dark spots, because as much as the latter are diminished in appearance from their true [size] by the brightness of the Sun, so much will the former be enlarged.

No more effective is what Apelles adds in this same passage in his effort to keep Venus immeasurably larger than it is, and I indicated as much in my first letter. Contrary to what perception and experience show us, he pointlessly adduces the authority of men who are in other respects very great, but truly deluded in estimating to the visual diameter of Venus as a tenth that of the Sun.[22] They deserve to be partly, but not entirely, excused. The lack of a telescope exculpates them somewhat, for it has made a not insignificant contribution to the astronomical sciences, but in two particulars their diligence leaves something to be desired. The first is that the size of Venus should have been observed during the day, not at night, when its crown of rays makes it appear ten or more times larger than in daylight, when it is shorn; they would have then quickly realized that at such times the diameter of its very small globe does not equal the hundredth part of the solar diameter. Second, they should have distinguished one position from another, rather than indiscriminately proclaiming the visual diameter to be a tenth part of that of the Sun, because when the star is very close to the Earth, it is more than six times greater than when it is very far away. This difference, though it cannot be accurately observed except with the telescope, is

21. P. 72 above.

22. Paolo Palmieri, "Galileo and the Discovery of the Phases of Venus," *Journal for the History of Astronomy* 32 (2001): 109–121; A. Van Helden, *Measuring the Universe*, 65–76.

nonetheless quite perceptible to the naked eye. In this regard, the authority of the astronomers cited by Apelles and on whom he leans for support ceases here. And if it were conceded that some spot were visible on the solar disc—a spot whose length is not the hundredth part of the diameter, nor whose surface area the ten-thousandth part of the visible circle of the Sun—let Apelles not believe that he has conclusively proved from this the apparition of Venus [on the Sun's disc], for I reply yet again that during its morning [i.e., superior] conjunction its diameter does not equal the two-hundredth part of the Sun's diameter, nor its surface area the forty-thousandth part of the visible disc of the Sun.[23]

As for the second of his adversaries' refuges—that is, that Venus, being perhaps luminous itself, does not necessarily eclipse part of the Sun—this, in my opinion, is not condemned by what Apelles produces. For, as to the mere authority of the ancient and modern philosophers and mathematicians, I say that it has no force whatsoever in establishing knowledge of any conclusion concerning nature, and the most that it can do is to incline us to believe one thing more than another. Besides, I don't know how true it is that Plato had decided to put Venus above the Sun precisely because during conjunction he did not see it dark in appearance and below the solar disc.[24] I do know that what Ptolemy has to say on this matter differs significantly from what Apelles has attributed to him. It would have been too great an error for the Prince of astronomers to deny the direct[25] conjunction of Venus and the Sun. What Ptolemy says at the beginning of book 9 of his *Great Composition*[26]—where he investigates what must be the most probable order of the planets, disputing the logic of those who had put Venus and Mercury above the Sun simply because they had never seen it obscured by them—demonstrates the weakness of this argument. He says that not every star below the Sun will necessarily eclipse it, because they can be beneath it but not in any of the circles that pass through its center and our eye. Ptolemy does not, however, affirm here that this pertains to Venus; on the contrary, the fact that he added the example of the Moon—

23. See note 19, and p. 93, above.

24. Plato put Mercury and Venus "above" the Sun. See *Timaeus* 38d; Plato does not argue, however, that this is because no transits had ever been observed. In his *Planetary Hypotheses*, Ptolemy states "if a body of such small size (as a planet) were to occult a body of such large size and with so much light (as the sun), it would necessarily be imperceptible, because of the smallness of the occulting body and the state of the parts of the sun's body which remain uncovered." See G. J. Toomer in *Ptolemy's Almagest*, 419 n 2. See also Bernard R. Goldstein, *The Arabic Version of Ptolemy's "Planetary Hypotheses,"* American Philosophical Society, *Transactions* 57 (part 4), 1967, p. 6.

25. That is, a conjunction of Mercury or Venus with the Sun during which the planet appears on the Sun as a black spot.

26. *The Almagest*.

which in most of its conjunctions does not darken the Sun—clearly shows that he wanted to argue nothing else about Venus than that it can be below the Sun without obscuring it in all conjunctions.[27] Therefore it may very well have been the case that the conjunctions observed by such men were not conjunctions in the plane of the ecliptic. The Most Reverend Father Clavius speaks with great certainty, affirming that this shadow remains invisible to us because of its small size,[28] and although it appears from these authors' pronouncements that they inclined him to believe that Venus was dark, rather than luminous in itself, that mere opinion alone is not sufficient to convince adversaries, who lack no opportunity to produce the opposing viewpoints of others.

The other argument offered by Apelles, derived from the Moon's darkening in its passage below the Sun,[29] has no force unless he first shows that the deficiency [i.e. darkening] of the Sun is visible as soon as the Moon occupies less than a 1/40,000 part of its disc.[30] Without this, the comparison of the Moon to Venus is not valid, but it is clear to everyone how difficult this would be to demonstrate.

That Mercury has been seen under the Sun by various people is not only doubtful, but even unbelievable, as I pointed out to Your Lordship in the other letter. And as to citing Kepler in this connection, I have no doubt at all that, being of a most acute and free intellect, and being a greater friend to truth than to his own opinions, he would be entirely persuaded that such dark marks seen on the Sun were some of the spots and that the conjunction of Mercury merely provided an opportunity to make more regular and accurate observations during those hours. With such diligence these spots would also have been seen at other times, as they will frequently be seen in the future, and as I have already shown them to many people.

Consequently, the darkness of Venus has unquestionably been demonstrated by the single [sensory] experience I wrote about in the first letter (and which

27. *Almagest* IX, 1: "To us, however, such a criterion seems to have an element of uncertainty, since it possible that some planets might indeed be below the Sun, but nevertheless not always be in one of the planes through the sun and our viewpoint, but in another [plane], and hence might not be seen passing in front of it, just as in the case of the moon, when it passes below [the sun] at conjunction, no obscuration results in most cases." See *Ptolemy's Almagest*, 419. For a complete review of this question, see Bernard R. Goldstein, "The Pre-telescopic Treatment of the Phases and Apparent Size of Venus," *Journal for the History of Astronomy* 27 (1996): 1–12.

28. Christoph Clavius S.J., *In Sphaeram Ioannis de Sacro Bosco Commentarius* (Mainz: Reinhard Eltz, 1611), 45–46; see in this connection James M. Lattis, *Between Copernicus and Galileo: Christoph Clavius and the Collapse of Ptolemaic Cosmology* (Chicago: University of Chicago Press, 1994), 75.

29. P. 195, above.

30. That is, 1/40,000 in area. See Palmieri, "Galileo and the Discovery of the Phases of Venus," 115.

now Apelles puts in the third place),[31] that is, that Venus is seen to vary, as does the Moon, in her shape. Let this serve as a single, solid, and strong argument to establish its revolution about the Sun, such that no room whatsoever remains for doubt. And therefore Apelles ought to regard it as worth delineation as the central figure in the most conspicuous and noble part of his engraving, rather than in a corner in the guise of a pilaster, as if it were a prop and support of some [other] element that without it would seem to menace onlookers with collapse.

But I pass to some consideration of Apelles' replies and his additions to what he has written earlier on the subject of the solar spots. Here it appears to me in general that he is rather less resolute in these findings than he was earlier, although he seems willing at the same time to present them as if merely modified rather than wholly revised. And at the end he affirms that all the things said in the first letters remain unchanged. For all that, I have some hope of seeing a third tract from him[32] with opinions that are, in essence, very much in conformity with mine. I say this not on account of these letters—for given the difficulty of the language, they cannot be read by him—but because as he reconsiders these matters, those same observations, those same reasons, and those same solutions that persuaded me to write what I have written in the first and second letter and what I add in this one will come to him as well. And already one sees how many details he puts in this second tract that were not yet seen in the first. Earlier he thought that the solar spots were all of a spherical shape, saying that if they could be observed separated from the Sun, they would appear to us like so many miniature moons, some sickle-shaped, others hemispherical, and still others gibbous, and some, perhaps, entirely full. Now he writes with more truth that they are very rarely spherical and most frequently of an irregular shape. He has likewise observed that very few or none of them maintain the same shape for the entire time they remain visible, but rather that they change extravagantly, now growing, now diminishing. And, what is more, he has seen how some suddenly appear and others dissolve, even in the middle of the Sun, and how some split into two or more parts and how, by contrast, many unite into one, all details mentioned by me in the first letter. He once thought that they were wandering stars and located at different distances from the Sun, such that some were less and others more remote, and thus that very

31. Scheiner gives the phases of Venus as discovered by Galileo as the *third* reason why he believes "Venus, moving under the Sun, makes a shadow that eclipses the Sun an extent proportional to Venus moving below it." See p. 196, above.

32. Scheiner never wrote a third tract. He did however publish the results of two decades of research on sunspot in his *Rosa Ursina* in 1630. See pp. 321–322, below.

many of them went wandering about between the Sun and Mercury, or again, between Mercury and Venus, at established distances, making themselves visible only upon encountering the Sun. But now I do not hear such a vast distance confirmed, and it seems to me that he is satisfied with showing that they are neither within the solar body nor contiguous to its surface but only a small distance beyond it, as can be gathered from the reasons that he uses in demonstrating his opinion.

I would readily agree with Apelles that the spots are not in the Sun, that is—immersed within its substance—but I would not in fact affirm this on the basis of the reasons he adduces, the first of which is an assumption that certainly will be denied him by anyone who wanted to defend the contrary. For there is no one so simple as to grant that the Sun is hard and immutable after arguing that the spots are immersed inside the solar substance and admitting moreover their continuous mutability of shape, mass, separation, and mingling; he will instead resolutely deny this assumption and the proof that Apelles produces from it, which is based on the opinion that the latter ascribes to all philosophers and mathematicians. And [such a thinker] will have no small reason to deny it, both because in matters of knowledge the authority of a thousand people's opinion is not equal to the spark of reason of a single individual, and because these modern telescopic observations strip away the force of those decrees of the writers of times past, who, if they had seen these things, would have judged them differently. Moreover, those same authors who thought that the Sun was neither yielding nor mutable believed still less that it was dappled with dark spots, and therefore where the assumption of its immaculacy is contradicted by experience, it would be fruitless to seek recourse in the presumption of its hardness and immutability, for where what appeared more solid gives way, those less sturdy ideas will provide still less support. In fact, these adversaries, gaining strength, will deny that the Sun is hard or immutable because not conjecture alone, but rather actual experience shows them that it is spotted. And as to the mathematicians, it is not known that any of them ever discussed the hardness and immutability of the solar body, or that mathematical science would suffice to fashion demonstrations of similar properties.[33]

Apelles' second reason is based on the observation of some spots as darker toward the circumference of the Sun than toward the central region, where they appear to become brighter. It does not seem that this fact forces Apelles' adversary to suppose them beyond the Sun, both because experience of the matter for the most part, if not always, shows the contrary, and because rarefac-

33. On the roles of the technical, mathematical astronomer and the philosopher/cosmologist, see above, pp. 95–96, and Westman, "The Astronomer's Role in the Sixteenth Century."

tion and condensation—characteristics not denied the spots—are enough to account for that effect. Indeed rarefaction and condensation may explain it no less well than what Apelles offers us here in saying that the stronger and more direct irradiation made when the spot is near the middle of the disc rather than when it approaches the circumference produces this less intense blackness. For returning to his illustration, and rereading his demonstration,[34] I say that it is not true that the rays coming from the surface AG are very weak because of the spherical inclination of the Sun in those parts. On the contrary, because there is not a single ray but an immense sphere of light diffused from every point in the solar surface, there is no point at all on the upper surface of either spot D or IK away from the eye that does not receive an equal number of rays, such that these spots are equally illuminated. Likewise, it is not true that the rays from the sloping [declining] surface AG arrive at the eye more enfeebled than do those from the middle, as experience shows. And therefore in my opinion it would perhaps be better to say (whenever one does not want to rely on arguments about rarity and density)[35] that the same spot appears less dark around the central regions than toward the limb on account of the fact that here it is seen edgewise and there straight on, just as happens when a sheet of glass seen edgewise seems dark and very opaque, but appears clear and transparent when viewed full on. And this would serve as evidence to demonstrate that the breadth of these spots is much greater than their depth.

All that Apelles has added to prove that the spots are neither pools nor deep chasms in the solar body can be freely conceded, for I do not believe that anyone would ever put forward such an opinion as true.[36] But because neither I nor, as far as I know, others have contended that the spots are immersed in the substance of the Sun—rather, I have repeatedly written to Your Lordship and, if I am not mistaken, have necessarily demonstrated, that they are either contiguous to the Sun or separated from it by a distance that is imperceptible to us—it is a good idea for me to examine the reasons that Apelles proffers as irrefutable arguments by which their appreciable distance from the solar surface is made manifest.

34. See the diagram on p. 201 above.

35. This is an echo of the traditional argument to explain the spots visible on the Moon with the naked eye. They were said to be caused by "rarer and denser parts." Christoph Clavius also used this argument against the spots seen by Galileo through the telescope. See *OGG*, 11: 93; Roger Ariew, "Galileo's Lunar Observations in the Context of Medieval Lunar Theory," *Studies in the History and Philosophy of Science* 15 (1984): 213–226.

36. One of the first to observe sunspots in Rome, in the autumn of 1611, was Domenico Cresti Passignani, a friend of Lodovico Cardi da Cigoli. As Cigoli reported to Galileo, at that time, Passignano thought sunspots to be depressions in the Sun's surface. See *OGG*, 11: 209, 212.

Apelles bases his reasoning on the alleged fact that the spots appear to spend unequal intervals of time below the face of the Sun. Those that traverse it along the longest line, passing through the center, seem to stay longer than those that cross it along lines remote from the center. And he offers observations of two spots, one of which stayed for sixteen days on the diameter, while the other, passing somewhat farther from the center, ran its course in fourteen days. Now, here I wish I could find words with which to deny that observation without offending Apelles, whom I intend always to honor. For having made a great number of most diligent observations of this particular, I have, on the contrary, not met with a single one from which anything can be concluded but that all spots without exception remain below the solar disc for equal intervals, which in my judgment are a bit more than fourteen days.[37] And I affirm this all the more resolutely because from now on it will be in everyone's power to make thousands and thousands of observations without difficulty. As to the particular experiment that Apelles proposes, I have a few objections to his having chosen in the first observation not the passage of a single spot, but rather of a squadron composed of a very large number of spots, and ones that changed their relative positions. From this it follows that this observation, since it is subject to many accidental variations, is not sufficiently certain to sustain such an important conclusion alone. Rather, the individual irregular motions of these spots render the observations subject to so many changes that no conclusion can be drawn except from an accumulation of a great many particulars. This I have done on the basis of more than 100 large and accurate drawings, and I have in fact found some small differences in the intervals of the crossings, but I have also discovered, in turn, that the spots on the circles closer to the center of the disc are sometimes no less slow than those [moving] along the more distant circles are at other times.

But even if we were in no position to compare the drawings already made with those that will be made, it nevertheless seems to me that we can detect a certain contradiction in the very things proposed and admitted by Apelles, for which reason one might quite legitimately doubt the truth of the observation he chose and, in consequence, that of the conclusion drawn from it. For first I notice that because he has to use the inequality of the times of the spots' transit as a conclusive argument for their appreciable distance from the surface of the Sun, Apelles is forced to believe that they are in a single sphere that revolves with a motion common to all. For if he wanted each spot to have its own proper motion, nothing could be deduced from this that would relate to evidence of

37. Galileo was not, in fact, quite correct about variable rotation. Scheiner would treat the matter fully in *RU*, Book 4, Part 2, pp. 553–570.

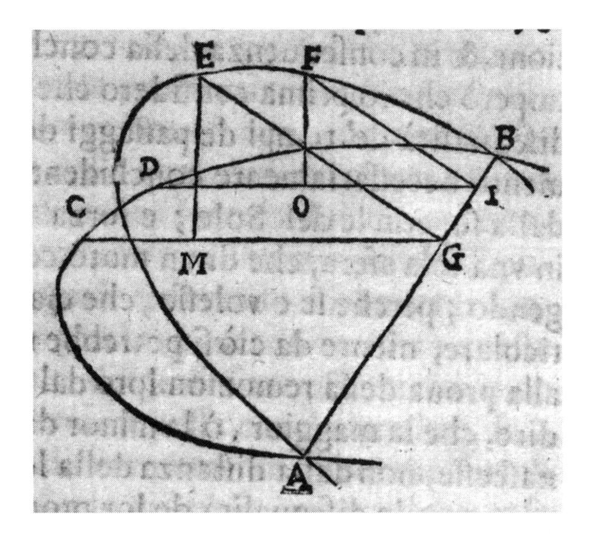

their distance from the Sun, for one could always say that the greater or lesser delay of this spot or that arose not from the remoteness of their sphere from the Sun, but from the true and real inequality of their proper motions. I note next that the lines described on the solar disc by the spots do not diverge from the ecliptic—the greatest circle of their revolution—either toward the North or the South beyond certain limited distances, which are at most 28, 29, or in rare cases 30 degrees. Now, having established these things, I think that with the very obvious internal contradictions among Apelles' pronouncements I can render ineffective what he proposes here as an argument for the separation of the spots from the surface of the Sun. For, granting his assumptions, even to the extent most favorable to his conclusion—that is, that the first spots traversed the longest line, the diameter of the Sun, in at least sixteen days and that the other one traversed it in at most fourteen days on a parallel no less than 30 degrees from the diameter—I will show that it follows from this that their distance from the Sun would have to be so great that many other manifest details simply cannot be valid. And first, for the fullest understanding of this fact, I will demonstrate that if two spots travel over the solar disc, one along the diameter and another along a lesser circle, the times of their passages always have a lesser proportion among themselves than these circles do, however large the sphere that carries them around. For this demonstration, I propose the following lemma.

Let the half circle ACDB be rotated around its diameter AB. On its circumference two points, C and D, are taken, and from these the perpendiculars CG and DI are drawn to the diameter AB; and let it be understood that in the rotation the half circle ACB is moved to AEB, so that point E is the same as point C, and F the same as D, and the line EG is the same as GC, and IF the same as ID; and from the higher points E and F drop the perpendiculars EM and FO to the lower plane so that these fall on the first lines, GC and ID. It is evident that if the circle AEFB were rotated 90 degrees and in consequence were perpendicular to the plane of the other circle ACDB, the perpendiculars falling from points E and F would be these same lines EG and FI, but being elevated here less than 90 degrees they fall, as mentioned, to M and O. I say that the lines CG and DI are proportionally cut by points M and O. For in the triangles EGM and FIO the two angles EGM and FIO are equal, the inclination of the two planes ACB and AEB being the same, and the angles EMG and FOI are right angles; therefore because triangles EMG and FOI are similar; and since EG is to GM as FI is to IO; and since the two lines EG and FI are equal to CG and DI, therefore CG is to GM, as DI is to IO; and by subtraction [*dividendo*], CM is to MG as DO is to OI.[38]

This being demonstrated, consider the circle HBT, intersecting the solar globe along the diameter HT, which is the axis of revolution of the spots; and from the center A, draw AB perpendicular to the axis HT, such that in revolving, line AB describes the greatest circle; and with point L being any other point taken on the circumference TBH, draw the line LD parallel to BA, and LD will be the semidiameter of the circle whose circumference is described by the revolution of point L. Now it is evident that if the Sun turned on itself and there were two spots at points B and L, both would traverse the solar disc in the same interval. This would be seen by an eye positioned at an immense distance on the line produced from the center A perpendicularly to the plane HBT, which would be the circle of the [solar] disc, and the lines BA and LD would appear half as long as those that the spots B and L would describe in their movements. But if the spots were not contiguous to the Sun, but were rather in a sphere that surrounded it and were noticeably larger than it, there is no doubt that that spot which appeared to cross the solar disc along the diameter BA would take more time than another that crossed it along the shorter line LD, and the larger we imagine the orb carrying these spots to be, the greater the gap between these two intervals will become. But it could never, in fact, happen that the difference between these times was as great as the difference between the lines BA and LD; but it would always be the case that the proportion of the time of transit along the longest line, AB, to the time of transit by any other

38. See appendix 5.2. The term *dividendo* is a technical term in proportions meaning "by subtraction" or "by separation" or "by division." It means:

If $\frac{a}{b} = \frac{c}{d}$, then $\frac{a-b}{b} = \frac{c-d}{d}$.

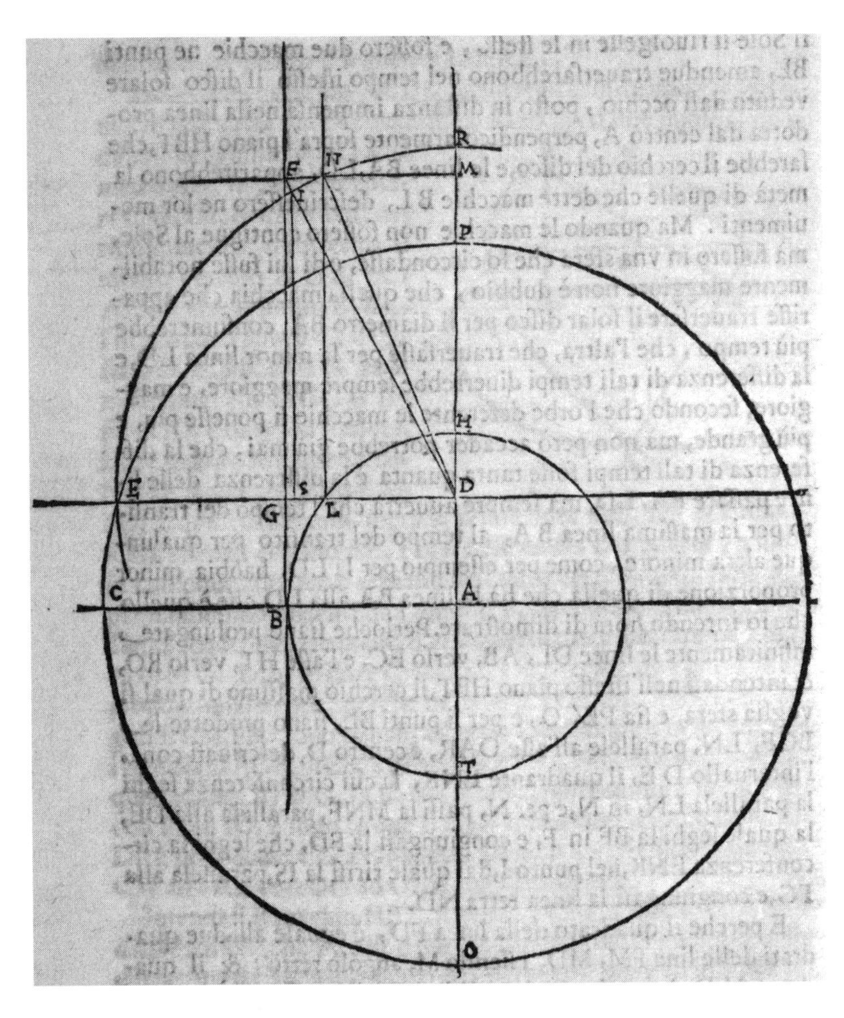

smaller line, as, for example, along LD, would be less than that of the line BA to LD, which is what I intend now to demonstrate.

For let the lines DL and AB be extended infinitely toward E and C, and the axis HT toward R and O, and take in the same plane, HBT, the great circle of any sphere, and let this circle be PECO; and through points B and L let the lines BGF and LN be produced parallel to the axis OAR. And with center D, describe with the segment DE the quadrant ENR, the circumference of which intersects the parallel line LN at N; and through N draw MNF parallel to DE, which intersects BF at F; and join to it FD, which intersects the circumference ENR at point I, from which draw IS parallel to FG, and connect the segment ND. And because the square of the line FD is equal to the combined squares of lines FM and MD, M being a right angle, and the square of ND is equal to the two squares of NM and MD, the excess of the square of FD over the square of ND will be equal to the excess of the two squares of FM and MD over the two squares of NM and MD, which (subtracting the common square of MD) is the same as the excess of the square of FM over the square of MN. But because FM is equal to BA, as opposite sides of a parallelogram, and NM is equal to LD, and the excess of the square of BA over the square of LD is the square of DA, therefore the excess of the square of FD over the square of ND is equal to the square of DA; and therefore the square of FD is equal to the two squares of lines ND and DA, that is of the two lines ED and DA. But the square of the semidiameter CA is also equal to these same two squares, and therefore line FD is equal to line CA.[39] Because in the triangle FGD the line IS is parallel to FG, FD will be to DG, or CA to AB, as ID or ED is to DS; and, dividing, CB is to BA as ES is to SD. Thus if we understand the sphere to rotate about the axis PO and the half circle PCO to be raised up until the perpendicular dropped from the elevated point C falls upon point B, it is evident from the converse of the preceding lemma that the perpendicular from point E will fall on S; and therefore, when spot C begins to appear on the limb of the solar disc, that is, at point B, the distance the other spot, E, still has from the circumference of the disc will be the interval SL. And because after a 90-degree rotation the perpendiculars from the spots C and E will fall on points D and A at the same moment, it is clear that the time of passage along BA is equal to the time of passage of the other spot along all of SD, of which the time of the passage along LD is only a part.

We must now demonstrate that the time of the passage along BA to the time along LD has a smaller ratio than the line BA to LD; and since it has already been established that the time of passage along BA is equal to the time of passage along SD, if it is shown that the ratio of the time along SD to the time along DL is smaller than that of line BA to LD, then what is desired will be proved. But since the time of the passage along SD to the time of the passage along LD has the same ratio as arc IR to arc RN (because arc ENR is equal to the quadrant that point E would describe on the surface of the sphere as it turns around axis PO, in which circumference the perpendiculars erected from points S, L, and D would mark off arcs equal to IR and NR, and the lines SD and LD would be their sines as they are of the two arcs IR and NR), we must demonstrate that the ratio of the line BA to DL, or FM to MN, is greater than that of arc IR to arc RN. Now, because triangle FDN is larger than the sector IDN, so triangle FND will have a still greater ratio to sector NDR than sector

39. See appendix 5.3.

IND has to the same sector NDR, since triangle FDN has a still greater ratio to triangle NDM than to sector NDR, triangle NDM being smaller than sector NDR. Therefore triangle FDN will have a much greater ratio to triangle NDM than sector IDN will have to sector NDR, and by addition triangle FDM will have a greater ratio to triangle MDN, than sector IDR has to sector RDN. But triangle FDM is to triangle MDN as line FM is to line MN, and sector IDR is to sector RDN as arc IR is to arc RN. Therefore the ratio of line FM to MN, or BA to LD, is greater than that of arc IR to arc RN, that is, greater than that of the time of the passage along BA to the time of passage along LD.[40]

From this it is evident how close Apelles comes to an absolute impossibility in saying that he has observed one spot traverse the diameter of the solar disc in at least 16 days, and another a shorter line in 14 days at most. As I said above, even granting (as the most favorable instance of his assertion) that the second spot traversed a line 30 degrees from the diameter—something that is seen to occur very rarely or never with large spots, such as this one—if the ratio of 16 to 14 days, which he shows with great care to be a low estimate, were expanded by only $3\frac{1}{2}$ hours, so that one interval were 16 days and the other 13 days and $20\frac{1}{2}$ hours, the ratio would be absolutely false and impossible, because the ratio of these times would be greater than that which the diameter has to the chord of 120 degrees, which is as the time of 16 days to the time of 13 days and 20 hours 33 minutes.[41] Nevertheless, while he has avoided an absolute impossibility, he falls into one by his supposition, which is enough to show the weakness of the argument. For I will demonstrate how, assuming that one spot traverses the diameter of the Sun in a time interval $1\frac{1}{7}$ times as long as the passage of another spot moving along a parallel 30 degrees away, it necessarily follows that the sphere that carries those spots has a semidiameter more than double that of the semidiameter of the solar globe.

Of the great circle of the solar globe let PR be the axis and A the center; let line ABC be perpendicular to PR, assume that arc BL is 30 degrees, and let DLE be drawn parallel to AC; let FECH in the plane of the circle PBR be the great circle of a sphere that, revolving around the Sun, carries the spots that traverse lines BA and LD, the former in a time $1\frac{1}{7}$ times as long as the time of the latter. I say that the semidiameter of this sphere, that is, line CA, is by necessity more than double the semidiameter BA of the Sun. For if it is not greater than double, it will be either double or less than double, and let there be supposed through point B the [line] BG parallel to DA, and let it be that as CA is to

40. See appendix 5.4.
41. See appendix 5.5.

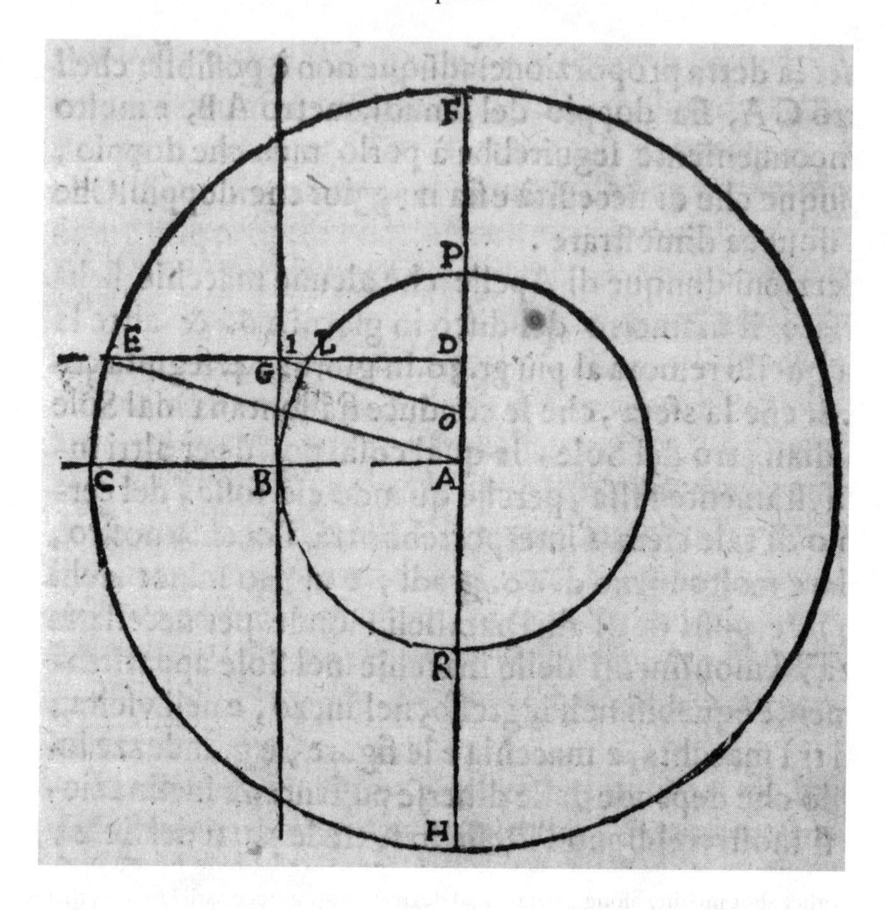

ED, so BA is to ID; and because CA is greater than ED, still more is BA greater than ID.[42] From the foregoing it is evident that when spot C appears at B, spot E will appear at I, and then both will appear at A and D at the same time. Thus the time of the apparent transit of spot C along AB will be equal to the time of transit of spot E along ID, and therefore the time along BA will have the same ratio to the time along LD as the time along ID to the time along LD. This ratio is that which arc-sine ID [*arco del sino*] has to the arc-sine [*arco del sino*] LD, taken in the circle whose semidiameter is the line DE. And because in triangle EAD, IO is parallel to EA, ED will be to DI as AD is to DO and AE to IO. But ED is double DI because CA has been assumed to be double AB. Therefore AD will be double DO and AE double IO; hence IO is equal to the semidiameter

42. Note that the figure is not altogether clear about the location of point I. It looks as though it coincides with point G, but it is actually on line ED between G and L, close to G, that is ID > GD = AB.

AB. And because arc BL is assumed to be 30 degrees, BA, that is, IO, will be double AD and thus four times OD. Then let IO be 1000; OD will be 250, DI 968, and its double, DE, 1936; but LD (sine of arc LP) is 866; therefore, if ED is 1000, ID will be 500 and DL 447, and the arc whose sine is ID would be 30°0′ and the arc whose sine is LD, 26°33′. But it would have to be [25°54′][43] to retain the $1\frac{1}{7}$ ratio of the one time interval to the other; therefore the arc-sine [*arco del sino*] LD is greater than what is needed to maintain the aforesaid ratio. Therefore it is not possible that semidiameter CA is double the semidiameter AB, and much greater inconvenience would follow from supposing it less than double. It follows, then, that it is of necessity more than double, which is what was to be demonstrated.

Therefore, from Apelles' assertions that some spots have traversed the diameter of the disc in sixteen days and others a parallel at most 30 degrees away from it in fourteen days, it follows, as Your Lordship sees, that the sphere that carries them is farther away from the Sun than that body's semidiameter. But this is manifestly false for other reasons. For, if this were the case, much less than 60 degrees of the greatest circle of this sphere would interpose itself between our eye and the solar disc, and much smaller arcs of the other parallels would be interposed. Hence, as a necessary consequence, the motions of the spots on the Sun would appear entirely uniform from ingress, through the middle, to egress. As regards what depends upon the different positions and angles of the observer, the distances between one spot and another, the shapes, and the sizes would always appear the same on all the parts of the Sun. Let Apelles himself bear witness to how inconsistent this is with the truth, for he has in fact observed the apparent slowness, the union or clustering, and the thinness of those spots near the circumference, and their speed, separation, and very noticeable growth around the middle parts. Because of such a contradiction I will not hesitate to say that it is entirely impossible if one spot traverses the solar diameter in sixteen days, that another crosses the above-mentioned parallel in fourteen. But I will also point out to Apelles that in refashioning the argument and noting with greater accuracy that the movements of the spots along any line whatsoever of the disc take place in equal times—as I have found from many observations and as anyone will be able to observe in the future—one must necessarily conclude that they are, as I have always said, either contiguous or separated from the surface of the Sun by an imperceptible distance. And, in order not to pass over a matter that might confirm and establish a conclusion so fundamental to this material, I add that Apelles, too, could realize this—let Your Lordship note how great the power of the truth is—from two other

43. Galileo's text has 25°45′, which is a misprint. It should be 25°54′. See appendix 5.6, p. 366.

necessary conjectures, which I would like to derive from his drawings, in order to remove every occasion for the suspicion that I, as if more intent upon disguising my errors than upon discovering the truth, had perhaps adjusted my sketches to my own conclusions. I could, of course, deduce them more precisely from some of mine, which are perhaps more accurately drawn, at least because of their larger format.

Therefore, let Your Lordship take the drawings[44] of the two days, 29 December at 2:00 and 30 December likewise at 2:00, in which spot μ,[45] which stands out from the others, begins to appear. This spot, as the author himself reports, emerged the first day in the shape of a thin black line separated from the edge of the Sun by a bright band no broader than its own width; but as the drawings show, the next day at the same hour its distance [from the edge] had tripled and the size of the spot had likewise grown considerably. In addition, he reports in regard to this spot—a rather constant one amidst the variability of the others—that its visual diameter was about the eighteenth part of the diameter of the solar disc. And because it grew into the shape of a half circle and was at its first appearance parallel with its entire diameter to the circumference of the disc, it follows necessarily that the apparent dilation of its shape took place not along the length of its entire diameter, but along the semidiameter perpendicular to it. This is shown by the drawing, so that the dimension of this spot, which at first appearance was very thin, toward the middle of the disc grew so much that it occupied about the thirty-sixth part of the diameter of the Sun, namely as much as the chord of $3\frac{1}{3}$ degrees. Now these two observations being established, I say that it is not possible that this spot was separated from the surface of the Sun by a noticeable distance.

For let ABD be the circle on the solar globe on whose circumference the spot appears to move, and assume that the eye is placed in the same plane but at an immense distance, so that the rays drawn from it to the diameter of this circle would be parallel lines. Now assume the spot μ occupies a width of 3°20′: its sine or chord—between which there is a most minute difference in such a small angle—will be 5814 parts of which the semidiameter AM itself contains 100,000 such parts. Next assume that arc AB is 8° and arc BD 3°20′, that is, the same as what is proposed as the width of the spot; and through points B and D pass the perpendiculars to the diameter AM, which are CBG and ODQ; ACO,

44. Galileo is referring here to Scheiner's demonstration and diagrams of his sunspot observations on pp. 196–200, above. For a full account of Galileo's demonstration in modern notation, see appendix 5.7.

45. This refers to a sunspot labeled η in the figures in *Accuratior Disquisitio*, although it is called μ in the accompanying text; see p. 197 above.

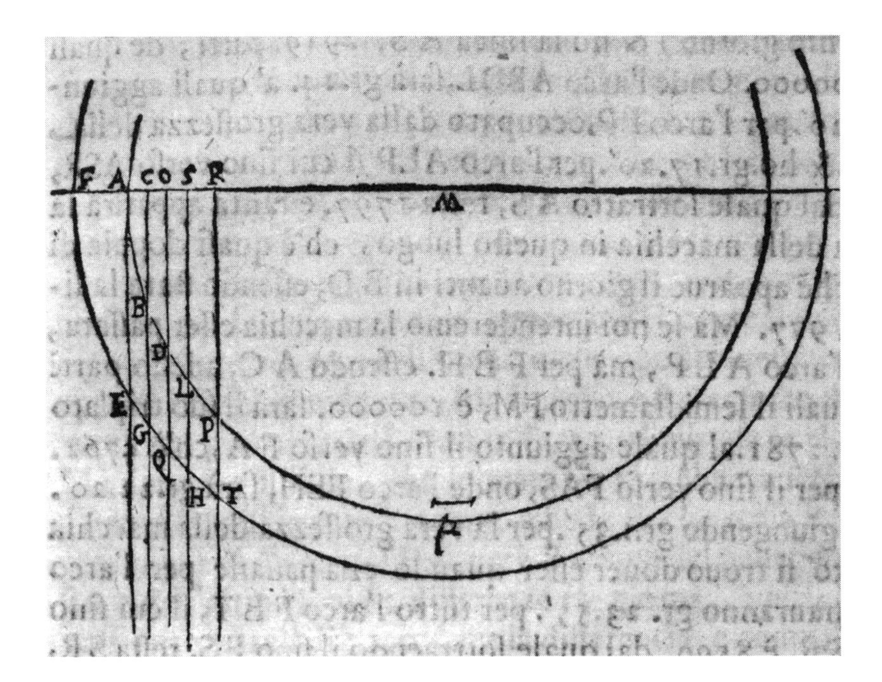

the versed sine of arc ABD, will be 1950, and AC, the versed sine of arc AB, 973, and the remainder, CO, will be 977. From this we have, first, that spot μ, located at BD, will appear very thin to us, that is, only the sixth part of that width, which it will manifest near the middle of the disc, at the location μ; and thus at BD it will seem to us equal to CO, or 977, and at μ it will seem to us 5814, or about six times as much. Further, we have the bright band AC equal to the apparent breadth of the spot, AC being 973 and CO 977. These requisite particulars correspond suitably to Apelles' observations. Now, let us see if these particulars would concur if we were to assume that the spots revolve at a distance of only one twentieth part of the semidiameter from the globe of the Sun.

Assume, therefore, that in a sphere with a semidiameter MF, segment AF will be 5,000 parts, the semidiameter AM 100,000 parts, such that the whole of FM will thus be 105,000 parts. But if MF is reset at 100,000, FA will be 4762, AC 927, CO 930, FAC 5689, and FACO 6619, and describing the circle FEGQ and drawing the parallel AE, we find arc FE to be 17°40′, FEG 19°25′, EG 1°45′, FEGQ 21°, and GQ 1°35′. Its chord when it is at μ would be 2765, while at GQ it had been equal to CO, i.e., 930, which does not amount to one-third of 2765. If, therefore, spot μ moves at such a distance from the Sun, it could never appear enlarged [in breadth] by more than three times, which is quite inconsistent with Apelles' observations and with mine. And let Your

Lordship take note that I make the present inference supposing that spot μ appeared to move along the diameter of the Sun and not, as it did, along a shorter line. Had we used this shorter one, the inconsistency would have been even greater, just as it would have been much more apparent if we had used even thinner spots. And whoever wanted to locate the spots at a semidiameter or more from the Sun would find this discrepancy immense and most manifest, for in that case absolutely no difference [in breadth] would be visible in their entire passage [across the solar disc].

I come now to another conjecture derived, according to Apelles' notes, from the growth of the bright band and of the spot's breadth in the course of only one day. Turning again to the same figure, and first supposing the spot contiguous to the Sun, I triple the versed sine of the bright band AC (which appears to have grown so much on the following day), and I find line AS to be 2919 parts of which AM is 100,000. From this, arc ABDL will be 14°, to which I add 3°20′ for arc LP, which is occupied by the spot's actual breadth, and I have 17°20′ for arc ALP, whose versed sine, ASR, is 4716, and when we subtract AS from this, the remainder will be 1797. And this is how large the spot will appear in this place, which is almost double what it appeared the previous day at BD, the line CO being 977. But if we were to suppose that the spot had moved not through arc ALP but rather through FEH, AC being 927 parts of which semidiameter FM is 100,000, its triple, ACOS, will be 2781, and when the versed sine FA, which is 4762, is added to this, it yields 7543 for the versed sine FAS, and thus arc FEH will be 22°20′. If we add 1°35′ to this for the real size of the spot—for we find that it must be that much if it passed along arc FEH—the entire arc FET will be 23°55′, and its versed sine is 8590. Subtracting the versed sine of FS from this, the remainder SR will be 1047, and this is the apparent size of the spot when located at HT, which is larger than that of the previous day, CO, by less than an eighth part. Thus, if the spot's movement were to occur along a circle distant from the Sun by only one twentieth part of its diameter, its visible size would not have grown the following day by one eighth part, but because it actually grew by more than seven-eighths, it necessarily skims over the solar surface.

And because this is one of the main points treated in this matter, I must not omit some other observations that Apelles produces on folios 43 and 44,[46] observations from which he attempts once again to persuade the reader of the distance of the spots from the Sun, relying upon those same arguments about the inequality of their sojourns under the solar disc. If this phenomenon took place as Apelles describes, it would necessarily follow that the spots were not

46. Pp. 221–223, above.

only not on the Sun, but not even anywhere near it. And what is more, taking their motions to be in general equitable and uniform—as the greater part of these most scrupulous observations suggest to me—it is absolutely impossible, as I have demonstrated above, that anyone will ever encounter differences in time such as those that Apelles postulates here, unless some of the spots progressed along lines whose distance from the center of the disc was not merely the 30° maximum that I observed, but 50° or 60° or more. This last conflicts not only with my observations but also with those that Apelles himself produces: there, spot G passes through the very center, as is seen in the drawing of 30 March; spot E, as is shown in the drawing of 25 March, does not pass further than 30° away, nor even 24°; and the same happens with spot H, as is seen in the drawing of the 30th of the same month. These things being established, Apelles adds next that spot E was below the Sun for at least twelve whole days, but G at most eleven, and H at most nine. But how is it possible that spot G, which crosses the entire diameter, would pass in less time than E, which travels more than 20° distant from the center? And that between the time of this spot's passage and that of the other, H, there would be an interval of three days or more, even though they move along parallels with little or no difference? And how is it that Apelles has forgotten those things of which earlier, on folio 18,[47] in the tenth item to be noted, he had written with such conviction, as for example, that it is certain that the spots traversing the middle of the Sun stay longer under it than do those passing nearer the limb? These things are absolutely impossible, unless one wants to maintain that the motions of the spots are all of different periods, which is neither true nor proposed by Apelles. And were such a proposition true, the entire discussion of such movements—from which he wants to infer the location of the spots relative to the Sun—would be wholly without vigor.

But because the force of the truth is too invincible, let us take up once more these same drawings, and let us consider them divested of all other aims except for that of arriving at the truth, and we will find that the intervals of these passages are equal, and all about fourteen days. And first, spot G,[48] which appeared on 26 March and was not seen earlier, is as distant from the circumference as the motion of three and perhaps four days' motion from the circumference. Clear testimony of this is not far afield; consider, on the same folio, spot B of 4 April. This spot is less distant from the circumference than was spot G on 26 March, and yet it had already traveled three days or more, as Apelles' two preceding drawings show us. The moment of its egress was thus not 3 April,

47. P. 199, above.

48. P. 222, above.

but two or three days later because its distance from the circumference was still so great. For, to turn once again to the same drawings, we will see what I am saying exemplified in spot E, which on 29 March is no farther from the circumference than G is on 3 April, and still it remains visible for two days, if not more. If, therefore, to the eight days of spot G represented on the engraving we add four earlier and two later, we will have fourteen days. The fact that it was observed neither before nor after those eight days must be attributed to its having been neither generated earlier nor preserved afterwards. And I am saying this because I assume that the observations were accurate; if they were not, one might be able to attribute their apparent absence not to the nonexistence of the spots but to something short of diligence on the part of the observer. But it appears to me that there is a flaw in the choice of observations, which should have been of those spots observed entering and leaving at the extreme circumference, and not of spots that appeared and disappeared so far from it. And so as not to call into doubt whether the spot that returned was the same one that had disappeared, those of constant duration throughout their transit should likewise have been observed. Spot E likewise shows that it, too, spent another fourteen days crossing the Sun because at its first observation on 20 March it was located as far from the circumference as could reasonably be accounted for by the motion of three days, which time, together with the eleven [other days] noted, comes to the total for which I am arguing. As to spot H, I will say, by Apelles' leave, that I hold it suspect in this demonstration, and I believe that the H on 1, 2, and 3 April is absolutely not the H of 28 and 30 March. Indeed, I still have doubts that these two last observations involve the same spot, seeing that the interval between H and G on the 28th is much greater than it is on the 30th, though this is just where it should, on the contrary, be much smaller, the spots being so close to the circumference. Apart from this, the fact that it was not seen on the intervening day, the 29th, is a very strong argument that the two cannot be the same spot. And the same doubt crops up about the H of 30 March and the H of 1 April, it not having been seen on the day in between, 31 March. But certain proof of such a mistake also derives from their different locations, for the H of 28 and 30 March appears to move along the same parallel as does G, which is distant from it in longitude, or the direction of the movement. But the H of 1, 2, and 3 April is at the side of G, and distant from it only in latitude. It is thus absolutely not the same spot as the first, and therefore it has no role in this issue.

And because, as I have said before, this is a crucial point in this matter, and the disagreement between Apelles and me is great—the revolutions of the spots all appear equal to me, and seem to traverse the solar disc in about $14\frac{1}{2}$ days, and to him seem so unequal that some take 16 days or more in their passage and others only 9—I believe that it is essential to begin anew with a repeated investigation to search for the exact character of these particulars. We must remember that Nature, unheeding of and indifferent to our entreaties, will neither alter nor change the course of her effects, and that those things that we endeavor to investigate now and later to promote to others did not exist but once and then vanish, but rather still subsist and will continue to subsist in this fashion for a long time, such that they will be seen and observed by a great many people. This should serve as a significant restraint, making us all the more circumspect in pronouncing our propositions, and protecting us from any excessive regard, either for ourselves, or for others, that would turn us away from the goal of pure truth. And in this connection I cannot hide from Your Lordship a bit of concern born of the use that Apelles desired to make of those two sunspots and their changes, the drawings of which I sent to Your Lordship in my first letter.[49] And I understand perfectly well that this intention arose from his courteous affection, for he was eager to obtain credit for them by saying that they agreed very much with his, and to create an opportunity to show that he had a pleasant memory of me. And yet I would have preferred that he not have gone so far as to risk losing something of the reader's respect by saying that because of such a precise agreement between my drawings and his, and especially between those of the second spot, he was certain of the absence of parallax, and in consequence of their great distance from us. There would be good reason for casting doubt upon this conclusion of his, since the figures I sent were of the sunspots drawn in isolation without reference to anything else, or to their place on the Sun, whose circle I did not even draw. This last detail leaves me likewise somewhat confused: exactly how was he able to judge whether or not I had distributed and arranged them accurately?

I hope that Apelles will be satisfied with what I have said up to now, and especially with what I have written in the second letter, and I believe that he will raise difficulties neither about the great nearness of the spots to the solar globe, nor even about its own revolution on itself. In confirmation of this revolution, I can add to the reasons that I wrote in the second letter to Your Lordship, that on the very face of the Sun, one sometimes sees certain small areas that are brighter than the rest, and which careful observation reveals to have the same motions as do the spots. That these are on the surface itself of the Sun, I do not believe anyone can doubt, for it is in no way credible that there is some substance brighter than the Sun outside of it. If this is correct, I don't think that there remains any room to doubt the revolution of the solar globe upon its own axis. And such is the chain of truths that the contiguity of the

49. Pp. 100 and 225, above.

spots to the surface of the Sun follows from this, as does their being conducted in turn by its rotation, for there appears to be no probable reason why these spots, were they at any great distance from the Sun, would be required to follow its revolution.

It remains for me to consider one of the conclusions that Apelles has drawn from the topics in dispute, the general tendency of which is, it seems, to maintain what he believes that he has established in his first letter. This conclusion, in the end, is that these spots are none other than stars wandering about the Sun, for not only does he return to calling them "solar stars," but he goes on to adjust certain connections and conditions shared by these and the other stars, such that every discrepancy and reason for separating the sunspots from the true stars is removed. In this connection, as well as in approbation for my theory of lunar peaks—a courtesy for which I thank him—Apelles says that my opinion is not improbable, because the same phenomenon is also observed in the majority of these spots. And he reasons that when my demonstrations are added to my hypothesis, it should satisfy everyone.

I not only concur with Apelles, but believe that I can demonstrate with compelling logic that the opinions of those who imagine inhabitants on Jupiter, Venus, Saturn, and the Moon are false and damnable, if what is meant by "inhabitants" is living beings like ours, and humans in particular.[50] For my part,

50. Though we know of no extant evidence for the claim, David Fabricius was identified as a brazen supporter of extraterrestrial life sometime in the seventeenth century. The suggestion appeared in Andrea Argoli's *Pandosion Sphericum*, first published in 1644; according to this astronomer and astrologer, "since both the Earth and the Moon are thought to be dense, opaque, and solid bodies, David Fabricius believed that the Moon was an earthly body, and that it had inhabitants, and that the *Lincei*, who observed the lunar globe, had discovered living creatures walking and moving about." In his *Lexicon Mathematicum Astronomicum Geometricum* (1668), Girolamo Vitali wrote that "those who say that the Moon is an earthlike body and that inhabitants live there and that animate beings roam to and fro—as for example, David Fabricius, who shamelessly boasted to have seen them, according to Argoli—those people are clearly dreaming." The assertion was adopted by Giacomo Leopardi in his *Storia dell'astronomia* (1813–1815), who in referring to "the madman David Fabricius" and his claim to have seen the lunar beings with his own eyes, compared him to the ancient satirist Lucian, whose *True History* involves a trip to the Moon. Leopardi returned to Fabricius in his "Dialogo della Terra e della Luna" (1824): here the Earth tells the Moon, "I am unable to make out a single inhabitant on you, though I hear that one David Fabricius, who saw better than a Lynx, once discovered some who were hanging their laundry out in the sun." Jules Verne, finally, noted in his *From the Earth to the Moon* (1865) that "in the seventeenth century, a certain David Fabricius boasted that he had seen the inhabitants of the Moon with his own eyes"; the Frisian astronomer's name is followed by those many who had written accounts of their visits to the populous lunar body. See in this connection Andrea Argoli, *Pandosion Sphericum* (Padua: Paolo Frambotti, 1653), 228; Girolamo Vitali, *Lexicon Mathematicum Astronomicum Geometricum* (Paris: Louis Billaine, 1668), 270; Giacomo Leopardi, *Storia dell'astronomia dalla sua origine fino all'anno MDCCCXIII*, ed. Armando Massarenti (Milan: Edizioni La Vita Felice, 1997), 138–139; Giacomo Leopardi, "Dialogo

I will neither affirm nor deny whether it can be judged probable that on the Moon or another planet there are living beings and plants that not only differ from the terrestrial ones but are also most remote from our every imagining: I will leave such determinations to those wiser than me. It is their views that I will follow, being certain that they will be sounder than what Apelles argues in this connection—that it would be absurd to put them on such bodies—as if, for example, one could not suppose that there were living beings on the Moon without also supposing them likewise on solar spots. Nor do I really understand Apelles' inference that the sunspots persuade us of the need to concede that some reflected light comes from the Earth. Rather, because the sunspots' reflection is not very conspicuous and what we see in them cannot be other than refracted light, if anything can be deduced from such a particular, it would sooner be that the Earth is of a substance that is transparent and permeable to the light of the Sun. And this does not appear to be true. I am not saying, however, that the Earth does *not* reflect light, but rather that I am absolutely certain for a variety of reasons and experiences that it shines no less than any other star, and that with this reflection it returns somewhat more light to the Moon than it receives from it.

But because Apelles is so unwilling to concede that such a powerful reflection of light comes from the terrestrial globe, and so ready to admit that the lunar body is transparent and can be penetrated by the rays of the Sun, as he argues here and even more openly toward the end of these discussions, I want to adduce one or two of the many reasons that persuade me that the former conclusion is true and the latter false. If these were by some chance resolved in Apelles' favor, they would be able to change my opinion. In the meantime, I will not be silent about the fact that I strongly suspect that this common notion—that the Earth, as dark, opaque, and rough as it is, is incapable of reflecting the light of the Sun, while on the contrary the Moon and the other planets reflect it well—is prevalent among the masses because we never have occasion to see it from some dim and remote place while it is illuminated by the Sun. By contrast, we often see the Moon when it is in the dark field of the sky, and we are encumbered by nocturnal gloom. And here it happens that when we lower our eyes to the Earth, after having fixed them on the wondrous splendor of the Moon and of the stars, we are in some fashion saddened by its darkness, and we form a dread of it as a thing repugnant by its very nature to all radiance. We fail to consider that the density, darkness, and roughness of the material in no way prevents it from receiving and reflecting the light of the Sun, and that illu-

della Terra e della Luna," in *Operette Morali*, ed. Giorgio Ficara (Milan: Arnaldo Mondadori, 1988), 81–88, at 82–83; Jules Verne, *De la terre à la lune* (Paris: Hachette, 1979), 22.

mination is a gift and quality of the Sun, and requires no excellence at all in the bodies that are to be illuminated by it. On the contrary, it is rather necessary to deprive them of certain more noble features such as transparency of substance and smoothness of surface, rendering the matter opaque and its exterior rough and rugged. And I am very certain, in contrast to popular opinion, that if the Moon were polished and smooth like a mirror, not only would it not reflect the light of the Sun to us as it does, but it would also be absolutely invisible, as if it were not there. And in its proper place I will make this manifest with clear demonstrations.

But not to stray from the matter at hand, I say that I find it easy to believe that if we had never happened to see the Moon by night, but only during the day, we would have formed the same impression of it as we have of the Earth. For, if we consider the Moon during the day, when on occasion it is more than a quarter illuminated and it happens to be in the wisps of some white clouds, or perhaps near the peak of a tower or a wall whose color is of a middling brightness, when these bodies are directly illuminated by the Sun such that one might compare their luster with the light of the Moon, their splendor will surely be found to be no less than that of the Moon. Therefore, if they were able to remain illuminated in this fashion even in the darkness of the night, they would appear to us no less luminous than the Moon. Nor would they illuminate the places around them less than it does, up to that distance from which their apparent size would be no smaller than the lunar face. But these same clouds and walls, stripped of the rays of the Sun, remain at night no less dark and black than the Earth. We should, moreover, derive great certainty about the Earth's ability to reflect light from seeing how much it spreads in a room deprived of all other illumination and lit only by the reflection from a wall facing it that happens to be struck by the Sun, even if that reflection passes through a hole so small that from the place in the room where it strikes its diameter does not appear to subtend an angle greater than that of the visual diameter of the Moon. Nevertheless, this secondary light is so powerful that when it has struck and passed from a first room into a second one, it will still be no less luminous than moonlight itself. A clear and available proof of this is that we will read a book more comfortably by the second reflection from the wall than we would by the primary reflection from the Moon.[51] Finally, I add that there will be few who, observing a flame on a distant mountain at night, have not wondered whether they were seeing a fire or a star grazing the horizon, the latter's light appearing no greater than that of a flame. From this, one would suppose not

51. On this assertion, see William Shea, *Galileo's Intellectual Revolution: Middle Period, 1610–1632* (New York: Science History Publications, 1972), 64–65.

only that were the Earth burning and wholly enflamed, it would appear no less bright than a star if seen from the dark part of the Moon, but also that every rock and every clod of dirt struck by sunlight is considerably brighter than if it were on fire. This will easily be observed if a lighted candle is brought near a stone or a piece of wood directly struck by a beam of sunlight, for the flame is by comparison invisible. And thus the Earth, struck by the Sun and seen from the dark part of the Moon, will appear as bright as all other stars, and as much more light will be reflected to the Moon [from the Earth] as the Earth exceeds the Moon in size, the surface area of our globe being, that is, about twelve times as large as the lunar body appears to be to us.[52] Besides, at New Moon, the Earth is closer to the Sun than is the Moon when it is full, and therefore the one is more strongly illuminated, that is, from nearer by; the light from the Earth will therefore reflect more forcefully to the Moon than will that from the Moon to the Earth.

For these and many other reasons and experiences that for the sake of brevity I will omit, the reflection of the Earth should, in my opinion, be considered sufficient to explain the secondary light of the Moon. There is no need to bring up any transparency, least of all the degree of transparency attributed to it by Apelles, in which I seem to see certain inexplicable contradictions. He writes that the transparency of the lunar body is so great that during solar eclipses, while one part was covered by the Moon, the disc of the Sun, perceptibly outlined and distinct, was clearly observed shining through the lunar depths. Now, I note that when an ordinary cloud, and not one of the densest sort, comes between us and the Sun, the cloud hides it so well that we would try in vain to determine its place in the sky within a few degrees, let alone be able to see its perimeter distinct and terminated. And very often the Sun is seen half-covered by a cloud, without the slightest trace of the circumference of its hidden part appearing, and yet we are sure that such a cloud will not be many tens or at most hundreds of *braccia*[53] in thickness. And besides this, if we are sometimes on the peak of a mountain, and happen to pass through such a cloud, we do not find it so dense and opaque that it does not permit sight at least for a few *braccia*, which the same thickness of glass or crystal would perhaps not do. Thus we gather of necessity that if it is true, as Apelles writes, that the transparency of the Moon is infinitely greater than that of a cloud—two thousand miles of the lunar substance impeding the passage of solar rays much less than a few *braccia* of a cloud—then the lunar substance must therefore be much more

52. The diameters of the Earth and Moon are to each other roughly as 7:2. Their surfaces will therefore be to each other roughly as 49:4, or 12:1.

53. 1 *braccio* = 58.4 cm. See p. 111, note 7, above.

transparent than glass or crystal, a statement that other considerations will condemn as impossible. For first of all, no reflection, or none but the faintest sort, would take place in a diaphanous body in which the solar rays were so deeply immersed, whereas, on the contrary, a very great reflection comes from the Moon. Second, the terminator or line separating the illuminated part of the Moon from the part not touched by the direct rays of the Sun would either be nonexistent or very indistinct. This can be seen in a large glass ball filled with turbid water or with another not entirely transparent liquid; were the water clear, this terminator would not appear at all. Third, the lunar substance being so perspicuous that it allowed sunlight to pass through 2000 miles of its depth, it cannot be doubted that 1/200 or 1/300 of the same material would be altogether transparent. This, however, is wholly contradicted by the lunar mountains: even though many appear very thin and narrow, all of them cast the darkest shadows on the surrounding lowlands, as is observed in countless places, and especially around the border between the illuminated and obscure areas where the lights abut the shadows with almost unimaginable sharpness and abruptness, an effect that cannot come about except in materials that are similar in ruggedness and opacity to our steepest mountains. Finally, were the light of the Sun to penetrate the entire depth of the Moon, the splendor of that hemisphere untouched by its rays would always have to appear unchanging, nor could it ever diminish, because that half of the Moon is always illuminated in the same way. Or if there were any variation at all to be seen there, then at New Moon the central region should appear darker than the rest, because here the depth of material to be penetrated is greater. And at quadrature there would have to be greater brightness near the terminator and less in the areas farther from it. These things, and many others that I omit for the sake of brevity, occasion considerable conflict between Apelles' hypotheses and actual appearances. Hence it is that the assumption of the opacity and roughness of the Moon, and the reflection of the light from the Sun by the Earth—hypotheses at once true and perceptible—solve every single problem with marvelous ease and thoroughness. But I will treat this more fully on another occasion.

Turning to Apelles' particular concerns, I begin to suspect that, carried away by the desire to maintain what he had originally said, and unable to accommodate the spots exactly to the properties once associated with the other stars, he has accommodated the stars to the properties that we know belong to the spots. This tendency, it appears, is most manifest in two other important points that he introduces. One of these is that it can probably be said that the other stars, too, are of various shapes but appear round because of the light and the distance, as occurs with the flame of a candle, and, one might add, horned Venus. In truth, this assertion could not be condemned as a manifest falsehood if the telescope had not resolved this uncertainty by showing us that all stars, fixed and wandering alike, are absolutely round in shape. Apelles' second point is that since it cannot be denied that the spots appear and disappear, he does not hesitate to assert—in order not to segregate them from the other stars on account of this feature—that the latter also dissolve and become whole again. To be precise, he considers as such those that I have observed moving around Jupiter, about which he repeats, yet again, what he wrote in the first letters, reaffirming it now as a well-established dictum, that, like the solar shadows, these stars suddenly appear and vanish, and, like those shadows, one follows upon another, but none ever returns unchanged. In confirmation of this, he derives no small argument from the difficulty and perhaps from the impossibility, as he sees it, of deriving their fixed periods from the many and most accurate observations, both his own and those of others, which he professes to have. Now here I really wish that Apelles would cease imagining me as someone so vain and frivolous as to have not only made public spots and shadows and to have offered them to the world as stars, but, more crucially, to have dedicated such fleeting, unstable, and transitory things to the glory of so great a Prince as My Lord the Most Serene Grand Duke, and to the eternity of such a royal House.[54] I answer him thus that the four Medicean Planets are true and real stars, permanent and perpetual like the others, and that they neither vanish nor hide except when they are in conjunction with each other or with Jupiter, or are obscured sometimes for a few hours in the shadow of that planet, as is the Moon in the shadow of the Earth. They have most regular motions and fixed periods; if he has not been able to investigate them perhaps he has not labored over this as much as I have, determining them at last after many vigils. As Your Lordship may have seen, I have already published them in the preface of my treatise *Discourse on Bodies in Water*,[55] and so that Apelles can dispense with even the slightest doubt, I am sending you the configurations for the next two months. They begin on the first of March 1613, with notes on their progress and hourly changes: Apelles will be able to compare them to the Medicean Stars themselves, and he will find that they correspond exactly, provided I have not made some inadvertent error in calculating them. I would like for him to set about counting them with new diligence, for he will find that they are no more than four in number, and that the fifth that he mentions was surely a

54. In his *Sidereus Nuncius*, Galileo dedicated the eternal moons of Jupiter to Grand Duke Cosimo II. See *OGG*, 3 (1): 55–58, and *SN*, 29–33.

55. *OGG*, 4: 63–64. We follow Drake's English version of the title here. See Galileo Galilei, *Discourse on Bodies in Water*, tr. Thomas Salusbury, with an introduction and notes by Stillman Drake (Urbana: University of Illinois Press, 1960). The periods are on 1.

fixed star, and he will also discover that the conjectures he permitted himself in supposing it to be wandering were based on various fallacies. To begin with, his observations are quite often erroneous, as I see from his drawings, because they leave out some stars that were conspicuous in those hours. Second, the distances among these stars and with respect to Jupiter are almost all wrong because of what I see as a lack of a method and of an instrument for measuring them. Third, there are gross errors in the movements of the stars, Apelles mistaking one for another more often than not and confusing the higher with the lower without recognizing them from one night to another. All of these things served to deceive him.

Star D, depicted in the diagram of 30 March,[56] was the one that describes the largest circle around Jupiter, and at that time was at its greatest digression, that is, at its mean longitude, and almost stationary, about fifteen minutes distant from Jupiter, for such is the semidiameter of its circle. It was not six minutes away, as Apelles estimated, having judged these intervals by sight, a procedure that offers ample opportunity for error. Supposing, then, the star's distance from Jupiter to be what it really was, and the star E having been observed a little to its west, it could very well be that because of Jupiter's retrogradation, star D appeared in conjunction in longitude with that planet on 8 April. Besides this, Apelles is badly mistaken in concluding that the motion of star E was faster than that of star D. First of all, he errs in saying that on 30 March the angle formed by star E, star D, and Jupiter was obtuse, when it follows from his own statements that it is of necessity acute, for the longitude of star D from Jupiter was then (as he says) six minutes, and that was also the southern latitude of star E, and its distance from Jupiter eight minutes. But in an isosceles triangle that has its sides equal to 6 and its base 8, the angle contained by these sides is necessarily acute, not obtuse, because the square of 8 is less than double the square of 6. Besides this, it is false that this configuration remained that way until 5 April; first because the star D depicted west of Jupiter on 5 April is not the star D of 30 March; rather this star D of March is subsequently the easternmost one, B, near the edge [of the circle] on 5 April, with which it does not form an acute angle, but rather a very obtuse one. And consequently what Apelles has concluded is false, that is, that the motion of star E is faster; rather, it is much slower than that of D. Moreover, if it were faster, I don't know how this would demonstrate that star E is wandering rather than fixed, for the cause of all inequality could be ascribed to the motion of D. This first reason is thus without force and proves the very opposite of what Apelles intended. And, moreover, what inconstancy must possess Apelles for him to allege here, in support of one

56. See p. 208.

of his whims, that the stars noted in his observations and marked with the same letters remain unchanged, and then to say a little later that he firmly believes that they continuously appear and disappear, never returning the same? And if this is so, what can he, or anyone else, gather from his discussions?

As to the other reason that Apelles proffers in confirmation of the true existence of the fifth Jovian star, my faith in him and the prestige he enjoys with me not permitting me to cast doubt on the question of whether it is real [*an sit*], I can say nothing except that I do not understand how it could happen that a star whose telescopic appearance is of a size and splendor equal to one of the first magnitude can become in less than ten days and, more puzzling, without moving more than a quarter or an eighth of a degree, or in truth, without changing its place at all, so much smaller that it vanishes entirely. Apart from the so-called "New Stars" of 1572 in Cassiopeia and of 1604 in Serpentarius, I don't know that a similar portent has ever been seen in the heavens. And if this fifth Jovian star was one of those—or a second-rate one, being so much dimmer and short-lived—it was indeed provident of Apelles to have secured the longevity and splendor of the Most Illustrious House of Welser for it.

Therefore, neither the Jovian nor [the fixed] stars are spots and shadows, and the solar shadows and spots are not stars. In truth, I am not insisting on nomenclature, for I know that everyone is free to adopt it as he sees fit. As long as people did not believe that this name conferred on them certain intrinsic and essential conditions, I would care little about calling them "stars," for it was thus that those of 1572 and 1604 mentioned earlier were also named. Meteorologists use the name "star" for those with hair, those that fall, and those that race through the air;[57] and finally, lovers and poets are permitted to call their ladies' eyes "stars:"

> When Astolfo's successor sees
> Those smiling stars appear above him.[58]

For like reasons one might also call solar spots "stars," but they have, in essence, characteristics that differ considerably from those of actual stars. It happens

57. Certain types of comets were described as "long-haired stars," while the latter two terms seem to refer to meteorites.

58. Lodovico Ariosto, *Orlando Furioso* vii, 27, 1–2. The allusion is to a famous scene, Alcina's seduction of Ruggiero, Astolfo's successor, in the work of one of Galileo's favorite authors. A double illusion is at stake here: unlike the lovers Astolfo and Ruggiero, the poet knows that the lady's eyes, elsewhere described as "dark, or rather as two brilliant suns," are not really stars; Ruggiero discovers after his dalliance that she's no lady either, but rather an ancient, shrunken, and toothless hag. See *Orlando Furioso* 7: 12, 2, and 7: 72–74.

that true stars always appear of one shape, the most regular among all shapes, while the sunspots are of infinite shapes, and all very irregular. The former are consistent, never changing in size or shape, and the latter are always unstable and changing; the former are always the same, and of a permanence that predates the memories of all past ages, and the latter are generated and dissolved from day to day. The former are never visible, unless full of light, and the latter are always dark and never brilliant. The former are either entirely without motion or else are self-moving with particular and regular motions; the latter move with one motion only, common to all, a movement regular only in general but altered by infinite and unequal particulars. The former are located at different distances from the Sun, and the latter are all touching or imperceptibly distant from its surface; the former are never visible unless they are very far from the Sun, and the latter are never seen except when they are in conjunction with it. The former are of a substance in all probability dense and opaque, and the latter, like fog or smoke, are rarified. Now I do not know why the sunspots ought to be classed with things with which they share not a single feature that is not also common to a hundred other phenomena, none of them stars; they should be grouped instead with those things that resemble them in every last detail, and I would compare them to clouds or smoke. Certainly if someone wanted to simulate them out of some of our materials, I do not believe he could manage a more apt imitation than by placing a few drops of incombustible bitumen on a red-hot iron plate, for this would mark the iron with a black spot from which source a dark smoke would rise spreading in bizarre and changing shapes.[59] And if someone were to argue that in order to sustain the immense light that constantly spreads from this great lamp through the furthermost reaches of the universe, it would have to be fed food and fuel without end, then he would indeed find not only one but a hundred experiences, each in agreement, in which we see that all materials set ablaze and converted to light at first turn to a dark and obscure color. Thus we see in wood, straw, paper, candles, and in sum in all that burns that the flame is implanted in and rises from those parts of this material that had first turned black. And I would add that perhaps more accurate observation of those small areas that I mentioned above that are brighter than the rest of the solar disc[60] might reveal that those are the very places where, a little earlier, some of the largest spots had dissolved. I do not intend, however,

59. On this analogy see Rivka Feldhay, "Producing Sunspots on an Iron Pan," in *Science, Reason, and Rhetoric*, ed. Henry Krips, J. E. McGuire, and Trevor Melia (Pittsburgh, PA: University of Pittsburgh Press, 1995), 119–143.

60. See p. 281, above.

to assert any of these things for certain, nor to oblige myself to defend them, for I do not like to mix dubious matters with certain and established ones.

On this side of the Alps, as I understand it, a certain opinion is current among a not negligible number of Peripatetic thinkers, men for whom philosophizing does not bear the burden of the desire for truth and for its causes, since it is with anything but objectivity that they deny these new developments, and joke amongst themselves, judging them to be illusions. Now is the moment for us to jest about these men, and for them to keep both quiet and out of sight. As I was saying, there is an opinion circulating among them that defends the inalterability of the heavens, something that Aristotle himself might have abandoned in this day and age. This opinion conforms to that of Apelles in all but this: whereas he supposes only one star for each spot, they imagine each spot a mass of many very small stars, which gather together with their different motions, at times in greater, and at times in lesser abundance, and then, in separating, form both larger and smaller spots, of irregular and quite varied shapes.[61] Having already gone beyond the bounds of brevity with Your Lordship, such that you will have to read the present letter in installments, I will take the liberty to touch on a few particulars on this point.

In this, the first thought that comes to mind is that those who subscribe to this opinion have not had the opportunity to make many scrupulous and sustained observations, because I am persuaded that certain difficulties would have made them not a little dubious and perplexed in accommodating such a thesis to the appearances. Although it is by and large true that many individual objects that would remain invisible because of their small size or distance when amassed together can form a body perceptible to our sight, we must nonetheless not stop at this generalization, but rather descend to the actual particulars of stars, and to those that are observed in sunspots, and carefully examine in what manner both the former and the latter phenomena might mingle together and aggregate. And so as not to be like that governor of a castle who, having to defend a fort with a small number of soldiers, in order to help that part which he sees assailed, rushes there with the entire force, while leaving other places open and defenseless, it is necessary that while we exert ourselves to defend the immutability of the heavens we do not forget the dangers to which other propositions, likewise necessary for the preservation of Peripatetic philosophy, might be exposed. And therefore, if the Peripatetic position is to remain intact and solid, for the preservation of its other propositions we must first say of

61. This view was mentioned by Cardinal Conti in his letter of 18 August 1612 to Galileo. See appendix 3.

the stars that some are fixed and others wander, calling those that are all in the same heaven, and all move with its motion, but remain immobile with respect to each other "fixed," but "wandering" those that have each their own individual motion. And we affirm also that the revolutions of the latter as well as of the former are each regular in themselves, and it is not necessary to give their motive intelligences the trouble of toiling now more and now less, which would be a condition incompatible with their nobility and inalterability and that of the spheres. If these propositions hold, it cannot be said, first of all, that these solar stars are fixed, because were they not shifting positions with respect to each other, it would be impossible to see the continuous changes that are really observed in the spots, for we would always find instead that the same configurations return. It must therefore be that they are mobile, each one moving in regular fashion by itself, but with differing motions among them. And the amassing and separation of some of them might occur in such fashion, but they would never be able to form spots, and we will understand this when we consider some particulars that are noticed in the latter. For example, because we see some very large spots appear and disappear, it must be admitted that they would have to be composed not of only two or four stars, but of fifty or a hundred, because we also see other little spots that are less than the fiftieth part of one of these bigger ones. If, therefore, one of these large spots were to dissolve, such that it vanished entirely from sight, it would have to disperse itself into more than fifty little stars, each one with its proper and individual motion, regular and differing from those of all others, for two with a common motion would never converge or would never separate on the face of the Sun. But if these things are true, who does not see that the formation of the sunspots would be absolutely impossible, above all because they last not just for many hours, but rather for many days? It is just as impossible for fifty boats, each moving at a different speed, to all come together and remain as a convoy for a long time. If the little stars were separated and therefore invisible, they could only be arranged in long rows, one after the other, all along their parallels, on which they all move toward the same part, as is clear in the case of the visible sunspots. It is very unlikely that forty or fifty or a hundred of them could draw together with any frequency and remain thus for a long time; on the contrary, with these varying movements, only extremely rarely would such a numerous assembly of stars converge in one place. But it would then be absolutely impossible for them not to disperse in a very short time, and yet, by contrast, we see many sunspots remain together for days on end with little change in their shape. He who wants, therefore, to maintain that the spots are a mass of minute stars must usher into the heavens and amongst these stars themselves countless motions, all riotous, discordant, and utterly removed from any regular pattern, which would not be consonant with any probable philosophical system. It will, moreover, be necessary to imagine them more numerous than all other visible stars, for if we consider the multitude and size of all the spots that have ever been seen under the hemisphere of the Sun, and we dissolve these into fragments of imperceptible size, we will find that there must necessarily be hundreds upon hundreds of them. And, moreover, since it is credible that there are others, not only above the other hemisphere but on the sides of the Sun as well, we cannot reasonably escape the conclusion that they number more than a thousand. Now what proportion can be maintained between the distances of the wandering stars and the periods of their revolutions, if in the descent from the immense orbit of Saturn to the smallest one of Mercury there are no more than ten or twelve stars, and no more than six different revolutions about the Sun, and yet hundreds and thousands must now be placed within such a small orb? For it would likewise be necessary to enclose them within the confines of the orbit of Mercury because they never make themselves visible in a bright appearance at some distance from the Sun.[62] But what am I saying about enclosing them in the orb of Mercury? Let us say also that since it is conclusively demonstrated that the spots are all contiguous to, or insensibly removed from the surface of the Sun, then anyone who wants to imagine them a mass of minute stars must first find a way of persuading us that above the solar surface hundreds and hundreds of dark dense globes go winding about at different speeds, constantly colliding with and blocking each other, such that the courses of the fastest are impeded for a few days by the slowest, and in many places squadrons of visible stars are formed from the confluence of a large number of them, until the throng of the approaching multitude, displacing at last those preceding them, opens a path, and the crowd disperses.

One must go to great lengths, and to maintain what? And with what effect? In order to preserve the celestial material from the condition of the sublunary elements, right down to every last little change. If what is called "corruption" were annihilation, Peripatetics would have some reason for being such enemies of it, but if it is nothing other than a modification, it hardly merits so much hatred. And it does not seem to me that anyone would reasonably quarrel about the "corruption" of the egg while a chick is being generated from it.[63] Besides,

62. Because its proximity to the Sun, Mercury is visible only when it is at its greatest elongation from that body. The solar planets postulated by Scheiner are never seen as bright objects any distance removed from the Sun. They must therefore move in extremely tight orbits around the Sun.

63. On this argument, see Biagioli, *Galileo's Instruments of Credit*, 138–141, 188, 216.

because what is called "generation" and "corruption" is only a little alteration in a small part of the elements, and one that would not even be perceived from the Moon, our nearest orb, why deny it to the heavens? Do [Peripatetics] perhaps imagine, arguing from the particular to the general, that the entire Earth is going to dissolve and decay, so that there will come a time when the universe includes the Sun, the Moon, and the other stars, but finds itself without our globe? I do not really believe that they entertain such doubts. And if the Earth's minor alterations do not threaten it with total destruction—for they are not its flaws, but rather its greatest ornaments—why deny them to the other bodies in the universe? Why fear that the heavens will vanish through changes no more hostile than these to the preservation of Nature? I suspect that our desire to evaluate everything according to our own meager measure makes us fall into strange fantasies, and that our particular hatred of death makes us detest frailty. Nevertheless, I don't know, on the other hand, if, in the pursuit of immutability, we would prize an encounter with the Medusa's head, so that she would turn us into marble or diamond, stripping us of our senses and of other movements that could not exist without bodily alterations. I want neither to continue this argument nor to scrutinize the merits of Peripatetic logic; I reserve these tasks for another time. I will add only this, that I think it is not the act of a true philosopher to persist—if I may say so—with such obstinacy in maintaining Peripatetic conclusions that have been found to be manifestly false, believing perhaps that if Aristotle were here today he would do likewise, as if defending what is false, rather than being persuaded by the truth, were the better index of perfect judgment and the most noble system of profound learning. And it seems to me that minds like these give others reason to suspect that they perhaps value an accurate examination of the strength of both Peripatetic and opposing arguments less than they do the preservation of Aristotle's absolute authority. For these minds, it appears sufficient to shield him from all manner of dissension with as little labor and trouble as finding texts and comparing passages, rather than investigating true conclusions and forming from them new and compelling proofs. And it seems to me, moreover, that we degrade our condition too much, not without offending Nature to a certain extent, and, I would almost say, divine Goodness—who, to help us understand His vast edifice, has granted to us 2000 more years of observations and vision twenty times as acute as what Aristotle had—by preferring to learn from that philosopher what he did not and could not know, rather than from our eyes and our own discussions. But not to stray further from my main point, I say that for the present it is enough for me to have demonstrated that the spots are neither stars, nor solid matter, nor located far from the Sun, but that they appear and disappear around it in a manner not dissimilar to that of clouds or other smoke-laden vapors around the Earth.

This is all that has occurred to me to tell Your Lordship at present about this matter, which I believe should be the emblem of all the new discoveries I have made in the heavens, and which should offer me the leisure to return to my other studies without interruptions, now that I have also fortunately succeeded in finding, after many vigils and much fatigue, the periodic times of all four Medicean Planets, and in preparing the tables and matter pertaining to the calculations and their other particular features. Though I will shortly publish these things along with the remaining considerations of other celestial novelties, my ideas are not yet in order because of the unexpected wonder with which Saturn has lately come to disturb me. I would like to give Your Lordship an account of this.

I have already written to you that about three years ago I discovered, to my great surprise, that Saturn is triple-bodied, that is, an aggregate of three stars arranged in a straight line parallel to the equinoctial. Of these stars, the middle one was much larger than the lateral two. I judged them to be immobile with respect to each other, and my belief was not unreasonable, for having seen them so close together in my first observation that they almost appeared to touch each other—a configuration in which they remained entirely unchanged for more than two years—I did well to conclude that they were totally immobile among themselves. In that period, only a second of arc, a motion incomparably slower than all the others, even than that of the greatest spheres, would have become perceptible, either separating or else completely uniting the three stars. I saw Saturn triple-bodied again this year around the summer solstice, and as someone who had no doubts about its constancy, having ceased to observe it for more than two months, I returned to it at last a few days ago and I found it solitary, without the company of the customary stars, and in sum perfectly round and shaped like Jupiter, and so it still remains. Now, what is there to say about such a strange metamorphosis? Were the two smaller stars perhaps consumed like solar spots? Did they vanish or suddenly flee? Has Saturn by chance devoured his own children? Or was the initial appearance an illusion and a mirage with which the telescope had managed to deceive me, and so many observers along with me, for such a long time? Has the time now come to revive the hopes, now on the point of withering, of those who, supported by the most profound contemplations, have discovered that the new observations are illusory and cannot be sustained? I do not have anything definite to relate about such an unusual, unexpected, and novel matter; lack of time, the unparalleled event, the weakness of my intellect, and the fear of error all render me quite confused. But permit me for once a bit of temerity, to which I confess as openly as Your Lordship will kindly excuse it. I declare that I do not intend to enroll what I am about to predict among those propositions derived from certain principles and firm conclusions, but merely from some probable conjectures of mine. I will present these when they are needed either to demonstrate the justifiable likelihood of the opinion toward which I am at present inclined, or to establish the certainty of the assumed conclusion in the event that my idea coincides with the truth. The propositions are these: the two smaller Saturnian stars, hidden from view at present, will perhaps appear a bit for two months around the summer solstice of the next year, 1613, and will then conceal themselves, remaining unseen until toward the winter solstice of the year 1614. Around this time, it may happen that they will reappear again for a few months, hiding once more until the following winter solstice approaches, at which time I believe firmly and with the greatest certainty that they will again appear and disappear no more. At the following summer solstice, which will be that of the year 1615, it will seem as if they want to disappear, but I do not believe that they will do so completely, but rather that, soon afterwards as they start to reveal themselves again, we will see them more distinct, brighter, and larger than ever. And I would almost venture to say with certainty that we will see them for many years without a single interruption. As few doubts as I have about their return, I am proceeding with restraint in regard to their particular features, for these are based at present on probable conjecture alone.[64] But whether these things take place precisely in this fashion or in another, I say to Your Lordship that this star, too, and perhaps no less than the emergence of horned Venus, agrees in a wondrous manner with the harmony of the great Copernican system, to whose universal revelation we see such favorable breezes and bright escorts directing us, that we now have little to fear from darkness and cross-winds.

I will now stop taking up Your Most Illustrious Lordship's time, but not without asking you to offer anew my friendship and aid to Apelles. And if you were to decide to let him see this letter, I ask you not to send it without the accompaniment of my excuses, in the event that it seems to him that I have disagreed too much with his opinions. For desiring nothing other than to find the truth, I have freely expounded my opinion, which I am also willing to modify

64. In modern terms, Galileo's conjectures were based on the motion of the Earth with respect to the ring plane during Saturn's retrograde loop. See Bruno M. Deiss and Volker Nebel, "On a Pretended Observation of Saturn by Galileo," *Journal for the History of Astronomy* 29 (1998): 215–220. See also A. Van Helden, "Saturn and His Anses," *Journal for the History of Astronomy* 5 (1974): 105–121.

whenever my errors are made known to me, and I will be particularly obliged to anyone who does me the favor of revealing and correcting them.

I kiss Your Most Illustrious Lordship's hands, and send you cordial greetings at the bidding of the Most Illustrious Signor Filippo Salviati, in whose pleasant villa and company I continue to make my celestial observations. May our Lord God grant you the fulfillment of your every wish.

From the Villa delle Selve, 1 December 1612.
Your Most Illustrious Lordship's
Most devoted servant
Galileo Galilei, Linceo

Postscript

The configurations of the Medicean [Planets] that I send to Your Most Illustrious Lordship are for the two months of March and April and up to the eighth of May. And I will be able to send you others soon [*alla giornata*]: they will perhaps be more accurate, and certainly more convenient for comparing to the apparent positions because of the more temperate season and the less inconvenient hours.[65] In the meantime, there are some considerations about these [tables] that should be pointed out to Your Lordship, and by you to Apelles or to others who might happen to make the comparisons.

And first, it should be noted that because of Jupiter's great splendor the stars closest to its body are not easily seen, save by the sharpest sight and with the best instrument, but as they move away, emerging from the radiation and thus becoming more visible, they show signs of having been quite close to Jupiter itself a little earlier. For instance, in the three configurations of the first night of March, the western star closest to Jupiter will not be seen during the first observation, three hours after sunset, being then almost contiguous to it, but because it moves away from it, at four hours [after sunset] it can be seen, and even better after five hours and throughout the rest of the night. The eastern star next to Jupiter on the ninth night of March will be seen with difficulty at the hour marked [in the table], but because it is moving away, in the following hours it will be seen very clearly. The contrary will happen to the eastern one

65. The sentence makes most sense if we assume that *alla giornata* means "in a few days." Those further tables will presumably be for May/June, when it is warmer and the Sun sets later, so the observations can be made without getting cold and don't have to be made at an inconvenient hour.

on the 15th day of the same month, for at the hour marked [in the table] if great care is taken, it can be seen, while not much later, moving toward Jupiter, it will be lost among its rays. When the air is very clear, which matters a lot in this affair, it is true that one of these four, being somewhat bigger than the other three, can be made out almost up to the moment it touches Jupiter, as can be observed in the one nearest to the western side on the 22nd of March, which will approach [Jupiter] and will be visible until it is very close.

But a more astonishing cause of the occultation of some of them is occasioned by the diverse eclipses to which they are variously subject because of the different inclinations of the shadow cone of the body of Jupiter itself. I confess to Your Lordship that this phenomenon worried me not a little until the reason for it came to mind.[66] These eclipses are sometimes long, and sometimes short, and on occasion invisible to us; these differences arise from the Earth's annual motion, from Jupiter's different latitudes, and from the eclipsed body's proximity to or distance from Jupiter, as Your Lordship will see more clearly in due time. During this and the next two years we will not have great eclipses;[67] nevertheless, what we will see is the following. Of the two stars to the east on the night of 24 April, the one further from Jupiter will be seen in the manner and time described [in the table], but the other, closer one, while separated from it, will not appear, being immersed in its shadow. But around the fifth hour of the night, emerging from the darkness, it will suddenly appear about two Jovian diameters from it. On the 27th of that same month, the eastern planet closest to Jupiter will not be seen until the fourth hour of the night, remaining in the shadow until that time; it will then suddenly emerge, and be visible already about one-and-a-half diameters distant from Jupiter. If you observe with diligence on the evening of the 1st of May, you will see the eastern star closest to Jupiter, but not before it has traveled a semidiameter away from it, having been in darkness earlier. And a similar effect will be seen on the 8th of the same month. With the next configurations I send you, I will include the other more notable and longer[68] eclipses that will follow upon these.

I want finally to submit to your most considerate judgment that you should not marvel—before I make my excuses—if what I offer you does not agree pre-

66. For the importance of the eclipses of Jupiter's satellites in determining the longitudes of places, see Albert Van Helden, "Longitude and the Satellites of Jupiter," in *The Quest for Longitude*, ed. William J. H. Andrewes (Cambridge, MA: Harvard University Press, 1996), 85–100.

67. "Great eclipses" most likely refers to their length. During the next two years, presumably Jupiter's ecliptic latitude is small.

68. See note 67, above.

cisely with the experiments[69] and observations that you and others will make, because the occasions for error are many. An almost inevitable one is carelessness in the calculations; besides this, the minuteness of these planets, and their being observed with the telescope, which so enlarges each and every object, have the effect that an error of only one second in the conjunctions and distances of these stars will be more apparent and noticeable than an error a thousand times greater in the aspects of other stars. But what is more important, the practitioners of this art will excuse me on account of the novelty of the matter, the brevity of the moment [of these observations], and the possibility that there are in the motions of these stars other differences and anomalies besides those I have thus far observed. And because in many thousands of years a great number of men have not yet perfectly determined the periods and explained every difference among the other wandering stars, the case of a single individual who has not precisely explained the little Jovian system in two or three years will surely be pardonable and favorably reviewed. For it is to be believed that the workmanship of this system, designed by the greatest Artificer, is of an immensity that far outstrips mortal minds.

69. For a discussion of the problematical relationship betwen *experimentum* and *experientia* in the sixteenth and early seventeenth century, see Peter Dear, *Discipline and Experience: The Mathematical Way in the Scientific Revolution* (Chicago: University of Chicago Press, 1995), 11–31.

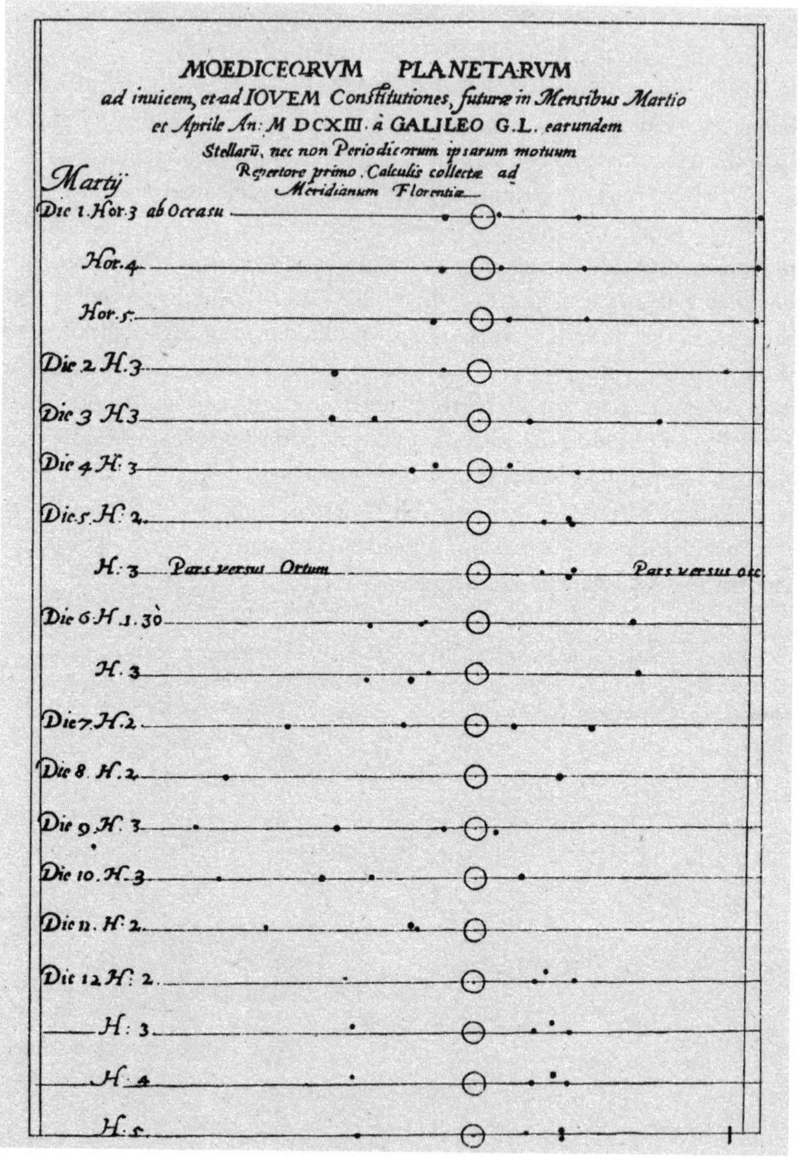

MOEDICEORVM PLANETARVM
ad inuicem, et ad IOVEM Constitutiones, futuræ in Mensibus Martio
et Aprile An: M DCXIII. à GALILEO G.L. earundem
Stellarū, nec non Periodicorum ipsarum motuum
Repertore primo. Calculis collectæ ad
Meridianum Florentiæ

Martij

Die 1. Hor. 3 ab Occasu

Hor. 4.

Hor. 5.

Die 2. H. 3

Die 3 H 3

Die 4 H: 3

Die 5. H: 2.

H: 3 Pars versus Ortum — Pars versus occ.

Die 6. H. 1. 30

H. 3

Die 7. H. 2

Die 8. H. 2

Die 9. H. 3

Die 10. H. 3

Die 11. H: 2

Die 12 H: 2

H: 3

H. 4

H. 5

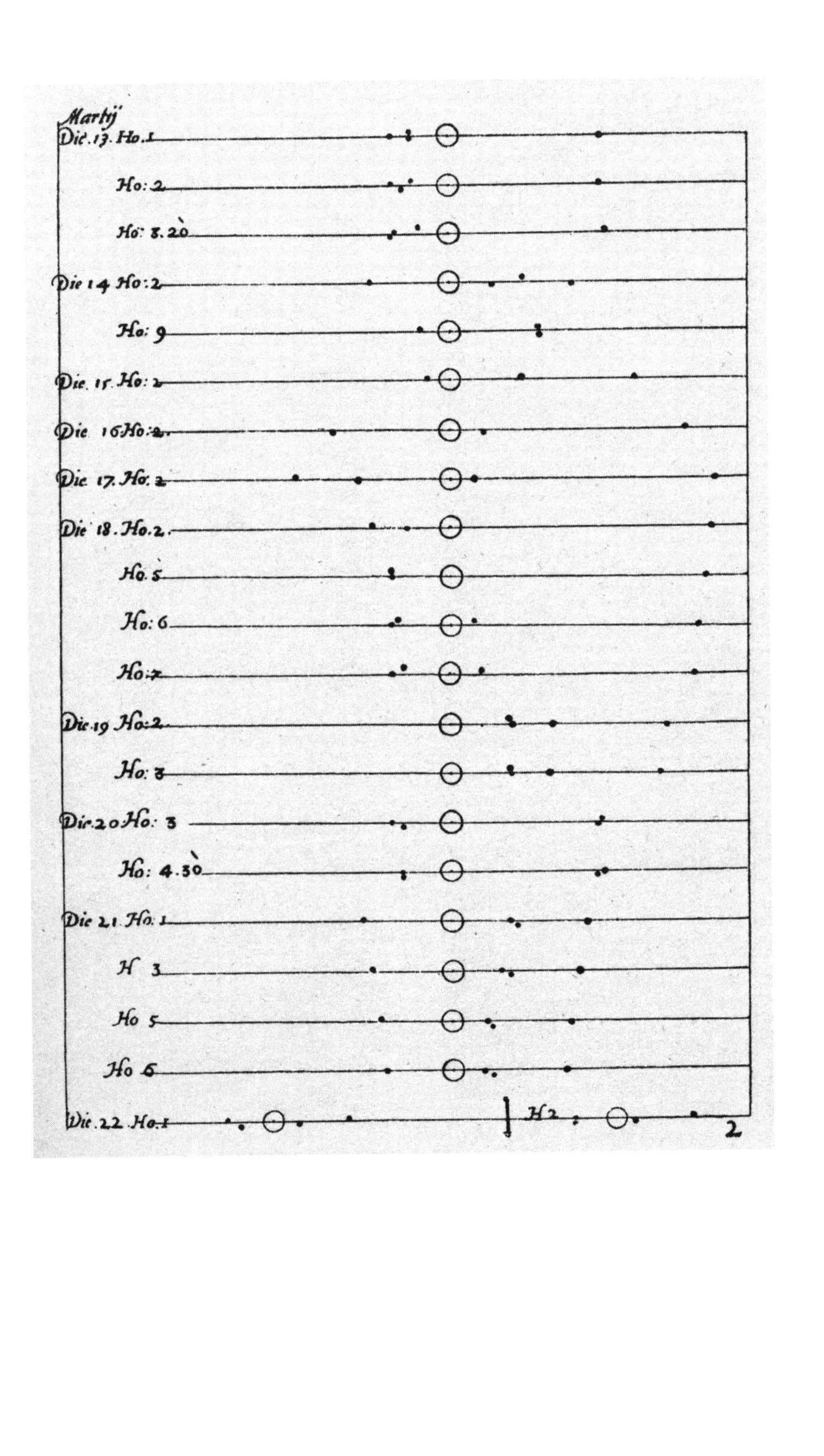

Martij
Die. 13. Ho. 1
Ho: 2
Ho: 3. 20
Die 14 Ho: 2
Ho: 9
Die. 15. Ho: 2
Die 16 Ho: 2
Die 17. Ho: 2
Die 18. Ho. 2.
Ho. 5
Ho: 6
Ho: 7
Die 19 Ho: 2
Ho: 3
Die 20 Ho: 3
Ho: 4. 30
Die 21 Ho: 1
H 3
Ho 5
Ho 6
Die. 22 Ho. 1
H 2
2

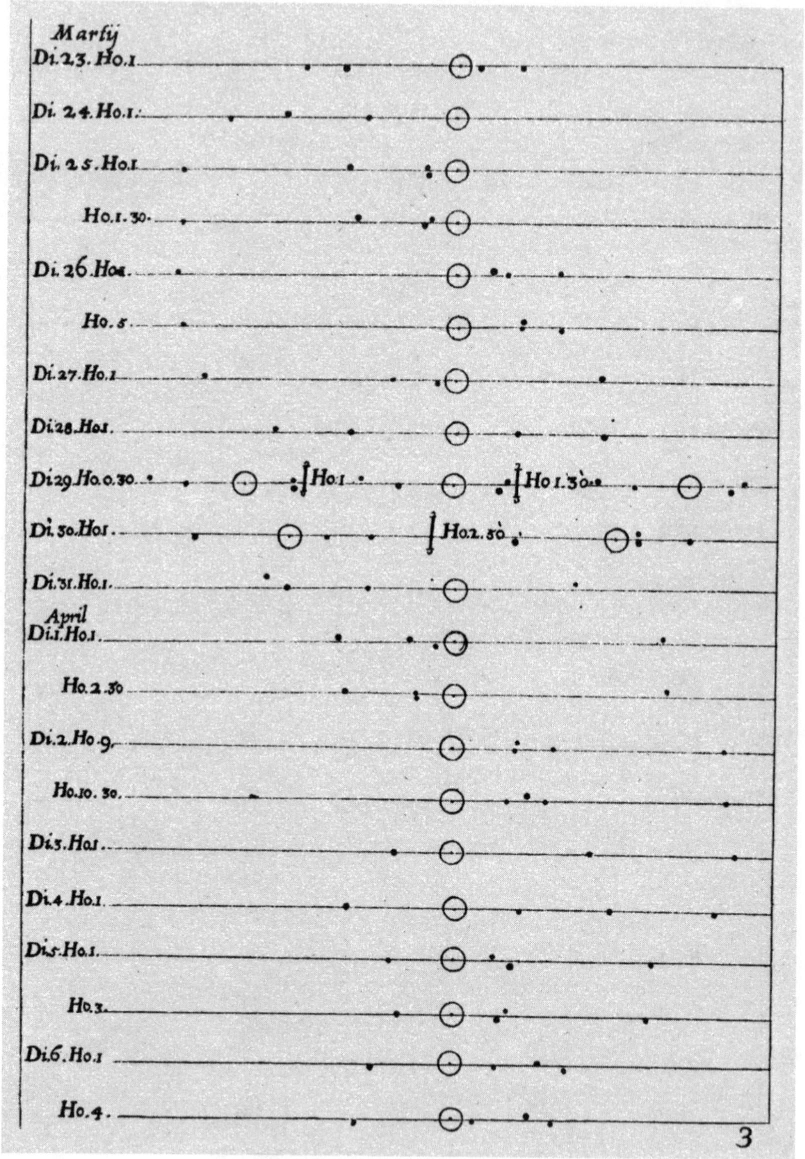

Martij

Di.23. Ho.1

Di. 24. Ho.1.

Di. 25. Ho.1

Ho.1. 30.

Di. 26. Ho.1

Ho. 5

Di. 27. Ho.1

Di. 28. Ho.1.

Di. 29. Ho.0.30 Ho.1 Ho.1.30

Di. 30. Ho.1 Ho.2.30

Di. 31. Ho.1.

April

Di.1. Ho.1

Ho.2.30

Di.2. Ho.9.

Ho.10. 30.

Di.3. Ho.1

Di.4. Ho.1

Di.5. Ho.1.

Ho.3.

Di.6. Ho.1

Ho.4.

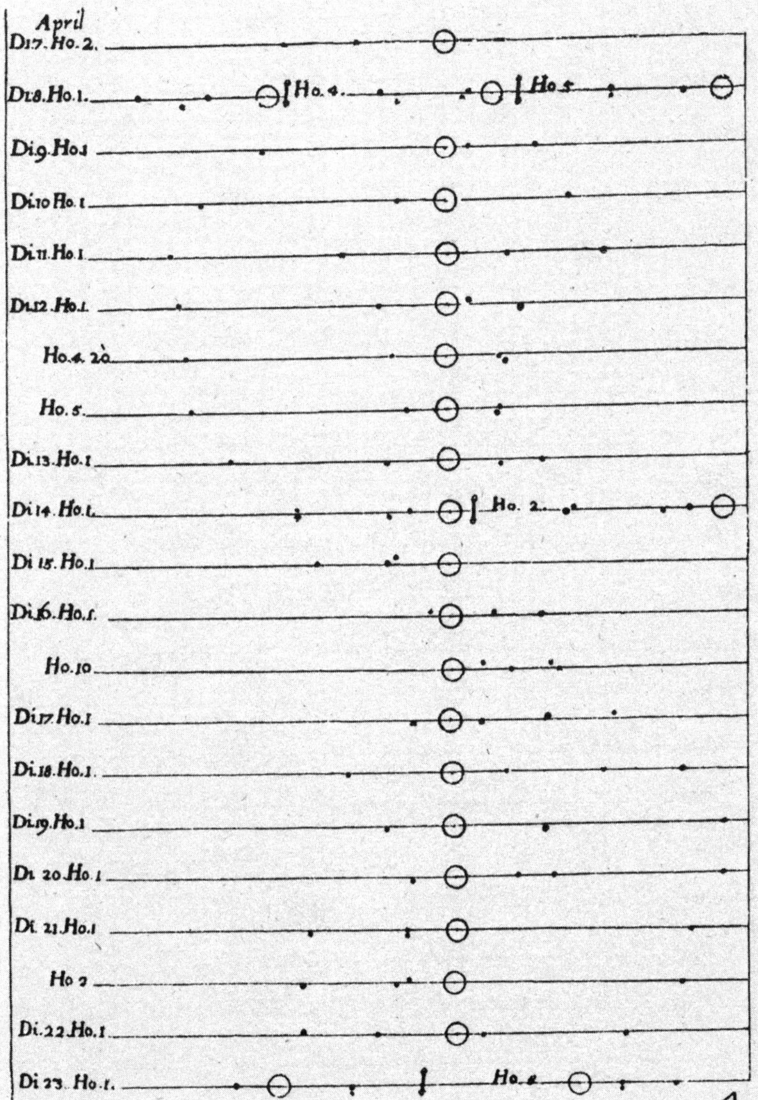

April

Di.7. Ho. 2.

Di.8. Ho. 1. Ho. 4. Ho. 5.

Di.9. Ho. 1

Di.10 Ho. 1

Di.11. Ho. 1

Di.12. Ho. 1.

Ho. 4. 20

Ho. 5.

Di.13. Ho. 1

Di.14. Ho. 1. Ho. 2.

Di.15. Ho. 1

Di.16. Ho. 1.

Ho. 10

Di.17. Ho. 1

Di.18. Ho. 1.

Di.19. Ho. 1

Di. 20 Ho. 1

Di. 21. Ho. 1

Ho 2

Di. 22 Ho. 1

Di 23. Ho. 1. Ho. 6

4

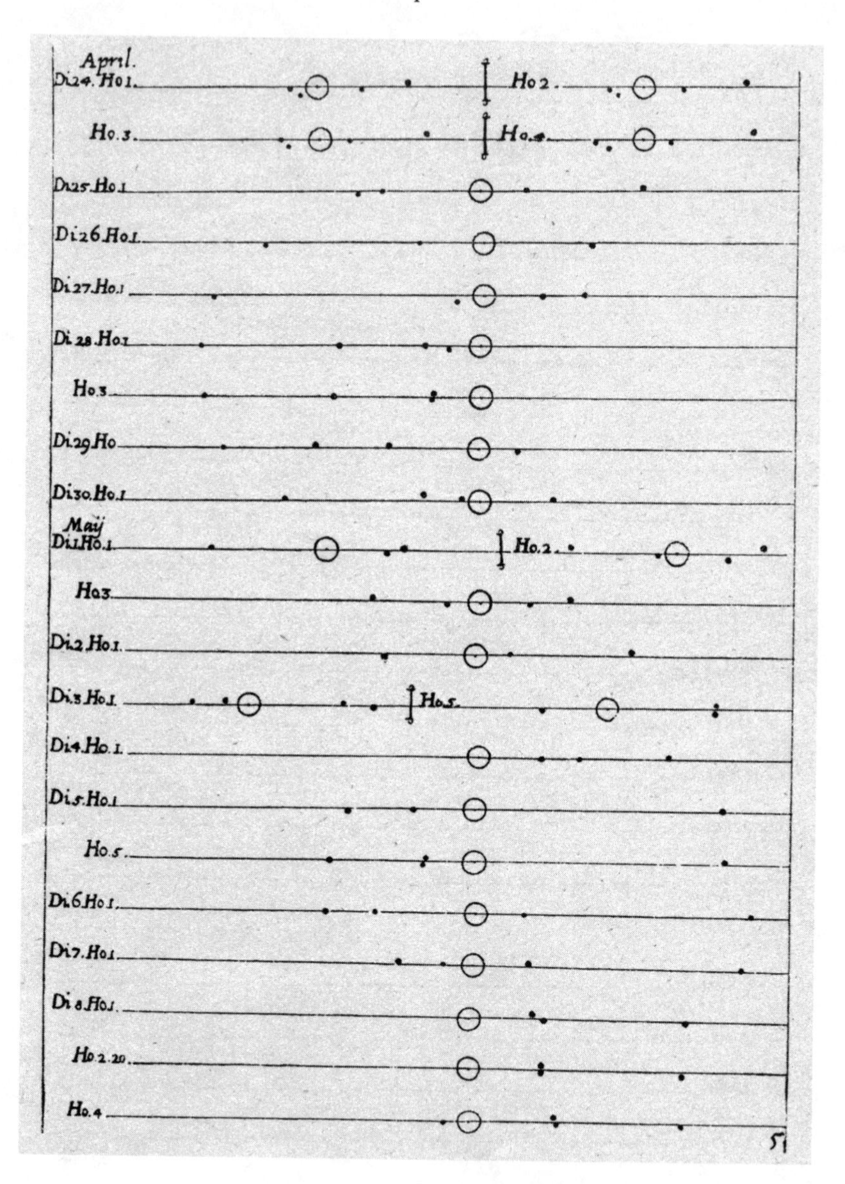
April.
Di24. Ho1.
Ho2.
Ho.3.
Di25. Ho1
Di26. Ho1.
Di27. Ho1
Di28. Ho1
Ho.3
Di29. Ho
Di30. Ho.1
Maij
Di1. Ho.1.
Ho.2.
Ho3.
Di2. Ho1.
Di3. Ho1.
Ho.5.
Di4. Ho.1.
Di5. Ho1
Ho.5.
Di6. Ho1.
Di7. Ho1.
Di8. Ho1.
Ho.2.20.
Ho.4.
51

13

Rematch, 1630–1632

A year after the emergence of *Accuratior Disquisitio*, Scheiner published a set of theses defended by one of his students, Johann Georg Locher, entitled *Disquisitiones mathematicae de controversis et novitatibus astronomicis*. This work offered a brief discussion of the sunspots:

> They are blackish bodies wandering about the Sun [*circa Solem erratica*] with various motions; neither their number nor their nature has yet been defined. They are so close to the Sun that the senses cannot separate them from it, for they have no parallax whatsoever with respect to it, and they set together with the Sun. From this it is evident that they are very close to the Sun and are absolutely not in the air. To put it another way, the notion that they do not remain on the Sun the entire day, or that they appear [only] for a little while when the Sun is over the horizon, or that they are not [seen] in the same place when observed simultaneously from widely separated places such as Italy, Germany, or Poland is contradicted by experience. Whether they are stars is still to be determined. Consult the paintings of Apelles, go to the *History* of Galileo, and expect more in time.[1]

1. Christoph Scheiner and Johann Georg Locher, *Disquisitiones mathematicae de controversiis et novitatibus astronomicis* (Ingolstadt, 1614), 66.

The thesis also alluded to another solar phenomenon, faculae, which had first been mentioned by Galileo in his third sunspot letter (p. 281, above):

> Faculae are small areas on the Sun that are brighter than the rest of its body. Where, in what way, and what they may be will be explained elsewhere soon. They have been observed with the spots from the beginning, and Signor Galileo wrote this in the third letter of his *History*, fol. 131.[2]

Whereas it was possible, even plausible, that the dark spots, *maculae*, were extrasolar planetary bodies—Galileo had entertained the possibility in his first announcement in print[3]—it was less feasible to assign the brighter faculae, which had the exactly same motions as the dark *maculae*, to orbits around the Sun. The faculae were very likely an important consideration in Scheiner's abandonment of the planetary nature of sunspots.

In the course of his student's thesis, Scheiner also offered a full review of the information available about Jupiter's satellites, beginning with a statement about their discovery:

> Jupiter's wonderful retinue was first discovered a few years ago by Signor Galilei, the distinguished and brilliant Italian mathematician (for in vain and belatedly some Calvinist this year for the first time tried very rudely to persuade us otherwise [see note 5, below]), and the entire community of astronomers was deservedly carried away in its admiration for him. For around this Lord, if you will, four attendants revolve, of different motions, sizes and distances.[4]

Early in 1615, Scheiner sent a copy of the *Disquisitiones* to Galileo, addressing him as follows:

> Noble, Excellent, and Most Esteemed Lord,
>
> Now that the occasion has arisen, I am gladly doing what I have often intended: to address Your Excellency with a letter and to interrupt you with a trifling present. One of my students recently defended the *Mathematical Disquisitions*, and I send Your Lordship a copy, not because I wish to teach [you] something, but rather to declare that my heart is well disposed [to you],

2. Ibid. Scheiner observed faculae in 1624, and later established a taxonomy of four basic types. On the importance of the faculae in his *Rosa Ursina*, see Luigi Ingaliso, *Filosofia e cosmologia in Christoph Scheiner* (Soveria Mannelli: Rubbettino, 2005), 194, 200–201, 205, 209.

3. In the first edition of his *Discourse on Bodies in Water*. See pp. 77–78, above.

4. Scheiner and Locher, *Disquisitiones mathematicae*, 78.

and to request in return some epistolary communication, if that is proper. It does not, however, escape me that the Copernican opinion and hypotheses are looked upon very favorably by Your Lordship, but my opinions, or rather those of my student, are such that they do not seek to escape the censure of the more learned [*doctiorum*]. Therefore, although I do not think that in this matter one should be violently deprived of his opinion, I judge that reason should not be spared in order to find the truth. For if Your Excellency puts forward something to the contrary, it will never offend us, but we will willingly read what is produced against [us], always in the hope that from it greater light will be shed on the truth.

In astronomical matters, there is hardly anything new. Simon Marius published *Mundus Iovialis*,[5] which, if Your Lordship does not have it yet, please let me know and I will see to it that you get it. You will be amazed by the man's arrogance and, if you wish, you can rightly correct his errors. I ask of you one thing at this time: if you have tables of the revolutions of the Medicean Stars—and I hardly doubt that you do—you deem it worthy to communicate them to me. I offer myself ready to return the favor. May Your Lordship fare well and pray to God for me.[6]

Two months later, Scheiner tried again,[7] this time sending Galileo a copy of his new book, *Sol Ellipticus*.[8] There is no record of Galileo's response to overtures from a colleague who was a member of an order that was, in this period, well disposed toward him.[9]

5. Simon Marius published *Mundus Iovialis* in 1614. In it he claimed to have discovered the satellites of Jupiter on the same night Galileo did, 7 January 1610 (Gregorian). In his *Disquisitiones Mathematicae* of that same year, Scheiner dismissed Marius's claim and called him a Calvinist. For examinations of Marius's claim, see Joseph Klug, "Simon Marius aus Gunzenhausen und Galileo Galilei," *Abhandlungen der II. Klasse der Königliche Akademie der Wissenschaften*, 22 (1906); 385–526; J. A. C. Oudemans and J. Bosscha, "Galilée et Marius," *Archives Néerlandaises des sciences exactes et naturelles, publiées par la Société Hollandaise des Sciences*, 2nd ser., 8 (1903): 115–189; J. Bosscha, "Simon Marius, réhabilitation d'un astronome calomnié," *Archives Néerlandaises des sciences exactes et naturelles, publiées par la Société Hollandaise des Sciences*, 2nd ser., 12 (1907): 258–307, 490–527. The Latin text with German translation can be found in *Simon Marius, Mundus Iovialis—Die Welt des Jupiter*, tr. Joachim Schlör, commentary by Alois Wilder (Gunzenhausen: Schenk Verlag, 1988). For an abridged English translation, see A. O. Prickard, "The *Mundus Jovialis* of Simon Marius," *The Observatory*, 39 (1916): 367–381, 403–412, 443–452, 498–503.

6. Scheiner to Galileo, 6 February 1615, *OGG*, 12: 137–138.

7. Scheiner to Galileo, 11 April 1615, *OGG*, 12: 170–171.

8. *Sol Ellipticus: hoc est novum et perpetuum Solis contrahi soliti Phaenomenon, quod noviter inventum Strenae loco* (Augsburg: Christopher Mangius, 1615).

9. Galileo had exchanged letters on scientific subjects with Christoph Clavius from time to time from 1588 to 1611, a year before Clavius's death, had enjoyed a rather favorable reception at the

Though Scheiner clearly considered himself a rival of Galileo's, his tone was polite, if awkward. His letter of February 1615 reflects the advice of the general of his order, Claudio Aquaviva, who had written to him two months earlier:

> In refuting Galileo, we consider . . . another method more suitable, and that is that after Your Reverence has put forward the arguments for the truth, and refuted the arguments of others without naming the author, you end with the conclusion that on the basis of these demonstrations the [Copernican world] system of Galileo collapses. Thus the confutation proceeds with greater benevolence and modesty.[10]

In the course of that same letter, however, Aquaviva made several points that illustrate Scheiner's dilemma. First, if his letters on the sunspots were to be reprinted, the pseudonym "Apelles" should be retained. Second, all further publications by Scheiner were to be sent on to the censors in Rome after approval in the province. And then Aquaviva laid down the law about what Scheiner could and could not write:

> I want to command Your Reverence this alone: insist on the solid doctrines of the ancients and do not teach the opinions of many moderns. Be assured that the latter do not please us, and that we will not allow our members to publish such things. Before all else, one should guard against maintaining anything about the fluidity of the heavens against the unanimous opinion of all Holy Fathers and Scholastics, or about stars that move like fish in the sea or like birds in the air. Against this opinion, we recommend that you reconcile yourself with the more solid and older doctrines.[11]

That same day Aquaviva wrote to Adam Tanner, Scheiner's superior at Ingolstadt, that "Father Scheiner has been warned by us to lay aside those new opinions about the heavens."[12]

Here, then, was Scheiner's dilemma. The arrangement of the planetary heav-

---

Collegio Romano in the spring of 1611, and carried on a scientific correspondence with Clavius's successor, Christoph Grienberger, for several years.

10. Archivium Romanum Societatis Jesus (ARSI), Germania Superior 4, f. 103r. We thank Ugo Baldini for transcribing this letter. For a German translation, see Franz Daxecker and Lav Subaric, "Briefe der Generaloberen P. Claudio Aquaviva SJ, P. Mutio Vitelleschi SJ und P. Vincenzo Carafa SJ an den Astronomen P. Christoph Scheiner SJ von 1614 bis 1649," *Sammelblatt des Historischen Vereins Ingolstadt* 111 (2002): 101–148, at 102–103.

11. ARSI, Germania Superior 4, f. 103r; Daxecker and Subaric, "Briefe," 103.

12. ARSI, Germania Superior 4, f. 103r; Daxecker and Subaric, "Briefe," 103.

ens implicit in his *Tres Epistolae* and *Accuratior Disquisitio* was that a host of planetary bodies orbited the Sun, that at least four of them orbited Jupiter, and that perhaps two more revolved about Saturn. This meant that there were at least three and possibly four centers of motion in the universe: the Earth, the Sun, and Jupiter, and perhaps Saturn as well. Realistically, only a fluid heaven, one in which these bodies moved about "like fish in the water or birds in the air," could accommodate this arrangement. The question had become more and more pressing since the new stars of 1572 and 1604 and the comets of 1577 and 1607. Johannes Lanz's plea for guidance from Christoph Clavius on this cosmological issue in 1607 (see above, p. 41) had not been answered by Clavius, who was, after all, a mathematician and neither a philosopher nor theologian. Four years later, after Galileo's discoveries with the telescope, Claudio Aquaviva had offered an emphatic reply, admonishing all the members of his order to follow the solid doctrine of traditional philosophy,[13] and he reiterated that point here personally to Scheiner.

Starting with *Sol Ellipticus*, Scheiner published on the strictly mathematical and noncontroversial subject of optics: atmospheric refraction and vision and the eye.[14] He could not restrain himself from sniping at Galileo; in the very book he offered him in 1615, *Sol Ellipticus*, Scheiner wrote that anyone who claims that at perigee Venus's apparent diameter is to the Sun's as 10″ is to 34′ (i.e., 10″ to 2040″) reasons badly.[15] In his first sunspot letter, Galileo had written "the apparent diameter of Venus [at the conjunction] was then not even the 6th part of a minute, and its surface was less than one 40,000th part of the surface of the Sun, as I know from sensate experience and in due time will make known to everyone."[16] Here Scheiner assigned Galileo's estimates of Venus's apparent diameter during a *superior* conjunction to an *inferior* conjunction. Such an argument could not have raised Galileo's estimate of Scheiner.[17]

It was in this period that Scheiner laid the foundation of his greatest scientific work, *Rosa Ursina*. He had adopted the method of projecting sunspots through a telescope developed by Benedetto Castelli and used to such good

13. See Richard Blackwell, *Galileo, Bellarmine, and the Bible* (Notre Dame: University of Notre Dame Press, 1991), 139–142.

14. *Sol Ellipticus* (1615); *Oculus hoc est fundamentum opticum* (1619).

15. Daxecker and Subaric, *Christoph Scheiners "Sol Ellipticus,"* 41–42.

16. Pp. 93 and 264, above.

17. There is the possibility that Scheiner thought the December 1611 conjunction he treats so exhaustively in his *Tres Epistolae* and *Accuratior Disquisitio* was an inferior conjunction, in which case his comments in *Sol Ellipticus* would be justified. But this seems highly unlikely, because it would mean that he was ignorant of the basic principles of astronomy. For the problem of estimates and measurements of the apparent diameters of planets, see Van Helden, *Measuring the Universe*, esp. 65–76.

effect by Galileo, and in his continuing study of sunspots and demonstrating them to others, he made successive improvements. Following the Sun with one's telescope in order to keep the solar image centered on the paper was very difficult. The first problem was that, as Galileo had already explained in his second letter (p. 127, above), the form of telescope he used projected an inverted image. In following the motion of the Sun, therefore, one has to turn the telescope in the direction contrary to the motion of the solar image. Practiced observers soon overcame the counterintuitive aspect of the procedure, but for demonstrations to laymen, this difficulty remained a crucial obstacle.

Scheiner had studied Kepler's *Dioptrice*, and he knew that there was more than one combination of lenses to achieve the telescopic effect. Replacing the concave ocular by a convex one would produce an inverted direct image but an erect *projected* image, making manipulation of the telescope much easier. In his *Rosa Ursina* Scheiner would recall that he had used this improved method more than a decade earlier to right "the projected images for the Most Serene Maximilian, Archduke of Austria, and later for His Holy Imperial Highness."[18] Since this combination of lenses presents an inverted image if one looks through it, one would expect that for terrestrial purposes it would be useless, and neutral, at best, for astronomical purposes. But when Scheiner looked through the combination, he found something unexpected:

> If you fit two like [convex] lenses in a tube . . . and apply your eye to it in the proper way, you will see any terrestrial object whatsoever in an inverted position but with an incredible magnitude, clarity, and width. But also, you will compel any stars you wish to submit to your sight; for since they are all round, the inversion of the position of the total view is not confusing with respect to the visual configuration.[19]

The astronomical telescope, as an instrument with a convex ocular is called, has a much larger field of view and brighter image than the Galilean form of the instrument. The replacement of the Galilean telescope by the astronomical telescope can, in fact, be dated from the publication of *Rosa Ursina* in 1630.

Archduke Maximilian III had called Scheiner to Innsbruck in 1616, and upon Maximilian's death in 1618 Scheiner stayed on there under his successor, Arch-

18. *RU*, f. 129v–130r.

19. *RU*, f. 130r. See also Van Helden, "The 'Astronomical Telescope,' 1611–1650," *Annali dell'Istituto e Museo di Storia della Scienza di Firenze*, 1 (1976): 13–35, at 23–24.

Figure 13.1. Scheiner's apparatus for projecting sunspots. *RU*, 77.

duke Leopold V. During this period Scheiner continued his studies of sunspots with several associates or students, especially Johann Baptist Cysat and Georg Schönberger, who oversaw these observations during his frequent absences, for he hoped one day to publish the completed study. Besides using a telescope with a convex ocular for projection, Scheiner also provided the entire apparatus with a convenient mounting (fig. 13.1). The main axis of the mounting is made parallel to the axis of rotation of the Earth, so that an object in the sky can be followed merely by turning the telescope around this axis. This means that on the pre-drawn circle on which the image of the Sun is projected, the Sun's path (the ecliptic) is always represented by a horizontal line. Scheiner systematically taught his students and associates to draw the perpendicular to this horizontal line, in order not to make errors in the complicated motions of the spots.[20]

But other duties increasingly occupied Scheiner, and it was not until he had settled in Rome in 1624 that he could return to sunspots. Obtaining the observations that he and others had made in the German region was complicated by the campaigns of the Thirty Years War, and many of those which he did manage to procure from various observers were useless because no perpendicular had been marked. His student Georg Schönberger did, however, send observations with the perpendicular line, and Scheiner was able to use them to demonstrate the curved motions of the spots. He had several examples en-

20. *RU*, 158–159. (Note that the pagination of *Rosa Ursina* is highly irregular and at times has to be indicated by recto and verso additions.)

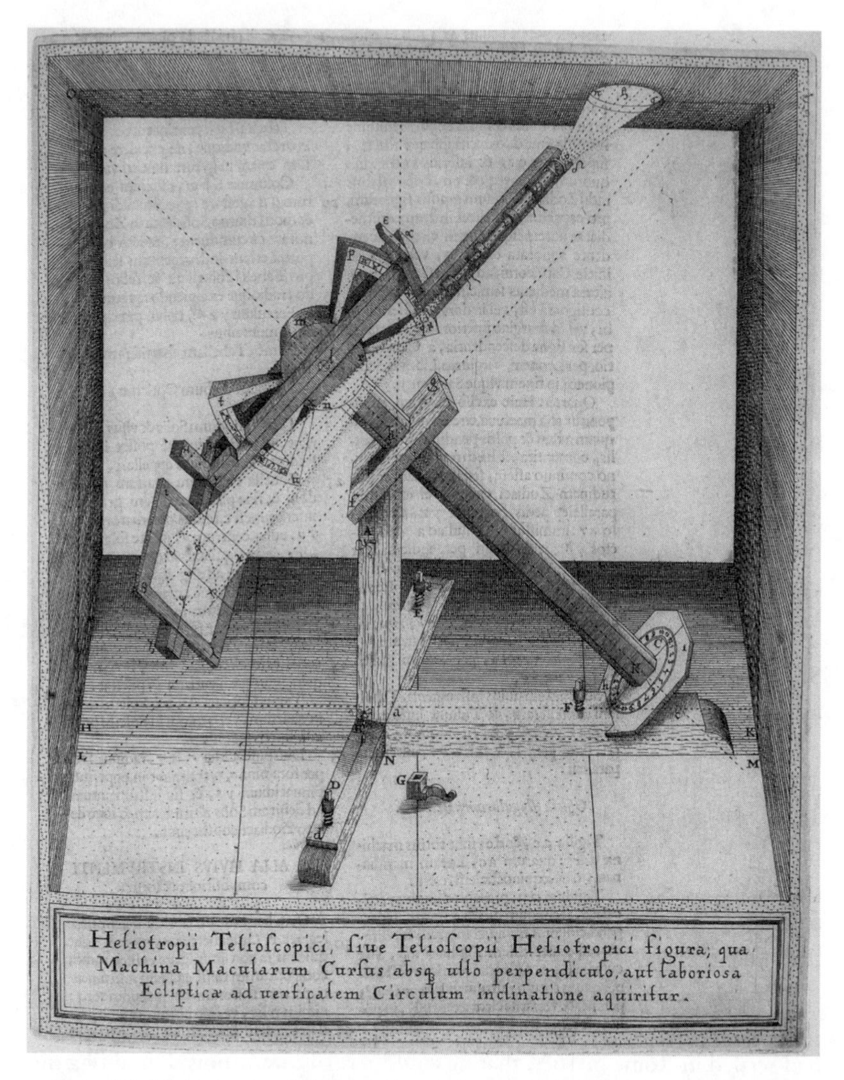

Figure 13.2. Scheiner's equatorially mounted projection apparatus. *RU*, 349. The caption reads: "Illustration of the Telescopic Heliotrope or Heliotropic Telescope, by which the course of spots is acquired without any perpendicular or laborious inclination of the ecliptic to the vertical circle."

graved, and returned these to Schönberger,[21] who would publish a little tract on the subject in 1626.[22]

In Rome, Scheiner made a large number of excellent observations of sun-

21. *RU*, 159.

22. *Sol Illustratus* (Innsbruck, 1626).

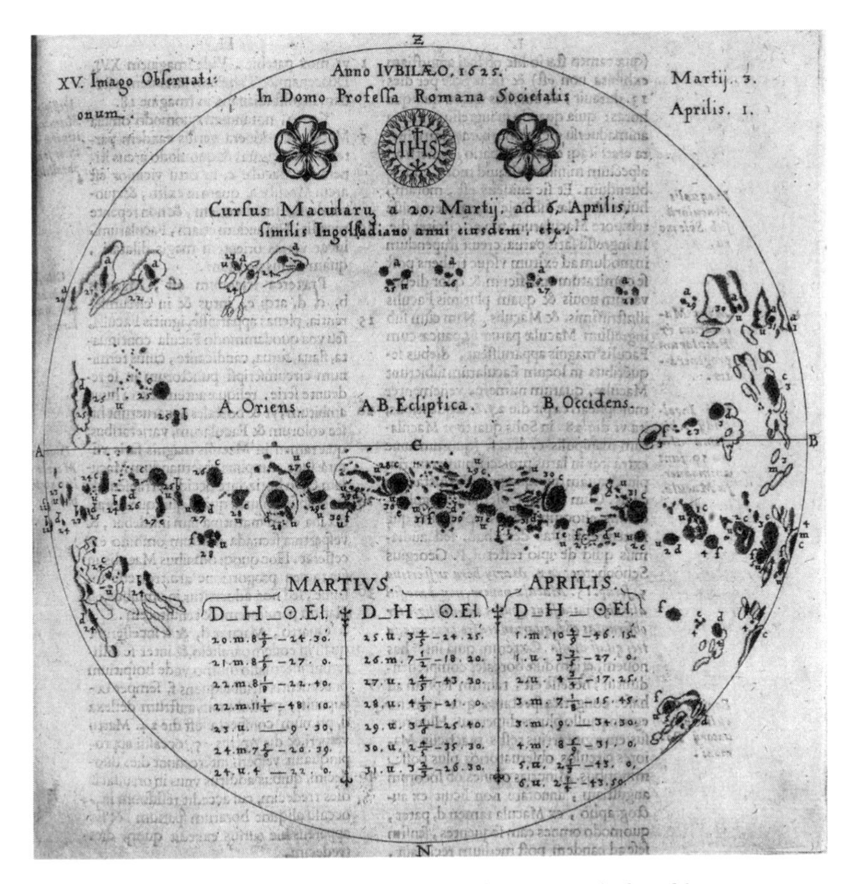

Figure 13.3. Observations of the course of a spot across the face of the Sun, April–May 1625. *RU*, 207.

spots, using this time an equatorial mounting for his projection apparatus (figs. 13.2 and 13.3), designed by his colleague Christoph Grienberger and, in 1626, began the publication process. The central argument of the *Rosa Ursina* was the demonstration that Galileo had erred on the path of the spots and had concluded that the Sun's axis of rotation was perpendicular to the ecliptic in his letters on the sunspots of 1612–1613. Scheiner determined that this axis is, in fact, inclined to that perpendicular by 7°15′.[23] Schönberger's publication threatened the secrecy, and Scheiner was especially eager to keep that information from Galileo before unveiling it in the *Rosa Ursina*. He even asked Archduke Leopold V of Austria, to whom Schönberger had dedicated his tract, to

23. For discussion and illustration of the sunspots' annual path, see Ingaliso, *Filosofia e cosmologia*, 209–213.

see to it that no copy reached Italy before the publication of the *Rosa Ursina*, lest "it expose my work to danger from Galileo."[24] The printing of *Rosa Ursina* began in 1626 and was finished in 1630. It was a magisterial work that was to remain the definitive study of sunspots for over a century, but it brought Scheiner little happiness.

But had Galileo missed the tilt of the Sun's axis of rotation? During his intensive observations of sunspots in 1612, his concern had been not the precise motions of the spots, but rather their nature—what they were not and what they might be. Moreover, an important part of his observations were made around that time of the year when the day-to-day motions appeared as straight lines on the Sun's disc. Galileo perhaps recognized that more sustained observations would reveal the inclination of the Sun's axis when he wrote, in the course of his second letter to Welser, "from the observations I have been able to make up until now I do not judge that the revolution of the spots is oblique to the plane of the ecliptic, in which the Earth lies" (see p. 113). But even if he had in fact been aware of the curved paths of the spots in this early phase of his investigations, it is unlikely that he attached any great significance to the phenomenon. Thus while the narrative of discovery aired in the *Dialogue Concerning the Two Chief World Systems*—that soon after the publication of the *History and Demonstrations* Galileo, in the company of Salviati, perceived the slightly curved path of one very large sunspot solely by charting its appearance each day at noon—seems overly simple (see below), certain details in the correspondence of 1613–1614 make it at least possible that his awareness of the issue dates to this period.[25]

The first suggestion that the sunspots' annual path might be used as evidence for or against the Copernican world system had emerged in a short treatise known to both Galileo and Cesi. In mid-February 1612, an author now identified as Willebrord Snellius published the *De Maculis in Sole Animadversis, &, tanquam ab Apelle, in tabulâ spectandum in publicâ luce expositis, Batavi Dissertatiuncula* (A Dutchman's brief discourse on the spots observed in the Sun and exposed by Apelles to public scrutiny, so to speak, in a painting on view).[26] The treatise summarized Apelles' methods and findings, expressed skepticism about his attempted observation of Venus, and ended with the con-

24. Scheiner to Archduke Leopold V, 19 December 1626. In Franz Daxecker, *Briefe des Naturwissenschaftlers Christoph Scheiner S.J. an Erzherzog Leopold V. von Österreich-Tirol 1620–1632*, 109–110; quotation on 110.

25. *OGG*, 7: 373–374.

26. The identity of the author has been definitively established through manuscript evidence by Rienk Vermij in *The Calvinist Copernicans: The Reception of the New Astronomy in the Dutch Republic, 1575–1750* (Amsterdam: Edita, Royal Netherlands Academy of Arts and Sciences, 2002), 44–45.

jecture that in a heliocentric universe sunspots would appear to undergo a shift with respect to the viewer over the course of the year.[27]

> You see then, sharp-sighted Apelles, what might finally be accomplished here. For if with those Lynx-like eyes of yours you were to see any long-lasting spot on the Sun, which is either not moved at all or only very slowly over the space of half a year, it could quickly be decided whether the Earth stands still or moves. And as this principle arises on account of this latest news, the motion of the Earth might be demonstrated from the motion of that spot. And do you know how? Imagine the Earth turning in its annual orbit around the Sun, which is the eye of the world because it is positioned at or near the center, and that the Earth moves in the direction of the zodiacal signs. Then both the Sun and the spot will have a single motion, such that they will appear to us to be moved from east to west.[28] And when we hasten from the summer solstice to the winter solstice, as the Sun appears to decline, that conspicuous spot will likewise appear to move in the same direction, from north to south. The contrary will happen while we progress from the winter solstice to the summer solstice.[29]

This suggestion, which of course does not take into account the tilt of the solar axis to the plane of the ecliptic, ended on an inconclusive note.

> You had indeed given me the greatest elation, since I might have observed the spots recorded by you move from the north to the south, in the month of October, but right away the speed of their motion in the opposite direction dashed that hope of mine. For everything you mention about this motion has to obey this rule very precisely, such that [those spots] must continuously proceed from west to east and from north to south, or in the opposite direction corresponding to the time of year . . . But if that perpetual spot does not move from its place, then it follows that the same face of the Sun looks upon us, so that the Earth is at rest and, on the contrary, the Sun moves, just as you now see it, although I foresee nothing of this sort.[30]

The powerful Dutch diplomat to whom this treatise was addressed, Cornelis van der Myle, intended to send it to Welser, who would have presumably re-

27. Snellius, *De Maculis in Sole Animadversis* (Leiden: Plantijn, 1612), 7–10, 17–18.

28. West to east. Galileo had pointed this same mistake out to Scheiner in his first letter; see pp. 90–91, above.

29. Snellius, *De Maculis in Sole Animadversis*, 17.

30. Snellius, *De Maculis in Sole Animadversis*, 17–18.

layed it to Scheiner and perhaps to Galileo as well.[31] Neither author referred in explicit fashion to the work, though it is worth noting that in the *Accuratior Disquisitio*, when Scheiner referred to those who had accepted the existence of the sunspots, and mistakenly listed among these "ear witnesses" "the very learned Simon Stevin, in the Netherlands," he was in all likelihood thinking instead of this treatise, which had included no new observations, but had focused entirely, if somewhat skeptically, on the ramifications of *Tres Epistolae*.[32] We also know that in October 1612 Galileo received a manuscript copy of the work from Cesi, who retained the printed version. Cesi criticized the treatise for its lack of new solar observations, its absence of quantitative data, and above all for its failure to mention Galileo by name when alluding to recent astronomical discoveries made with the *Batavica Dioptra*, or "Dutch telescope."[33]

Despite its negligible impact in Italy, however, it is likely that the treatise, or its conjecture concerning seasonal shifts in the sunspots' apparent path, influenced solar observers in Paris, for both Snellius and van der Myle had connections to Jacques Alleaume, then a prominent engineer and astronomer at the French court who followed Galileo's work.[34] We will return shortly to the conjectures of these observers.

Galileo certainly realized sometime in late 1613 or early 1614 that the complex but regular variations in latitude of Jupiter's satellites, which made them appear at times to move in a straight line, and other times to deviate a little to the north or to the south, were caused by the inclination of Jupiter's orbit to the ecliptic.[35] As we will discuss below, Galileo's discovery and explanation of this phenomenon might have prompted him to consider an analogous effect in the case of the Sun. Further, in July 1613, shortly after the publication of the *History and Demonstrations*, Orazio Morandi, the abbot of Santa Prassede in Rome, forwarded to Galileo a letter written that spring in Paris by Francesco

31. Vermij, *The Calvinist Copernicans*, 45.

32. See p. 220.

33. Federico Cesi to Galileo, 29 September 1612, in *OGG*, 11: 404; see also Cesi to Galileo, 6 October 1612, in *OGG*, 11: 409–410.

34. In his chapter on finding the altitude of the pole in his *Eratosthenes Batavus*, Snellius refers specifically to the compass developed "first by that brilliant and most ingenious man, our friend Jacques Alleaume"; see *Eratosthenes Batavus* (Leiden: Iodocum à Colster, 1617), 103. Prior to his employment at the court of Henri IV of France, the Huguenot Alleaume had been an engineer and *déchiffreur* in the Netherlands for Maurice of Nassau. On this important figure see Cornelis de Waard, s.v. "Alleaume," in *Nieuw Nederlandsch Biografisch Woordenboek*, ed. P. C. Molhuysen and P. J. Blok, 10 vols. (Leiden: A. W. Sijthoff's Uitgevers-Maatschappij, 1912), 2: cols. 17–19. On Alleaume's regard for Galileo's work, see Francesco Sizzi to Orazio Morandi, 10 April 1613, *OGG*, 11: 492–493.

35. See Noel Swerdlow, *The Renaissance of Astronomy in the Age of Humanism*: *Regiomontanus, Copernicus, Tycho, Kepler, Galileo* (forthcoming).

Sizzi.[36] Sizzi, working in the orbit of Alleaume, had informed Morandi that he and other Parisian observers had noticed, after almost a year's study of the Sun, that the sunspots' paths varied in their tilt, moving from one sort of appearance at the equinoxes to another at the solstices, and that the latter sort further differed from each other, the angle described at the summer solstice being in the opposite quadrant of the solar surface at the winter solstice. The letter bears no annotations by Galileo, and there is no record of a reaction to these observations, for which Sizzi himself offered no explanation. It appears that Galileo had moved on to other, more serious concerns and would not return to sustained study of the sunspots for fifteen years.

This did not mean, however, that he did not keep an eye on Scheiner. In early March 1614, Galileo learned from Federico Cesi that François Aguilon S.J. had proudly revealed Apelles' identity in print,[37] and had supported the latter's irritating claim to priority in this matter. "To what purpose this Apelles has revealed himself, I surely have no idea," Cesi wrote, "and I am amazed that the [Jesuit] Fathers assert his precedence in this observation, since they know how much earlier Your Lordship treated and showed these things."[38] But Scheiner's purposeful emergence as an author seemed confirmed later in the year, when Paolo Gualdo told Galileo that the Jesuit would soon be publishing another treatise on the Sun, the *Sol Ellipticus*. Galileo lost no time in urging his informant, as well as Cesi, to send the work as soon as either acquired it. The latter evidently greeted the prospect of a new publication as no more than an occasion for mockery: "Your Lordship can well imagine how I long to see what sort of foundation his solar ellipses or contractions might have."[39]

By mid-February 1615, Galileo had also requested Scheiner's work from Gianfrancesco Sagredo in Venice.[40] As mentioned above, Scheiner himself sent a copy of the treatise, as well as the thesis of a student, Johann Georg Locher, to Galileo in April.[41] While the *Sol Ellipticus* mentioned sunspots with some

36. On Morandi's traffic in books, see *OGG*, 11: 369, and especially Brendan Dooley, *Morandi's Last Prophecy* (Princeton, NJ: Princeton University Press, 2002). For the letters of Sizzi and Morandi, see *OGG*, 11: 491–493, 530, as well as Stillman Drake, "A Kind Word for Sizzi," *Isis* 49.2 (1958): 155–165.

37. François Aguilon, *Opticorum Libri Sex* (Antwerp, 1613), 421. In the final sentence of the preface to the reader, Aguilon compared his own position as an author to that of Apelles behind the canvas, a rhetorical flourish that might explain Gianfrancesco Sagredo's inference that Aguilon, rather than Scheiner, was the author of *Tres Epistolae* and *Accuratior Disquisitio*. See *Opticorum Libri Sex*, fol. **4 v and *OGG*, 12: 51.

38. *OGG*, 12: 29.

39. *OGG*, 12: 112, 115, 118, 122, 136.

40. *OGG*, 12: 158.

41. *OGG*, 12: 137–138, 170–171.

frequency in its discussion of a novel aspect of the Sun's appearance—the atmospheric causes of its oblong shape as it neared the horizon in the evening—it did not allude to any long-term variation in the spots' apparent path. Indeed, Scheiner's suggestion that projections made in the evening, unlike those in the morning or near midday, be plotted onto a pre-sketched ellipse rather than onto a circle, while in direct contradiction to the technique outlined by Galileo in the second of his letters to Marc Welser, would perhaps have been one more impediment in the detection of an annual pattern in the sunspots' path.[42]

Scheiner does not figure again in Galileo's writings until 1619, when Mario Guiducci's *Discourse on the Comets* made an unfavorable allusion to a "masked Apelles" who had claimed priority in the matter of the sunspots but who had discredited himself with "badly colored and poorly drawn images."[43] Because Guiducci's work was (rightly) assumed to be Galileo's, and because the *Discourse* was dedicated to Archduke Leopold of Austria, Scheiner's patron, the insult was of some significance.[44] Leopold's physician and mathematician, Johannes Remus Quietanus, mentioned it to Kepler, speculated about the possibility that "Apelles" would produce better images of the comet than those of the sunspots, and reported to Galileo that the Jesuit, upon reading the *Discourse*, had vowed "to repay [him] in the same coin."[45]

That the two astronomers were very distrustful of each other at this point is clear: Guiducci's *Letter to Tarquinio Galluzzi* (1620) and Galileo's *Assayer* (1623) derided a scholar who had claimed the discovery of the sunspots for himself, and while the allusion probably concerned the Frenchman Jean Tarde, Guiducci's deployment of the name "Apelles" suggested to Scheiner, and perhaps to many other readers as well, that he was the real target of such accusations. For his part, Scheiner alleged that the *Assayer* itself contained appropriations of his work on his *Sol Ellipticus*.[46]

Galileo and his associates were certainly aware of Scheiner's presence in

42. Christoph Scheiner S.J., *Sol Ellipticus* (Augsburg: Christopher Mangius, 1615), 28–29. The *Sol Ellipticus* received a favorable assessment in a letter read in the Royal Society in the 1670s; see Thomas Birch, *The History of the Royal Society of London*, 4 vols. (London: A. Millar, 1760), 3: 183–184. See also Daxecker and Subaric, *Christoph Scheiners "Sol Ellipticus."*

43. *OGG*, 6: 48.

44. On Scheiner's relationship with this patron, see Franz Daxecker, *The Physicist and Astronomer Christopher Scheiner: Biography, Letters, Works* (Innsbruck: Leopold Franzens University of Innsbruck, 2004), 15–19, 60–71. For Scheiner's letters to his patron, see Franz Daxecker, ed., *Briefe des Naturwissenschaftlers Christoph Scheiner S.J. an Erzherzog Leopold V von Österreich-Tirol, 1620–1632* (Innsbruck: Publikationsstelle der Universität Innsbruck, 1995).

45. *OGG*, 12: 469, 484, 489.

46. *OGG*, 6: 188, 214; Drake, "A Kind Word for Sizzi," 163–165; and Daxecker, *The Physicist and Astronomer*, 128.

Rome in these years, and they commented occasionally on both his forthcoming work on the sunspots and on his relationship with the powerful Archduke Leopold and with Cardinal Francesco Barberini, nephew to Pope Urban VIII. Guiducci, for example, reported in 1625 that Scheiner had been seen in the cardinal's *anticamera*, that the cardinal was at once greatly attached to Galileo and much influenced by "Apelles," and that the excellent telescope that the archduke and the Jesuit astronomer used had probably once belonged to Galileo himself.[47] In early 1626 Francesco Stelluti related that Scheiner was printing his sunspot observations, and that he had asked if it was true that Galileo was engaged in publishing a treatise called "On the Tides."[48] Scheiner appeared tolerably well-informed, for this was in fact the subject of the eventual fourth day of the *Dialogue*, the original title of the work, and a question that Galileo had been investigating for its evidence of a Copernican world system about a year earlier.[49] The Jesuit astronomer evidently added that he was eager to see such a work, and that he concurred with Galileo's opinion about the world system.

It is certain that the exchanges in 1625–1626 between Scheiner and Galileo's friends in Rome were guarded and less than candid, as if both sides correctly sensed that the much anticipated works of the two rivals would involve open conflict. Over the next few years Galileo's friends urged him repeatedly to finish his *Dialogue*, and in early 1629 Castelli, writing from Rome, told him "soon we will have a big new book on sunspots from the masked Apelles. We shall see."[50] A month later Castelli promised to send a copy to Florence as soon as it became available, and as if to inspire Galileo to devote himself particularly to the issue of solar phenomena, also reported on the timely return of a vast sunspot that had passed from view fifteen days earlier.[51] Galileo, for his part, insisted upon his low expectations of his rival's work, telling another friend that spring that he was certain that wherever the *Rosa Ursina* diverged from what had earlier been established in the *History and Demonstrations*, Scheiner would simply be offering "nonsense and lies."[52]

The enormous work emerged a year later, in the spring of 1630 (fig. 13.4); Juan de Alvarado S.J. of the Collegio Romano noted on 28 May 1630 that

47. *OGG*, 13: 249, 255, 266.

48. *OGG*, 13: 300.

49. On the change of the title see *OGG*, 19: 327; on Orazio Grassi S.J.'s discussion of the implication of the tides with Guiducci, see Stillman Drake, *Galileo at Work* (Chicago: University of Chicago Press, 1978), 292.

50. *OGG*, 14: 19.

51. *OGG*, 14: 22.

52. *OGG*, 14: 36.

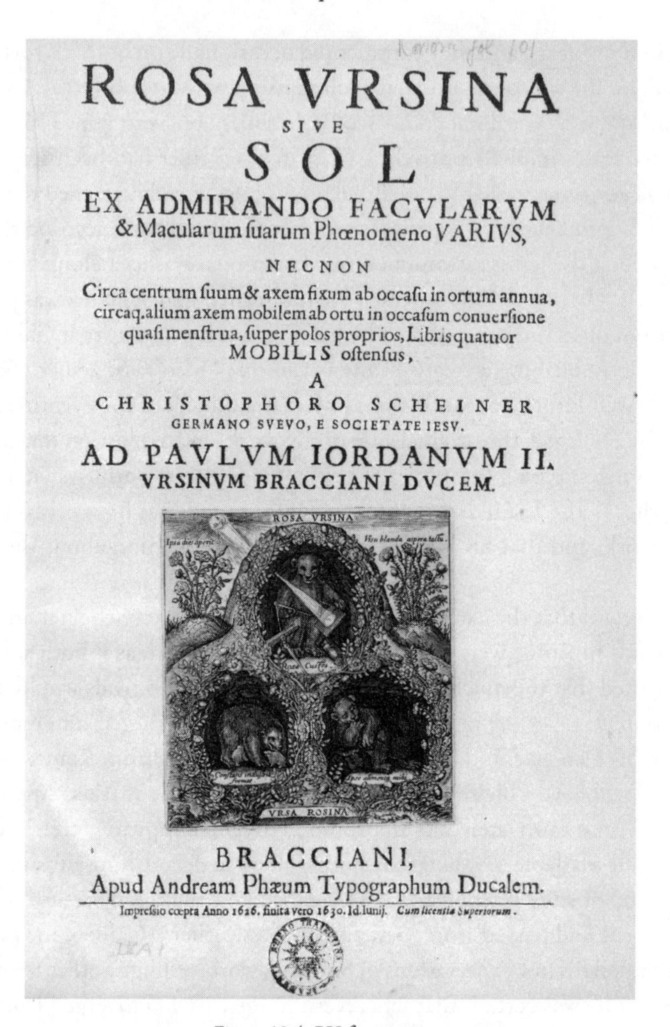

ROSA VRSINA
SIVE
SOL
EX ADMIRANDO FACVLARVM
& Macularum ſuarum Phœnomeno VARIVS,
NECNON
Circa centrum ſuum & axem fixum ab occaſu in ortum annua,
circaq. alium axem mobilem ab ortu in occaſum conuerſione
quaſi menſtrua, ſuper polos proprios, Libris quatuor
MOBILIS oſtenſus,
A
CHRISTOPHORO SCHEINER
GERMANO SVEVO, E SOCIETATE IESV.
AD PAVLVM IORDANVM II.
VRSINVM BRACCIANI DVCEM.

BRACCIANI,
Apud Andream Phæum Typographum Ducalem.
Impreſsio cœpta Anno 1626. finita vero 1630. Id. Iunij. Cum licentia Superiorum.

Figure 13.4. *RU*, frontispiece.

the *Rosa Ursina* had been licensed by Father Niccolò Riccardi, master of the Holy Palace.[53] Galileo was by then in Rome seeking permission for his recently completed *Dialogue*. While a rumor that Galileo's manuscript "attacked many of the opinions upheld by the Jesuits" was then circulating in Rome,[54] the astronomer and his friends were confident by the time he departed for Florence in late June that only minor adjustments needed to be made to his manuscript. The imprimatur was granted in mid-September 1630.

53. – *RU*, fol. IV v.
54. *OGG*, 14: 103.

Though Galileo heard in mid-April 1631 that Scheiner referred to his letters on sunspots with great frequency and hostility in the *Rosa Ursina*, he claimed not to have seen the treatise until the fall or winter of that year, when he expressed his displeasure to Paolo Giordano Orsini, duke of Bracciano, who now regretted, both for fiscal and for personal reasons, having agreed to finance the expensive publication.[55] The *Rosa Ursina*, for all its wealth of solar observations, was so openly rancorous that it would later be described to Galileo as a "big fat volume against Your Lordship, but one that could be divided into two basic headings: insults, and the assertion that Your Lordship's discoveries belong to others."[56] But if it seemed to some that Galileo was the target of the *Rosa Ursina*, Scheiner felt he was likewise the victim of the *Dialogue* when it emerged in the spring of 1632, becoming apoplectic, as the story goes, upon hearing the treatise praised in a Roman bookshop, and seeking to acquire it at all costs so as to begin work on a response.[57]

Scheiner had very accurately determined the solar axis of rotation. Accounting for this phenomenon by means of a geocentric construction was a straightforward astronomical exercise of the kind technical astronomers had done at least since Ptolemy. To the standard solar description of the Sun's diurnal and annual motions, Scheiner added a construction to make the Sun rotate in about a month on an axis that pointed to a place in the fixed stars 7° 15′ removed from the pole of the ecliptic. To keep this axis always pointing to the same spot (i.e., to keep it parallel to itself), Scheiner added a conical construction, as shown in fig. 13.5. (Copernicus had added a similar conical motion to the *Earth's* axis in order to keep it parallel to itself.) All of Scheiner's aims had thus been achieved: he had shown Galileo to be wrong about the motion of sunspots, had demonstrated the correct movement, and had supplied a mathematical model to account for that motion.

One can imagine Scheiner's reaction to Galileo's *Dialogue*. In the First Day, a discussion of sunspots formed part of Galileo's arguments against the perfection of the heavens, along with the location of the new stars of 1572 and 1604 and the superlunary location of comets. Salviati asked, "But you, Simplicio, what have you thought of to reply to the objection based on these annoying spots which have come to mess up the heavens and even more so the Peripatetic philosophy?"[58] Simplicio answered with Scheiner's original argument, that

55. *OGG*, 14: 255, 294–295, 322.

56. *OGG*, 16: 206.

57. *OGG*, 14: 360, 367.

58. *OGG*, 7: 77. We have taken the translation from Maurice A. Finocchiaro, *Galileo on the World Systems*, 100.

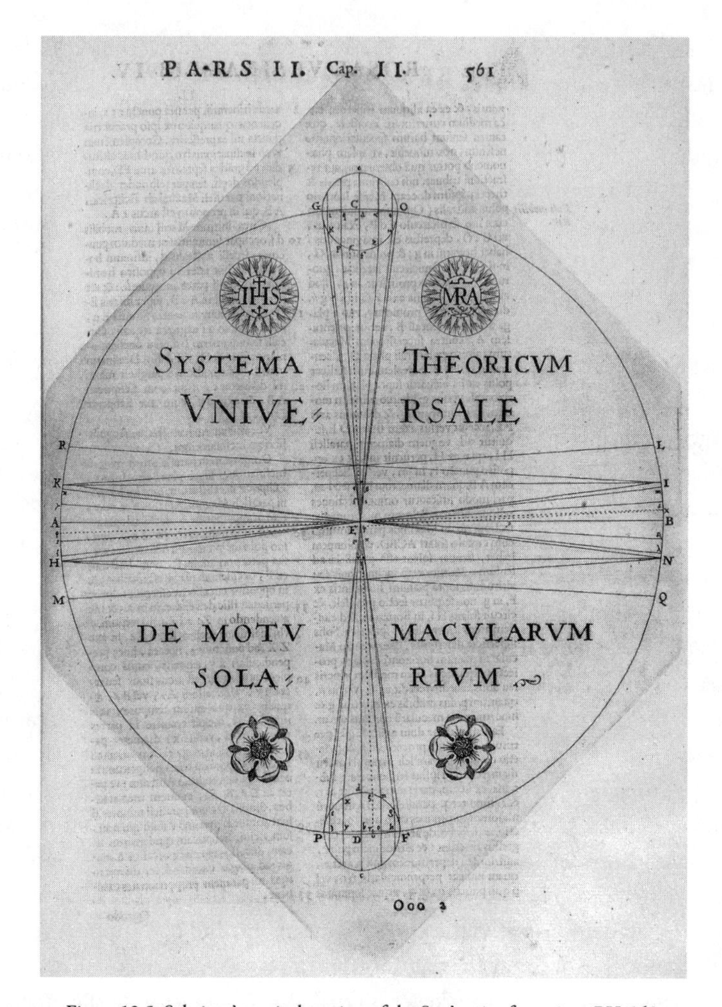

Figure 13.5. Scheiner's conical motion of the Sun's axis of rotation. *RU*, 565.

sunspots were dark bodies orbiting the Sun, and he concluded that "this seems to me to be the most convenient escape found so far to account for such a phenomenon and at the same time retain the indestructibility and ingenerability of the heavens; and, if this were not sufficient, there will be no lack of loftier intellects who will find better explanations."[59] And to compound the sarcasm, Galileo had Simplicio add, in reply to Salviati's argument that the spots were on the Sun's surface or very near it, "Frankly, I have not made the long and

59. *OGG*, 7: 78; Finocchiaro, *Galileo on the World Systems*, 101.

diligent observations needed to properly master the facts of this matter,"[60] as if echoing Cardinal Conti's words of 1612 (see appendix 3). It was obvious enough that this was Scheiner's argument; it was not necessary to mention his name.

In the Third Day of his *Dialogue*, in the course of arguments for the motion of the Earth and the Copernican arrangement of the planets, Galileo introduced a new argument: "But another effect, no less wonderful than [the explanation of retrograde motion in the Copernican system], and containing a knot perhaps even more difficult to untie, forces the human intellect to admit the annual rotation and to grant it to our terrestrial globe."[61] Galileo recounted his discovery of sunspots, his initial supposition that the Sun turned on an axis of rotation perpendicular to the ecliptic, and his eventual, but still timely, conclusion to the contrary. As Salviati explained it,

> Now several years later, being with me at my Villa delle Selve and being enticed by a particularly clear and protracted serenity of the heavens, he [i.e., Our Academician, Galileo] happened to find one of those solitary sunspots which are very large and thick, and at my request he made observations of its entire journey, carefully noting down its places from day to day when the Sun was on the meridian. We perceived that its passage was not exactly in a straight line, but a somewhat bent one; and it occurred to us to make other observations from time to time. We were strongly encouraged to do this by an idea which suddenly struck the mind of my guest and which he imparted to me in the following words:
>
> "Filippo, it seems to me that the road is open for us into a matter of great consequence. For if the axis around which the Sun revolves is not perpendicular to the plane of the ecliptic, but is somewhat inclined to this—as the curved path just observed suggests to me—then we shall have a more solid and convincing theory of the Sun and Earth than has ever yet been offered by anybody."[62]

Certain biographical details complicate this account. Filippo Salviati had left Florence and Tuscany late in 1613, and died in Barcelona in March 1614.[63]

60. *OGG*, 7: 80; Finocchiaro, *Galileo on the World Systems*, 101.

61. Galileo, *Dialogue Concerning the Two Chief World Systems*, 2nd rev. ed., trans., with revised notes, by Stillman Drake; foreword by Albert Einstein (Berkeley: University of California Press, 1967), 345.

62. Galileo, *Dialogue*, 346–347.

63. On Salviati, see Mario Biagioli, "Filippo Salviati: A Baroque Virtuoso," *Nuncius* 7, no. 2 (1992): 81–96.

These observations must therefore have been made before that time, and there is no evidence in Galileo's papers to support his claim. This and the subsequent explanation of the phenomena, recited as if by memory by the fictional Salviati, are the only two passages in the *Dialogue* that purport to be direct quotations of Galileo. Upon receiving his copy of the work late in May 1632, Benedetto Castelli wrote to Galileo, "When I got to that false attestation of the sunspots, I was beside myself with happiness in seeing how much light these dark marks shed on the matter; sunlight itself could not be brighter."[64] If Galileo had made this discovery in 1613 or even earlier, Castelli, who was working very closely with him and had developed their method of projecting sunspots, would surely have known about it and would have expressed himself differently about this newly introduced argument.

There is thus reason to doubt Salviati's "testimony" from beyond the grave. The explanation of the phenomenon itself, presented by Salviati as a direct quotation dating to the discovery made at Villa delle Selve, also merits further scrutiny:

> If the annual motion belongs to the earth—along the ecliptic and around the sun—and if the sun is situated in the center of this ecliptic, and if it turns upon itself not around the axis of the ecliptic (which would be the axis of the earth's annual motion) but around a tilted axis, then extraordinary changes would have to be seen by us in the apparent movements of the solar spots, provided we assume that the axis of the sun remains perpetually and unchangingly at the same tilt with the same orientation toward the same point in the universe.[65]

Galileo went on to argue that the assumption that the Sun rotated on such an inclined axis would produce the observed motions of the sunspots, and that the simplicity of this scenario, as opposed to the extremely complex motions one would have to ascribe to the Sun in a geostatic system, constituted proof that the Earth moved around the central Sun. He was somewhat vague on details, not specifying how great the inclination of the Sun's axis of rotation was or when during the year the spots appeared to pass over the Sun in a straight line or along curved paths.

Overall, the argument is imprecise, and has thus given rise to much specu-

64. *OGG*, 14: 358. "Ma quando gionsi a quel testimonio falso delle macchie del sole, hebbi a uscire di me stesso d'allegrezza, considerando quanta chiarezza davano in questa materia tali oscurità, che maggiore non ne può dare l'istessa luce del sole."

65. Galileo, *Dialogue*, 347.

lation about Galileo's acquaintance with some information about the results presented by Scheiner in *Rosa Ursina*. It is impossible, at this stage, to come to a firm conclusion. The facts are that Galileo himself never mentioned the inclination of the Sun's axis of rotation before the *Dialogue*, and that the passage in question was not written before 1629. But in view of Snellius' treatise of 1612, Sizzi's letter of 1613, and the analogy provided by the curved paths of Jupiter's satellites, it entirely possible, if not probable, that Galileo was familiar with the solar phenomenon, but long unaware of its potential as evidence for the motion of the Earth.[66] It has, however, also been argued on the basis of the same textual evidence that Galileo's knowledge of the annual paths of the sunspots derives wholly from the *Rosa Ursina*, and that his focus on this issue in Day Three of the *Dialogue* reflects changes made to his text after the imprimatur had been granted. His handling of the alternative hypothesis proposed by the Aristotelian Simplicio—that the appearance of the sunspots was compatible with an immobile Earth—would be, in this reading, a direct reply to Scheiner's *Rosa Ursina*, for here the point was to demonstrate the number of extra and unlikely movements that would have to be assigned to the Sun in such an arrangement.[67]

There are crucial differences between the two astronomers' deployment of these solar phenomena, and the relationship of the two texts has been and will surely continue to be the subject of controversy. Scheiner's knowledge of the variable appearance of the sunspots' paths was the product of painstaking observation, and it is an accurate depiction of the times when the sunspots will appear to have a rectilinear motion, and when they will seem instead to describe curves of slight convexity or concavity. Galileo's notion, by contrast, has none of the observational detail provided in Scheiner's account, and the ambiguously worded text itself appears to some readers to exchange the periods of rectilinear motion with those of maximum curvature.[68] As part of a larger presentation of evidence for a particular cosmological arrangement, however, Galileo's analysis of the apparent annual paths of the sunspots was by far more

66. The argument is described by Stillman Drake in "Sunspots, Sizzi, and Scheiner." In *Galileo at Work*, Drake states (335), "The fact that in order to know the annual dates of maximum and minimum curvature of the paths, observations for every day around those times are necessary, and since sunspots are not always to be seen this would normally mean the accumulation of observations from several different years. Galileo had had no time for that if, as I believe, the argument first occurred to him in 1629. A few observations in that year, combined with those he had recorded in 1612, would have sufficed to assure him that the hints of Sizzi's old letter were reliable, but not to supply him with the specific information in question, which is in fact lacking from his account."

67. See the commentary of Besomi and Helbing on Galilei, *Dialogo*, 2: 720–760.

68. Drake, "A Kind Word," 162.

convincing, at least within the confines of this section of the *Dialogue*, for the relative simplicity of his explanation, based on the assumption of a mobile Earth, contrasted sharply with the four different motions around four different axes that the Aristotelian Simplicio needed to ascribe to the Sun.[69]

In any case, the rumors that Galileo's difficulties in Rome were orchestrated by the Jesuits in general, and by Scheiner in particular, were not long in coming.[70] While it is certain that Galileo counted many sympathizers among the Jesuits, it has recently been established that Scheiner exercised some degree of influence on the Special Commission whose task it was to examine the *Dialogue* in 1632–1633.[71] The report of Father Melchior Inchofer S.J., appointed to the commission by Niccolò Riccardi, the longest and most critical of the three texts prepared for this investigation, asserted that with this publication Galileo did, in fact, violate the strictures placed on him in 1616 by "teaching, defending, and holding" Copernican views. Inchofer noted that Galileo used phenomena such as the sunspots, the tides, and the Earth's magnetic poles to teach the Copernican doctrine "ad nauseam," and he argued that the astronomer's "principal goal was to attack Father Christoph Scheiner, who had most recently written against Copernicanism."[72] Inchofer was also composing a treatise on the conflict of heliocentrism with Scripture during the period of the trial, and this work, emerging in the fall of 1633, both insists upon Scheiner's primacy in the discovery of the sunspots, and notes that the variation in their apparent paths, used by Galileo as an argument for a rotating Earth, had already been attributed to other causes by his rival.[73]

Even those rare scholars on good terms with both Galileo and Scheiner were somewhat taken aback at the latter's uncompromising hostility after the

69. Galileo's argument for the mobility of the Earth from sunspots has recently been the subject of discussion. See A. Mark Smith, "Galileo's Proof for the Earth's Motion from the Movement of Sunspots," *Isis* 76 (1985): 543–551; Keith Hutchison, "Sunspots, Galileo, and the Orbit of the Earth," *Isis* 81 (1990): 68–74; David Topper, "Galileo, Sunspots, and the Motions of the Earth," *Isis* 90 (1999): 757–767; Paul R. Mueller, "Unblemished Success: Galileo's Sunspot Argument in the *Dialogue*," *Journal for the History of Astronomy* 31 (2000): 279–300; David Topper, "Colluding with Galileo: On Mueller's Critique of My Analysis of Galileo's Sunspots Argument," *Journal for the History of Astronomy* 34 (2003): 75–77; Owen Gingerich, "The Galileo Sunspot Controversy: Proof and Persuasion," *Journal for the History of Astronomy* 34 (2003): 77–78.

70. *OGG*, 15: 87–88, 141; on the historiographic tradition, see Michael John Gorman, "A Matter of Faith: Christoph Scheiner, Jesuit Censorship, and the Trial of Galileo," *Perspectives on Science* 4.3 (1996): 283–320, on 286–289.

71. Richard J. Blackwell, *Behind the Scenes at Galileo's Trial* (Notre Dame, IN: University of Notre Dame Press, 2006).

72. *OGG*, 19: 350, 351; and Blackwell, *Behind the Scenes*, 39–40.

73. Blackwell, *Behind the Scenes*, 134, 157; on the possibility of a coordinated approach by Scheiner and Inchofer, see 89–90.

abjuration.[74] Among the few amusing vignettes from this grim period is that offered to Galileo by Father Raffaello Maggiotti, a close acquaintance of Castelli and Torricelli:

> in the Collegio [Romano], there's a German father who is putting together a giant volume against the *Dialogue*, or rather a big ark to enclose all terrestrial and celestial beings, and perhaps the aquatic ones as well. The German is putting his name on it and the bear is doing the work . . .[75] I expect it will have great Patristic authors, lots of Scripture, long stories and images, whole slabs of the *Dialogue* translated into Latin, and big doings: God help us! This fellow is lying very low, and I for one believe that he has dragged a barrel into his room so as not to waste time even in going to the wine cellar.[76]

Latent in Maggiotti's somewhat ribald depiction of Scheiner's forthcoming work, the *Prodromus pro Sole Mobili*, is perhaps an impression of the latter part of the *Rosa Ursina*, which upheld the rather controversial doctrine of fluid heavens, where planets were said to move like "birds in the air, and fish in the sea," and which involved a barrage of Patristic sources and Scriptural citation.[77] But the *Prodromus* was devoted in the main to the disputed priority claims regarding the sunspots, and the evidence that their annual paths, in particular, provided for the hypothesis of a mobile Earth.[78] Though Scheiner was unable to publish the work during his lifetime, Galileo's associates continued to regard him as powerful enough to jeopardize the printing of the *Two New Sciences*.[79] And the issue of sunspots appeared so volatile for some of the Roman clergy—memorably described by Father Magiotti as "sacks of coal"—that the Jesuit astronomer's research, rather than the *History and Demonstrations* or the condemned *Dialogue*, needed to be invoked in their presence, and the note "all these things are treated *ad satietam* in the *Rosa Ursina*" added to discussions of the solar phenomena.[80]

In conclusion, we might add to this Galileo's own words about Scheiner and

74. Daxecker, *The Physicist and Astronomer*, 43–44.

75. The ursine motif of the title page of the *Rosa Ursina*, a form of homage to the duke of Orsini, was much derided. Scheiner had originally invoked the motif of the mother bear who licks her cubs into shape, a simile for Virgil's method of revising the *Georgics* and still present on the title page of the *Rosa Ursina*, in the proem of the *Accuratior Disquisitio*.

76. *OGG*, 15: 300.

77. See the summary provided in Daxecker, *The Physicist and Astronomer*, 141–145, and Ingaliso, *Filosofia e cosmologia*, 225–245.

78. Blackwell, *Behind the Scenes*, 88, 219–222.

79. Daxecker, *The Physicist and Astronomer*, 33; and *OGG*, 16: 189, 301, and 17: 130–131.

80. *OGG*, 16: 425.

Rosa Ursina. To Fulgenzio Micanzio, theologian and consultor of the Venetian senate, he wrote in 1636:

> I marvel at your impassivity in reading *Rosa Ursina*, which contains so much childishness. But you will say to me that it is rather their overwhelming presence that causes no small delight. And who will not be astonished in considering the wit of the device of the three bears in the three caves [see fig. 13.4], one of whom projects the sunspots through a telescope, the other licks her cubs, and the third sucks its paws? And alongside, two mottos that are so meaningful and so cleverly positioned: *Rosa Ursina / Ursa Rosina*. But why begin to catalog the absurdities of this brute if they are without number? This pig, this malicious ass makes a catalog of my errors, which derive as a consequence from one single one, equally unnoticed at the beginning by him and by me, which was the very slight inclination of the axis of rotation of the solar body to the plane of the ecliptic. I am convinced that I discovered it before him, but I did not have an occasion to speak of it except for the *Dialogo*. But then let the poor devil realize his bad fortune, for he derives nothing wonderful from this observation [of the inclined axis], and I have discovered the greatest secret of nature with it. And this great secret, which I discovered, and the extreme marvel of which he fathomed after my announcement, is what has mortally wounded him, and turned him against me like a rabid dog. For it was my destiny alone to observe so many and such great new things in the heavens, and to deduce from them such numerous and stupendous consequences of nature, of which this one, it can be said, is the greatest. And this wretched man, who for so long held such a precious jewel in his hands, was not able to recognize it. I have said enough.[81]

81. Galileo to Fulgenzio Micanzio, 9 February 1636, *OGG*, 16: 391.

Appendix 1

Scheiner and Galileo on the nature of the Moon

Welser to Galileo, 7 January 1611, *OGG* 11: 13–14

. . .

I sent the answer from Your Lordship to Signor Brengger right away, and we should soon hear if he is satisfied, as is a certain other friend of mine[1] to whom I showed it. This friend, however, maintains another fancy, one that would appear very plausible to me if it were confirmed by your calculations and those of your peers. He says: *"From what has been reported so far, for my part I judge that it is not yet certain that any of those mountains project beyond the outer surface of the Moon because up to now that outer surface of the Moon has been imagined [to extend] from the summits of the mountains rather than up from the lower parts. That there are chasms within the Moon is the sole point that is agreed upon; this in no way means that lunar mountains rise above the greatest circles.*[2] *And these phenomena have yet to turn the philosophers away from their general opinion, which is that the Moon is a perfect sphere. For they say that those unequal asperities are*

1. Antonio Favaro doubted that the "other friend" was Christoph Scheiner because neither in his *Tres Epistolae* nor in his *Accuratior Disquisitio* did Scheiner deal with the nature of the lunar surface. *OGG*, 11: 13n. But it was indeed Scheiner; see Lanz to Guldin, 14 May 1611, Graz, Universitätsbibliothek, MSS 159, part 17, no. 6. See pp. 43–44, above. The italics indicate the use of Latin.

2. Scheiner means here the imagined spherical surface that contains the highest lunar mountains.

within it, just like stones of various colors amassed in various shapes, and so forth, inside a glass or crystal globe.[3] *This opinion has thus far not yet been overthrown by these phenomena."* But perhaps Your Lordship's instrument would rescue us from these doubts in the twinkling of an eye, and I can tell you that the method of making telescopes is much desired in this region, and as you have publicly stated your intention of divulging its theoretical basis, it is assumed that you would feel obliged to do so.

Galileo to Welser, February 1611, *OGG* 11: 38–41

. . . .

I am eagerly awaiting the reply of Signor Brengger, especially because Your Lordship's report of your other friend's satisfaction has given me hope that [Brengger] will be content with what I replied, even though this other friend may not think what I have written about the inequality of the lunar surface is a necessary conclusion, for which he adduces some reasons that I do not entirely understand. Your Lordship must therefore forgive me if in my reply I do not entirely satisfy him. And where in the beginning he says "From what has been reported so far," and so forth, *he appears to consider many surfaces in the body of the Moon, for he appears in particular to introduce the contradistinction of a maximum to other, non-maximum surfaces, just as in spheres some circles are called greatest in distinction to smaller ones described on the same sphere.*[4] *But in my view it is a new and unheard of thing to consider many surfaces in a solid; thus, just as the terrestrial body is terminated by one surface that is rough rather than precisely spherical, so I say that the surface of the Moon is not precisely spherical but rough, and crowded with chasms and prominences. And therefore it seems that the statement that follows is inaccurate as well:* "because that outer surface of the Moon," and so forth, *for the surface of the Moon is taken at the same time from the tops of mountains, from the mountains themselves, from the lower parts, and altogether from all extreme and apparent parts.*

"The sole point that is agreed upon," and so forth comes next. *I overlook what is inaccurately stated,* "that lunar mountains rise above the greatest circles," *for all peaks are referred to the greatest circles, because they are measured by*

3. This is the explanation of the medieval Aristotelians, echoed by Christoph Clavius, that the naked-eye spots on the Moon were caused by rarer and denser parts of the Moon's interior. See Roger Ariew, "Galileo's Lunar Observations in the Context of Medieval Lunar Theory," *Studies in the History and Philosophy of Science* 15 (1984): 213–226.

4. For instance, the Equator is a great circle, while circles of greater latitudes are referred to as small circles.

perpendicular lines between which the surfaces of the greatest circles alone extend. I see that based on the phenomena reported by me, the author only allows chasms and gulfs to penetrate to the inside of the greatest circle, but does not permit mountains to rise from it, which, however, is an error. For if the surface of the Moon, in other respects smooth and polished, abounded only here and there with chasms,

certainly at the terminator only some dark parts would curve into the luminous part, as in the adjoining figure, and no bright peaks, entirely separated from the illuminated region, would shine among the darkness of the remaining part; experience teaches us otherwise. Furthermore, almost all the large and ancient spots, namely those that are seen with the unaided eye, are surrounded by very high ridges. From this it is established that when the edge of illumination passes over these spots, above and below them some promontories, so to speak, jut out, rising above the dark part and swelling in a long chain, as the second drawing shows.

This can on no account take place in a smooth surface pitted with only a few cavities. Moreover, if those numerous deep black spots visible beyond the terminator inside the illuminated part were only chasms extending below the lunar surface and not surrounded by ridges of mountains, their openings would certainly project no shadow from the side facing the Sun, but experience teaches the contrary. For as is seen in the adjoined figure, when the illumination comes from A the rim c shines

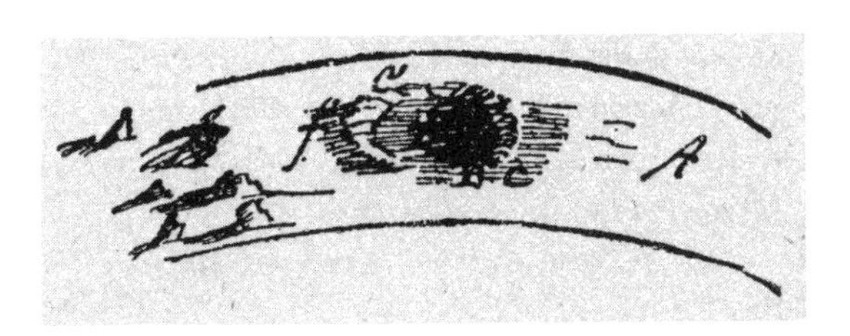

brightly, because just like a ridge of mountains it receives more light directly. Next is the very dark pool D, rendered shadowy by the wall of mountains c and by its own considerable depth. After this there is another ridge e, very bright, followed by a shadow f, which shadow, together with the brighter part c, would indeed not be there if the chasm D were a simple chasm extending beneath the surface. But you may see six hundred of such shapes on the Moon. Also, many cliffs appear such as those depicted near B; the side exposed to the Sun looks very bright, while the opposite one is dark, and their shadows stretch visibly into the plain.

He concludes, finally, that the phenomena described by me will not yet put to flight those philosophers who will maintain that my dark spots are instead chasms inside the solidity of the Moon, like numerous little stones of diverse colors in a crystalline globe, while in the meantime the outer surface of the Moon is transparent and smooth. Here I first advise him that those black spots of mine are in truth nothing other than shadows since they are increased, diminished, and entirely destroyed, and shift from right to left and vice versa, depending on how the Sun's light falls on the Moon, now obliquely, now perpendicularly, now from the west, and now from the east. No reasonable cause will be assigned to these effects except the unevenness of the Moon's surface. [The notion] that, finally, those chasms are filled with some diaphanous material so transparent that it does not in the slightest impede our view and the Sun's rays, such that the Sun cannot produce shadows and we cannot see them, I leave for the philosophers themselves to demonstrate. Until that point I maintain indeed that the surface of the Moon is rough like the surface of the Earth; by the Moon, I understand that body in itself dark and opaque which, since it is able to receive and reflect sunlight, is illuminated and exposed to our sight. And because of this, it is a matter of dispute whether it is surrounded by an entire heaven of transparent and invisible ether, and I assert this body, as it is seen by us, has innumerable prominences and cavities on its surface. And if someone wants to take as the Moon not only the body that we see but also some invisible and imagined material around it, that same person may, no less rationally, make the Earth also perfectly spherical, with its valleys and depths filled with the surrounding air and an aerial and imaginary surface extended to the heights of the highest mountains, limiting the terrestrial mass as he wishes. I would say that the Earth would be entirely similar to this philosophical Moon if at the time of Noah's flood the sea had been left bound up in ice. But then even the clearest waters are not so utterly limpid and transparent that they would allow our eyes a passage to sufficient depth to distinguish the shadows of the lower ridges. They lead us therefore to conclude that the visible Moon must be covered with some diaphanous substance—glass, crystal, diamond, something a great deal more pellucid than water itself—and reason teaches us that only ether is of this kind. But as soon as this is established, what else do we infer from it except that the lunar body is in fact bounded by a rough surface but located in the ether?

But perhaps I have gone on too long and, as you say so well, the excellent instrument might conceivably remove all doubt, as has happened with the Medicean Planets; after their very existence was firmly denied for a long time by the most eminent mathematicians, they have at last been acknowledged and recognized, once they were seen by these men. But what most amazes me is that after they have agreed that they are true planets, these men don't cite a single passage that I haven't already written and published. If they were to come up with some necessary point that I had left out, I would be able to believe that they had considered me truthful but unskillful in the art of observation, but as it stands I don't see how to escape the reputation—as hateful as it is to me—that they looked upon me as a liar.

Appendix 2

Galileo to Maffeo Cardinal Barberini

2 June 1612, *OGG,* 11: 304–311

Most Illustrious and Reverend Lordship and Most Honored Patron,

Among the many favors I have received from Your Most Illustrious and Reverend Lordship, there remains fixed in my memory the one you did for me at the table of My Lord, the Most Serene Grand Duke, when you passed through here recently, for during a debate about a certain philosophical question you took my side against the Most Illustrious Lord Cardinal Gonzaga and others who held opinions contrary to mine. And because it was suitable, on the orders of His Highness, that I lay out my reasons more clearly on paper and then publish them in print, as has now been done, it seemed to me that I should send Your Most Reverend Lordship a copy, and then ask you to do me the service, at your convenience, of seeing or listening to what I have proposed in this treatise, in which I believe that you will not fail to notice that you had the support of one of your servants as well as of Truth itself.

I believe that you must have heard the rumor making the rounds about the dark spots on the body of the Sun that are constantly being noticed and observed with the telescope, and because from your region someone has written me that highly esteemed men of your city[1] ridicule such reports as paradoxes and as great absurdities, I think it a good idea to relate briefly to Your Most Illustrious Lordship the state of affairs in this matter.

It was about eighteen months ago that, looking with the telescope at the body of the Sun when it was near its setting, I saw some very dark spots on it. Returning repeatedly to the same observation, I perceived that these spots kept changing their places, and that they did not always appear in either the same shapes or formations, and that sometimes there were many of them and at other times few, and sometimes none at all. I pointed out this peculiarity to some of my friends, and also this past year in Rome, I showed them to many prelates and to other men of letters. From there, the news spread to diverse parts of Europe, and for the past four months various observational drawings have been sent to me from different places, and in particular three letters written to My Lord Marc Welser in Augsburg, printed under the pseudonym of *Apelles hiding behind the canvas*. These letters were sent to me by Welser himself, who asked for my opinion about them, and also, in this connection, what I thought could be known about the essence of these spots. I have written him a letter of six folio pages on this subject, refuting the opinion of the masked Apelles and of those who thus far have spoken about them. And finally, after many and various ideas had paraded through my imagination, I came to conclude and to maintain as indubitable that these spots are contiguous to the surface of the solar body, and that they are continuously generated and dissolve there, some being of longer, and others of shorter duration; some are more dense and obscure, and others less so. For the most part, they change their shape—which is generally very irregular—from day to day. Frequently, some of them split into two, three, or more spots, while others that were at the outset separate unite into one. And finally, because of a motion universal and common to them all, I have arrived at an indubitable certainty that the Sun rotates about itself from the West to the East, that is, along with all the other revolutions of the planets, completing an entire revolution in about one lunar month. And from what I have observed, by far the greater number of these spots is generated between those circles of the solar globe that correspond to the Tropics, and beyond these circles I have almost never observed any of these spots, which in view of their generation, disintegration, rarefaction, condensation, and their separations and changes in shapes, and all other phenomena, if I had to compare them to some of the materials familiar to us, nothing would be found that imitated them better than do our clouds.

All that I say to Your Most Illustrious and Reverend Lord is so true, and

1. Bologna.

confirmed by me by so much proof, that I do not hesitate to propose it finally as entirely certain; and what I have heard as the scorn of many does not frighten me in the least, because we are dealing with matters that can always be observed by an infinite number of people and from all parts of the world, and little by little recognized as true by those endowed with better sense. And therefore I boldly dare to be the first to spread about conclusions that seem to be so strangely paradoxical. It only displeases me that those who scoff at these matters gamble, so to speak, in safety, certain of never losing, and standing to make great gains. For if what I am affirming and they are denying were found to be false, without having made the least effort, they would have the glory of having understood matters better than others who have made many laborious observations, whereas when it becomes certain that what I am saying is true, they will be excused for not having given their assent to such a surprising issue. If Your Most Illustrious Lordship has seen the three letters of the masked Apelles, I can send you a copy of the letter I wrote to My Lord Welser on the subject. In the meantime, I am sending you some drawings of the solar spots, made with great accuracy both with regard to their number as well as to their size, their shape, and their position from day to day on the solar disk [figs. A2.1–A2.9]. If Your Most Illustrious Lordship happens to discuss this solution of mine with the learned men of your city, I would be grateful to hear something of their opinion, and in particular that of the Peripatetic philosophers, because this new development appears to be the Last Judgment of their philosophy, "for there have already been signs in the Moon, the stars, and the Sun."[2] For this reason, this [Peripatetic] doctrine itself, along with the mutability, corruption, and generation of even the most excellent substance of the heavens, shows signs of deterioration and change, but not without the hope of regenerating into something better.

I have wearied Your Most Illustrious and Reverend Lordship enough. Please excuse me with your infinite kindness, and with the same kindness preserve the place you have allowed me in your favor. I bow down humbly.

From Florence, 2 June 1612
Your Most Illustrious and Reverend Lordship's
Most Devoted and Obliged Servant
Galileo Galilei

2. This is a garbled line of Luke 21: 25.

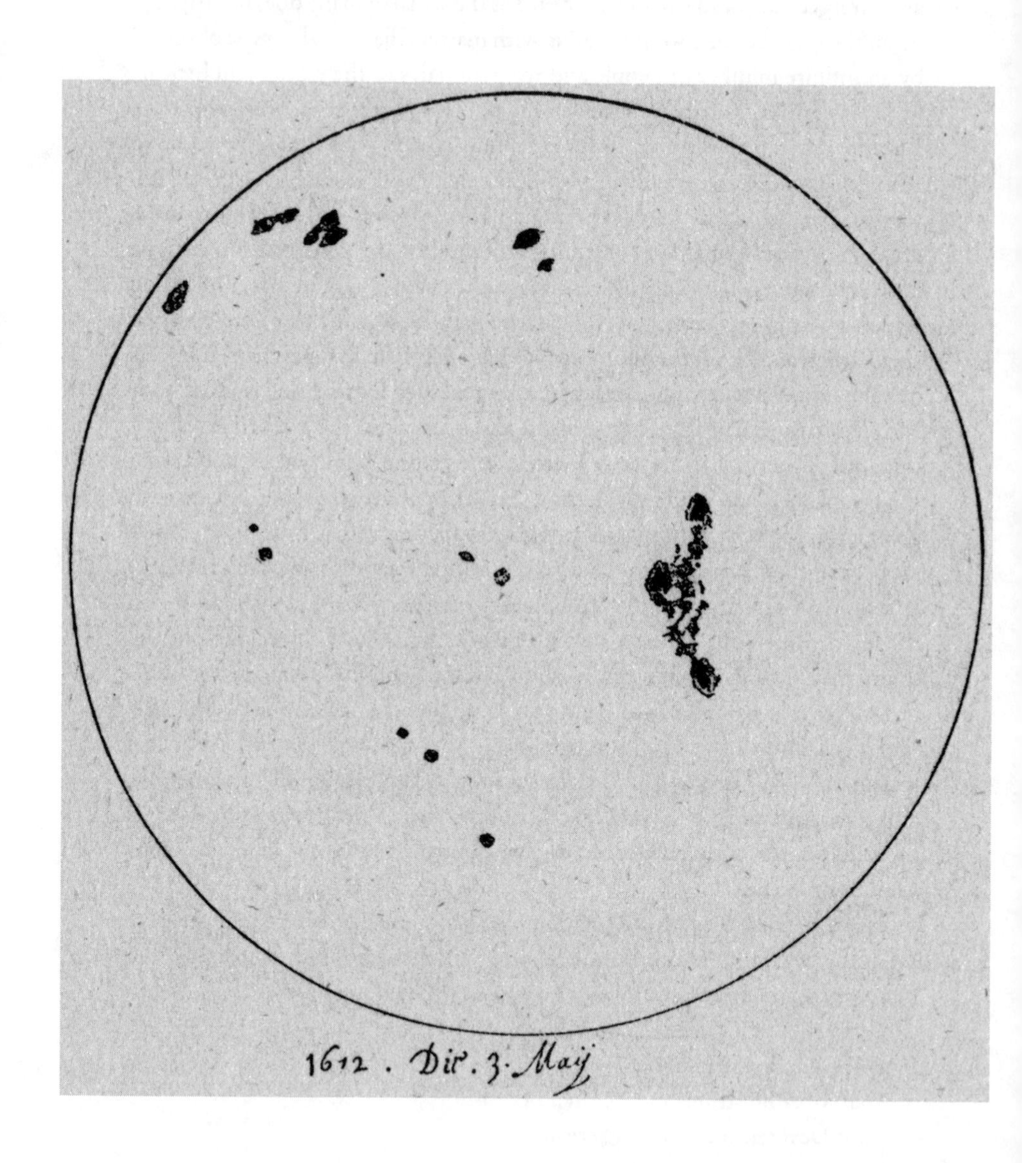
1612 . Dic. 3. Maij

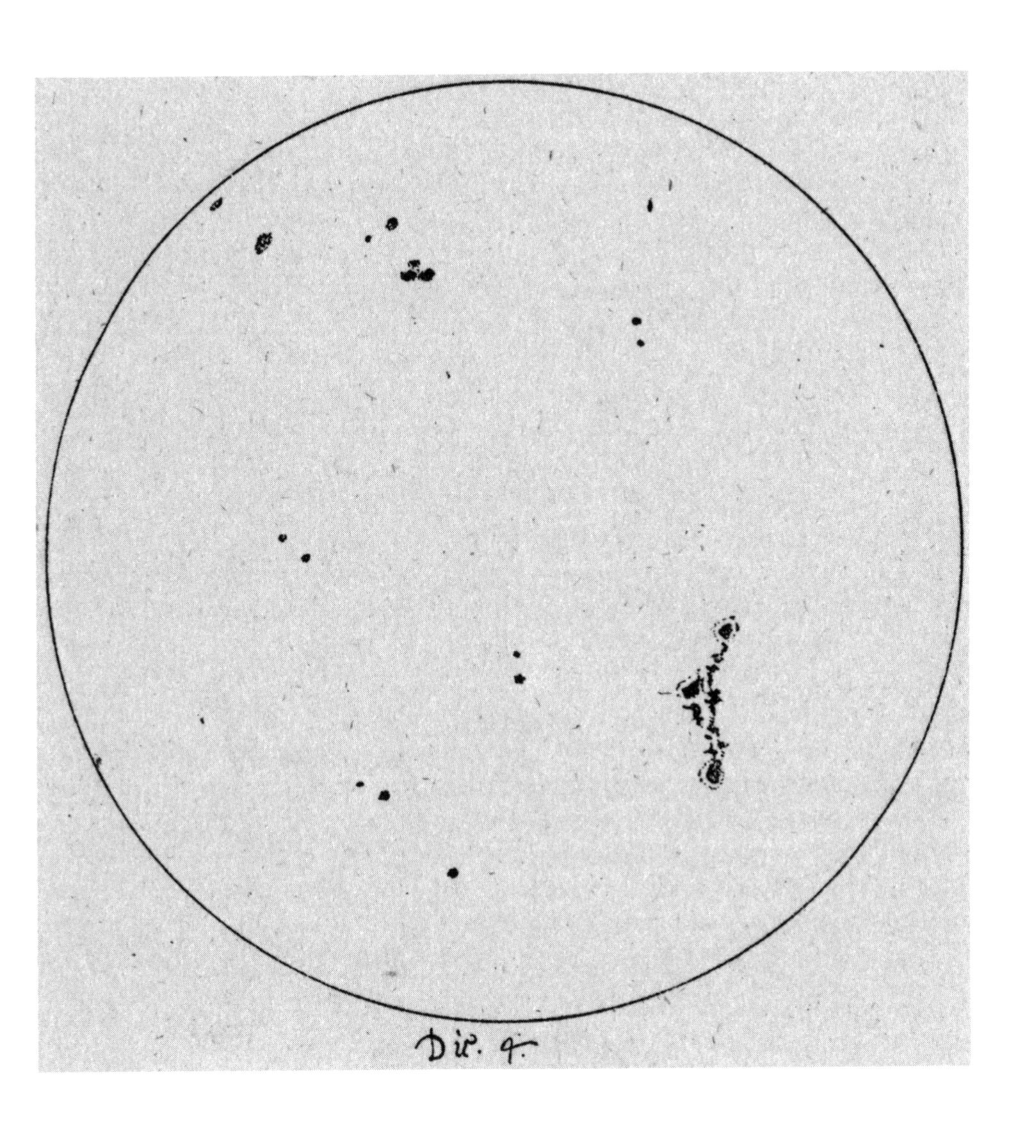
Dic. 4.

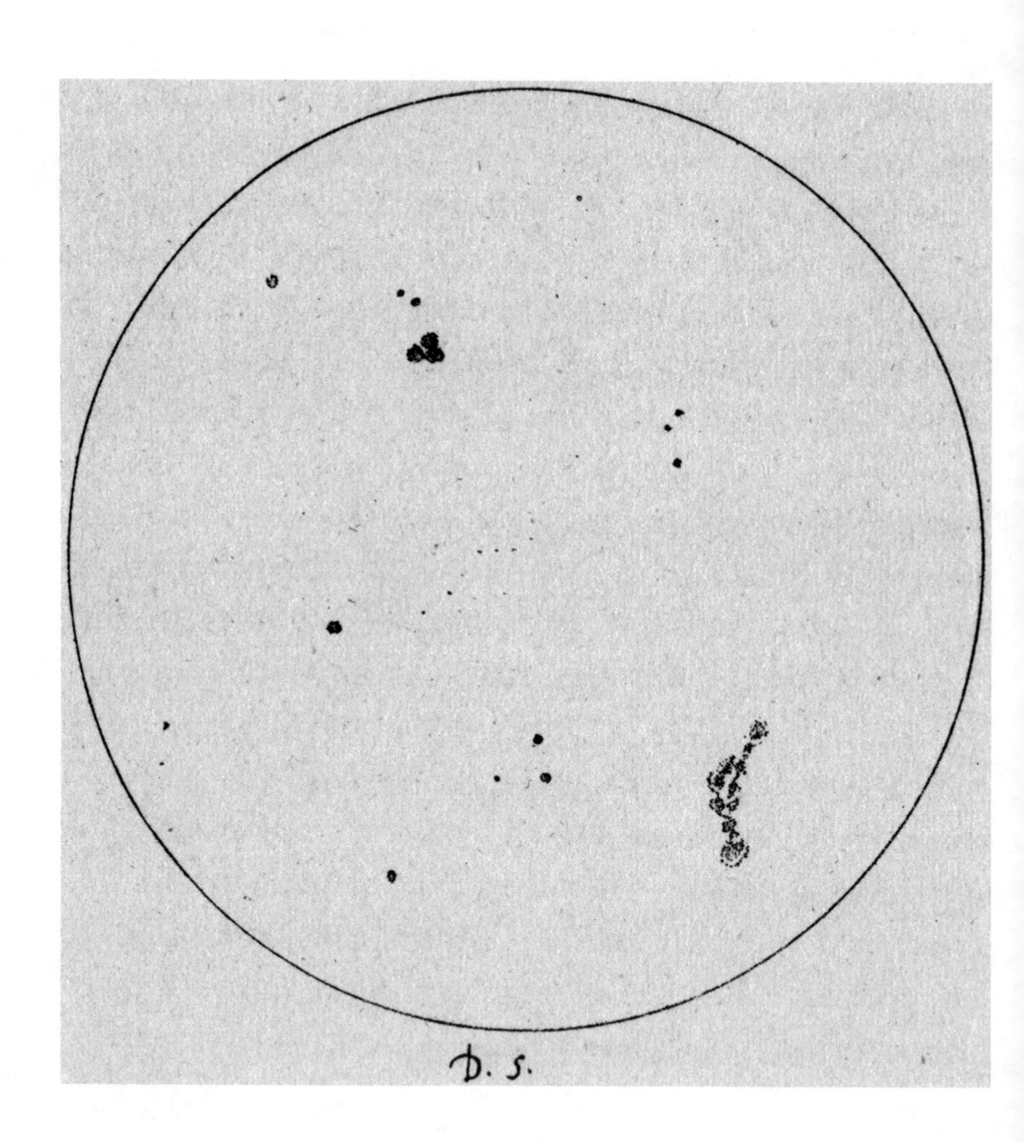
D. 5.

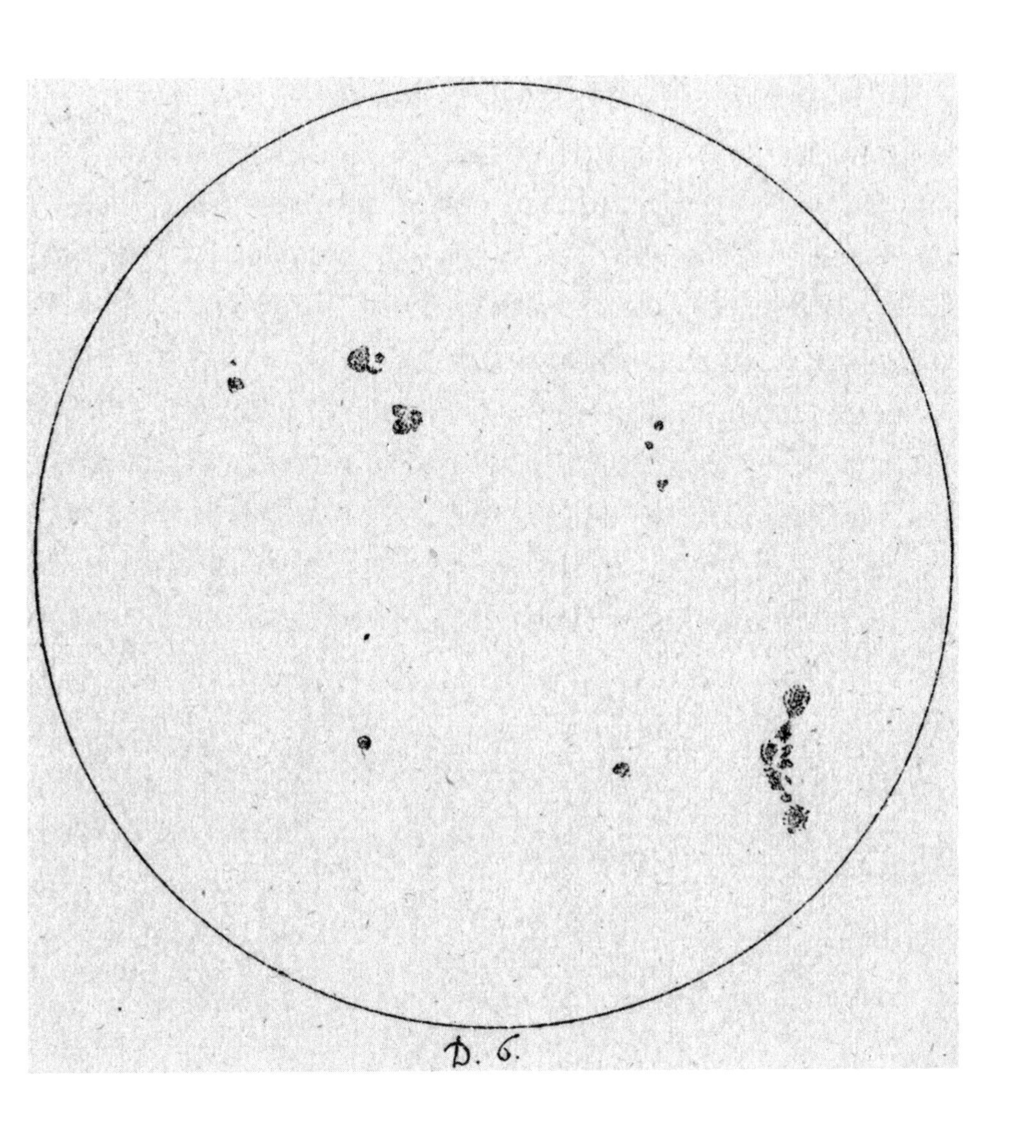
D. 6.

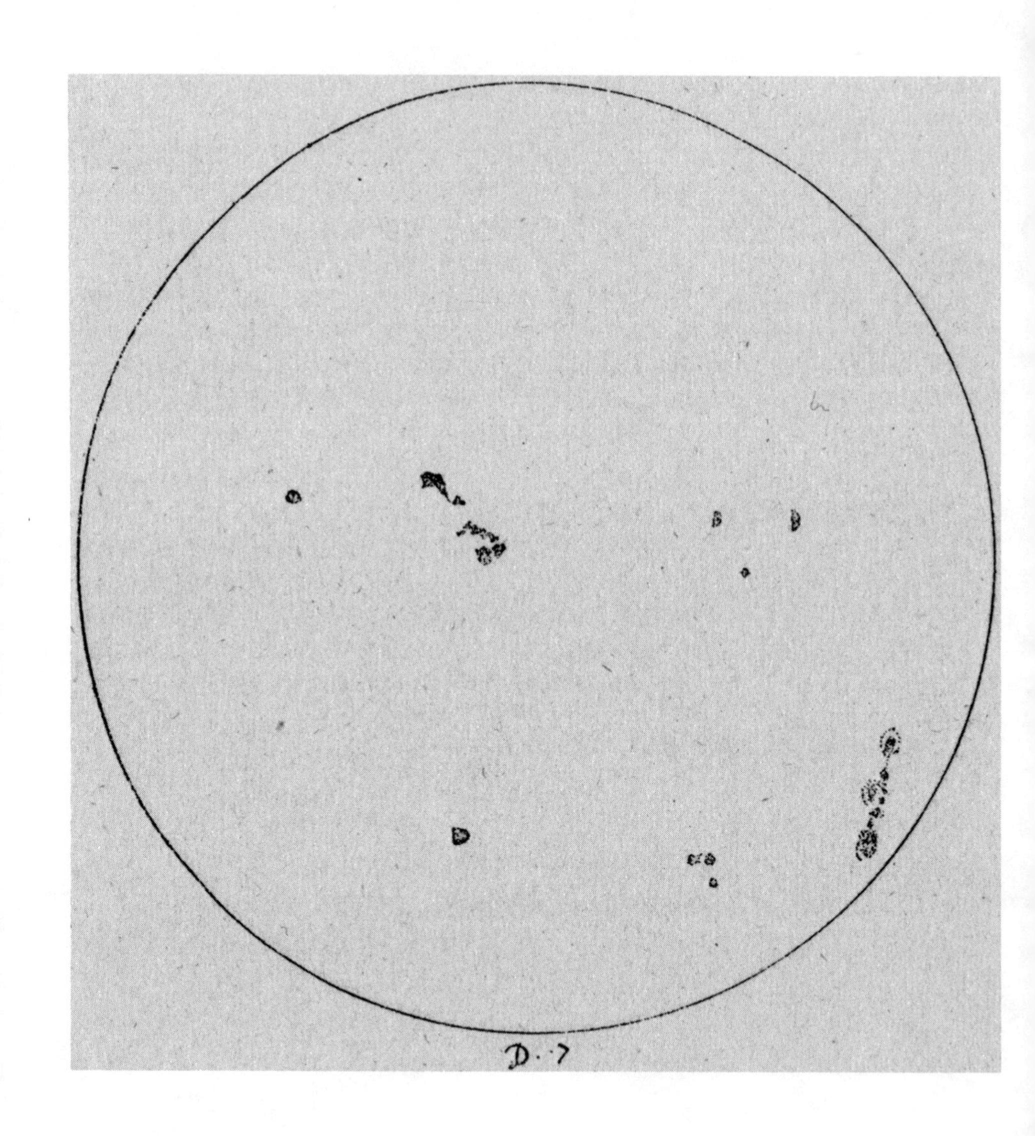
D. 7

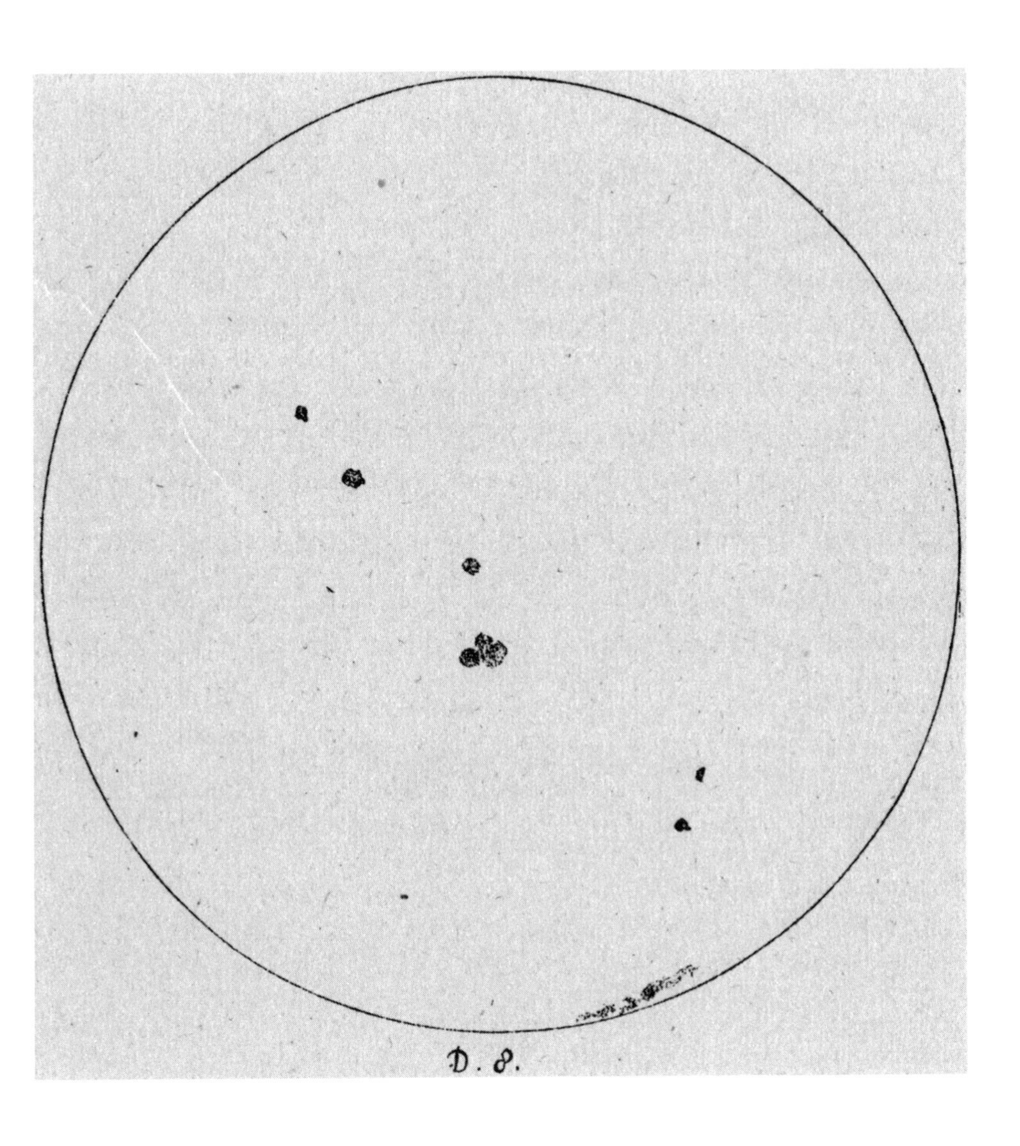
D. 8.

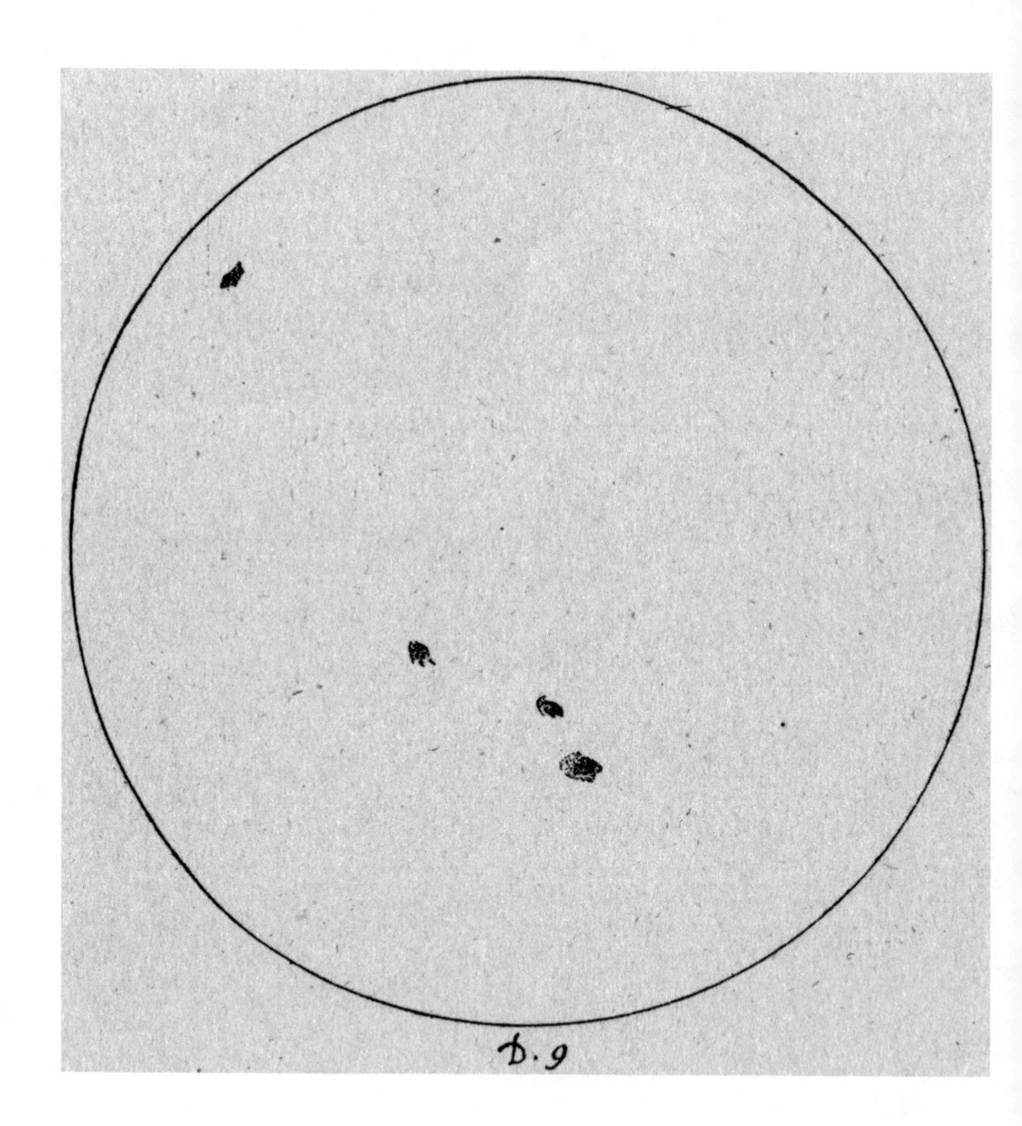
D. 9

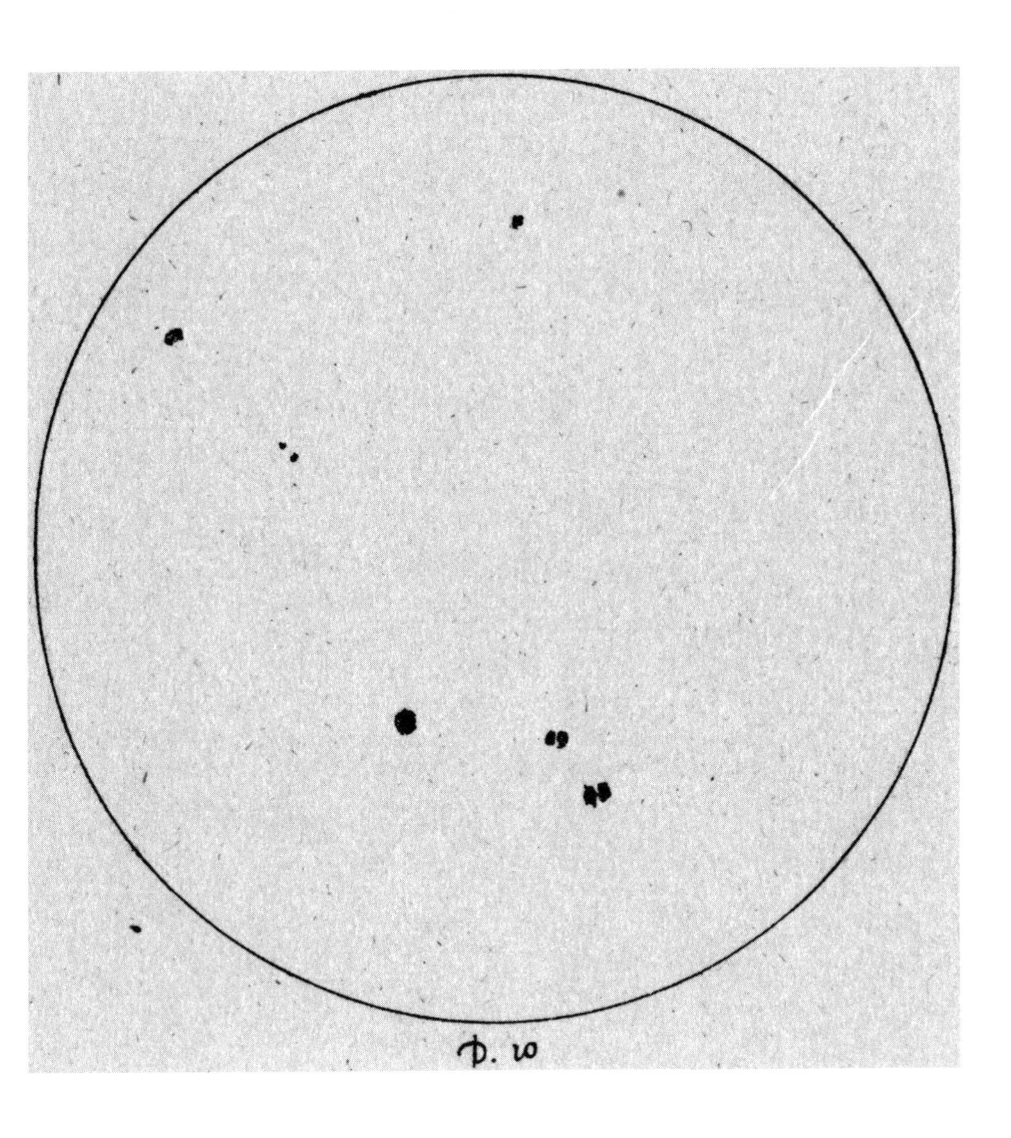
D. 10

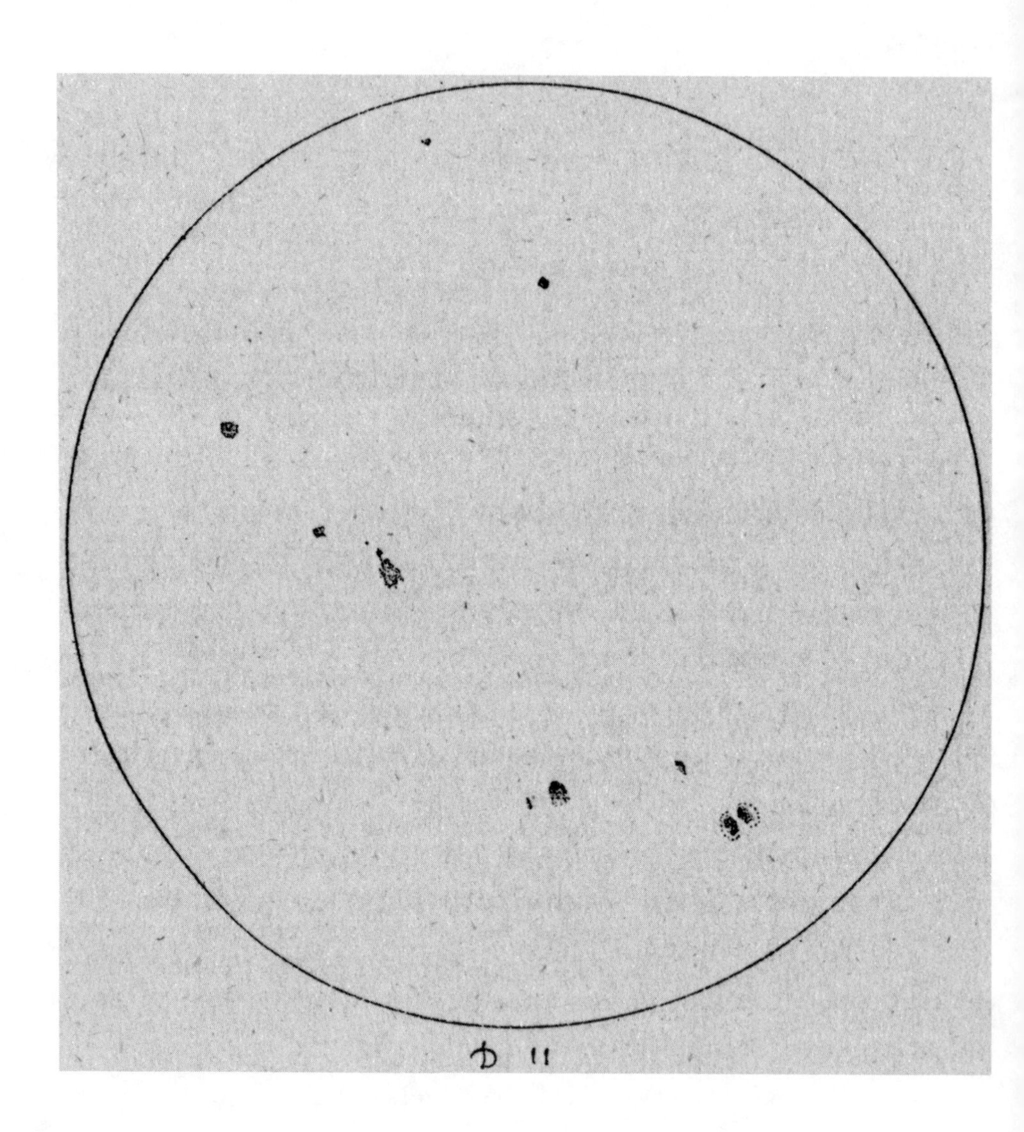
D 11

Appendix 3

Carlo Cardinal Conti to Galileo on Scripture

7 July 1612, *OGG*, 11: 354–355

Most Illustrious and Excellent Lord,

The questions raised by Your Lordship in your book[1] are very fine and curious, and based on very firm reason and certain experiments. But as these are new things, there will be no lack of opposition, which, I hope, will serve only to make your genius brighter and the truth more certain.

As to what you asked me, if Sacred Scripture supports the principles of Aristotle about the constitution of the universe: if you are speaking of the incorruptibility of the heavens, as it seems you hint in your letter, saying that each day new things are discovered in the heavens, I respond that there is no doubt whatsoever that Scripture does not support Aristotle, but rather, it even supports the contrary judgment, such that it was commonly believed by the Church Fathers that the heavens were corruptible. If then these things lately discovered in the heavens might prove such corruptibility, [that matter] will require long consideration. This is both because the sky, being so far from us, makes it difficult to affirm anything certain about it without lengthy observa-

1. Galileo's *Discorso intorno alle cose che stanno in su l'acqua o che in quella si muovono*, which had come off the press in April of that year. Galileo had not yet sent his first letter on sunspots to Conti.

tion, and also because if it is corruptible you need to have definite causes of these changes, which would have to appear at fixed and definite times. Nor can [these alterations] be accounted for without the heavens suffering corruption,[2] the way that some people will easily think they can account for the spots that appear on the Sun by the motion of some stars that turn below it. I think that these reasons and many others have been thoroughly considered and examined by you, and therefore I await more ample exposition of your observations, and judgments from you.

As to the motion of the Earth and the Sun, it is the case that one could be talking of two motions of the Earth. One of these is in a straight line and is due to the change of the center of gravity, and he who might propose such a motion would be saying nothing against Scripture because this is an accidental motion of the Earth; and so it was called by Lorini in his commentary on the first chapter of Ecclesiastes.[3] The other motion is circular, so that the heavens would stay unmoved and [merely] appear to us to move because of the motion of the Earth, as to sailors the shore appears to move.[4] This was the opinion of the Pythagoreans, followed later by Copernicus, Calcagnino,[5] and others, and it seems less in conformity with Scripture. For although those passages that say that the Earth is stable and firm can be understood as referring to the eternity of the Earth, as Lorini notes in the place cited, nevertheless, where it says that the

2. Conti was referring here to the Aristotelian notion of a phenomenon that could be adduced in a scientific demonstration, for which multiple repetition of the experience was necessary. See Peter Dear, *Discipline and Experience: The Mathematical Way in the Scientific Revolution* (Chicago: University of Chicago Press, 1995), 32–62.

3. Giovanni Lorini, *Commentarii in Ecclesiasten* (Lyon, 1606), 27. The commentary is on *Ecclesiastes* 1: 4: "One generation passeth away, and another generation cometh: but the Earth abideth for ever."

4. On the development of the nautical trope, adapted from *Aeneid* III:72 and used by Copernicus as well as by Galileo, see Ulrich Stadler, "Moving Objects, Moved Observers: On the Treatment of the Problem of Relativity in Poetic Texts and Scientific Prose," *Science in Context* 18.4 (2005): 607–627.

5. In his influential *Commentariorum in Iob libri tredecim* (Madrid, 1597–1601, and many subsequent editions), Juan de Pineda S.J., wrote, after stating that the Copernican opinion was plainly false: "Alii certe scientiam hanc deliram dicunt, nugatoriam, temerariam et in fide periculosam dicunt, atque ex ore antiquorum illorum philosophorum a Copernico et Caelio Calcagnino revocatam, potius ad ingenii specimen, quam ad philosophiam atque astrologiae bonum" ("Others rightly term this science insane, worthless, audacious, and dangerous to faith; they say it is an opinion of those ancient philosophers revived by Copernicus and Celio Calcagnino, and rather more proof of cleverness than something good and useful to philosophy and astrology"). Juan de Pineda, *Commentariorum in Iob libri tredecim* (Hispali: In Collegio D. Ermenegildi Societatis Iesu: Excudebat Ioannis Rene, 1598), cols. 461–462. This sentence was cited verbatim by Lodovico delle Colombe in his *Contro il Moto della Terra*, circulated in 1610/11 (see *OGG* 3: 290). See also Blackwell, *Galileo, Bellarmine, and the Bible*, 26–27.

Sun goes around and the heavens move, Scripture can then only be interpreted as speaking according to the common manner of the people, and *that* mode of interpretation is not to be admitted unless absolutely necessary. And yet, in a commentary on Job 9:6, Diego de Zuñiga says that it is more in conformity with Scripture to say that the Earth moves,[6] but his interpretation is not generally followed. And this is what has occurred to me on this proposition up until now; if Your Lordship wants to have further clarifications of other passages in Scripture, let me know so that I can send them.

And as for those black spots that Your Lordship sees on the Sun, I have wanted to send you a copy of what is written in an unusual book,[7] from which one learns that they are stars that revolve around it. And thanking Your Lordship for sharing your noble efforts with me, I end, and commend myself to You from the heart.

From Rome, 7 July 1612
My brother[8] is in Parma and is soon supposed to be in Rome, and I will share the book with him which, as the product of your genius and learning, will bring him much enjoyment.

At the pleasure of Your Lordship Lord Galileo Galilei
Cardinal Conti

18 August 1612, *OGG*, 11:376

Most Illustrious and Excellent Lord,

Your observations[9] are very diligent and beautiful; and whatever these spots are, they are matters beyond what has hitherto been believed. But because this is an issue of great consequence, and in places so distant from us, observations made over a lengthy period are needed, especially since someone seizing the opportunity of the Medicean Stars, which you observed, could maintain

6. Diego de Zuñiga's (Didacus a Stunica), *In Iob Commentaria* (Toledo, 1584), 205ff. For an English translation, see Blackwell, *Galileo, Bellarmine, and the Bible*, 185–186.

7. Conti is probably referring to Scheiner's *Tres Epistolae*.

8. Conte Conti.

9. Since Galileo's second letter to Scheiner is dated 14 August 1612, it is not probable that Conti was referring to this document. It is more likely that Galileo had sent him the first letter and some observations as well, like those he had sent earlier that summer to Maffeo Barberini (see appendix 2).

that these spots had been generated by stars, ones that are so minute as to be invisible when separate, but whose aggregation causes these spots. This person could also say that these stars are so numerous, and have such a variety of motions around the Sun, that in coming together in different patterns, they cause this diversity of spots. To convince people of this sort, extensive observations are needed, and all the more so for observing what else these spots might be, and whether they might render the heavens corruptible. For this reason, we hold that these spots are not on the solar body itself, but in another part of the heavens.[10] I hope that with your diligence and genius Your Lordship will be able to shed light on all this.

As to Sacred Scripture, I would like to learn more precisely in which matter Your Lordship seeks to know that it does not favor Aristotle. For if Your Lordship is speaking of the corruptibility of the heavens, there is no doubt that this is indicated in many passages. If Your Lordship is speaking of other dogmas, it is certain that it is against Aristotle, as for example about the eternity and control of the universe, but this has nothing to do with the present observations. But please let me understand what you want, because I will not fail to provide, so that you may be satisfied. And may God protect you.

Rome, 18 August 1612
At the pleasure of Your Lordship [Lord] Galileo Galilei
Cardinal Conti

10. "[E]t per convincer questi è necessaria lunga osservatione, come molto più per osservare che altra cosa siino queste macchie, et quando facciamo il cielo corrutibile, donde noi habbiamo che queste macchie non siino nell'istesso corpo solare, ma in altra parte del cielo." The phrase "et quando facciamo il cielo corrutibile" is problematical.

Appendix 4

Scheiner's calculation of the transit of Venus

1. Scheiner's calculation

Scheiner computed the time and duration of the transit at the superior conjunction of Venus and the Sun on the basis of Giovanni Antonio Magini's *Ephemerides* (Venice, 1582 and later editions) and *Tabulae secundorum mobilium coelestium* (Venice, 1585), calculated for the meridian of Venice. He did not correct for the difference in longitude between Venice and Ingolstadt, but the error thus introduced, about +4 minutes of time, is very small. The calculation, in which he appears to convert minutes to seconds for a number of steps, is as follows:

kl = 10 days · 15′1″ per day = 150′10″ = 9010″ (converting minutes to seconds)

$li = 17' = 1020''$

$ik = (kl^2 + li^2)^{1/2} = 151'7'' = 9067''$

$eh = mn = 7'18'' = 438''$

$$nk = \frac{li}{kl}\, mn = \frac{1020}{9010}\, 438'' = 49''$$

$$mk = \frac{ki}{kl}\, mn = \frac{9067}{9010}\, 438'' = 440'' = 7'20''$$

$pi = 1020'' + 49'' = 1069''$, $im = 9067'' + 440'' = 9507''$

$mp = 438'' + 9010'' = 9448''$

$hk = 9'$, $em = hk - kn = 9' - 49'' = 8'11'' = 491''$

$mo = \frac{pi}{im} em = \frac{1069}{9507} 491'' = 55''$

$eo = \frac{mp}{im} em = \frac{9448}{9507} 491'' = 487'' = 8'7''$

$eq = 17' + 1' = 18' = 1080''$

$oq = or = (eq^2 - eo^2)^{1/2} = (1080^2 - 487^2)^{1/2} = 963'' = 16'3''$

$qr = 2oq = 32'6''$

Duration: 32′6″/15′1″ per day = 2d 3h 18m 10s ≈ $51\frac{1}{3}$ h

$mq = eq - mo = 16'3'' - 55'' = 15'8''$

Incidence (immersion): 15′8″/ 15′1″ per day = 24h 11m 11s

$mr = eq + mo = 16'3'' + 55'' = 16'58''$

Egress (emersion): 16′58″/15′1″ per day = 1d 3h 6m 59s

Distance between Sun and Venus on 11 Dec noon (12 PM) is 7′18″

Time to mean (correctly “true”) conjunction:
7′18″/15′1″ per day = 11h 40m 3s

Time of mean (correctly “true”) conjunction:
11 Dec 12 PM + 11h 40m 3s = 11 Dec 11:40:3 PM

Beginning: 11 Dec 11:40:3 PM – 24h 11m 11s = 10 Dec 11:28:52 PM

End: 11 Dec 11:40:3 PM + 1d 3h 6m 59s = 12 Dec 14:47:2 PM =
13 Dec 2:47:2 AM.

2. Calculation of the transit from modern elements

Here we have arrived at the computation of transit using Alcyone Ephemeris. All times are mean times for the meridian of Venice, as in Magini's *Ephemerides*, as Scheiner made no adjustment for the meridian of Ingolstadt. Apparent times, by applying the equation of time, are all about six minutes later.

Time of true conjunction: 1611 Dec 11, 1:54 PM
Longitude of Sun and Venus: 259°13′28″
Latitude of Venus: –9′7″
Least separation between Sun and Venus: 8′59″
Time of least separation between Sun and Venus: 11 Dec 11:28 AM
Interval from conjunction to least separation: –2h 26m
Rate of elongation of Venus: 36.1″ per hour
Radius of Sun: 16′15″
Radius of Venus: 5″
Sum of radii: 16′20″

The beginning and end of the transit occur when the angular separations of the Sun and Venus are equal to the sum of the radii, 16′20″, at:

Beginning: 10 Dec 1:06 PM

End: 12 Dec 9:51AM

The total duration of the transit is thus 44h 45m. The middle of the transit is at

10 Dec 1:06 PM + 22h 22m 30s = 12 Dec 9:51 AM – 22h 22m 30s = 11 Dec 11:28:30 AM

This is consistent with the least separation at 11 Dec 11:28 AM, which is also the middle of the transit.

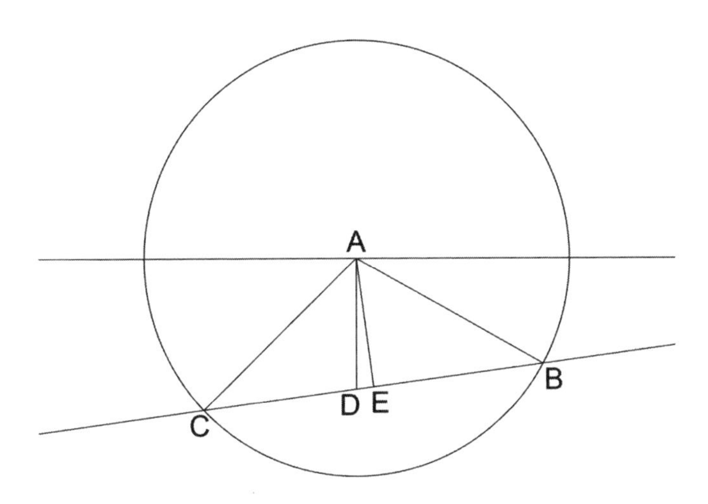

In the figure:

AB = AC is sum of radii: 16′20″
AD is latitude of Venus at true conjunction: 9′7″
AE is least separation: 8′59″
DE = 2h 26m · 36.1″/h = 1′27.8″
BE = EC = 22h 22m 30s · 36.1″/h = 13′27.7″
BD = BE + DE = 14′55.5″
Immersion: 14′55.5″/36.1″/h = 24h 48m 22s
DC = EC – DE = 11′59.9″
Emersion: 11′59.9″/36.1″/h = 19h 56m 31s
BC = BD + DC = 26′55.4″

Duration: 26′55.4″/36.1″ per hour = 44h 44m 52s ≈ 44h 45m, consistent with the previous calculation.

Circumstances of transit

	Date and time	Sun long.	Venus long.	Venus lat.	Separation
Start	10 Dec. 1:06 PM	258°10′21″	257°55′26″	−6′39″	16′20″
Middle	11 Dec. 11:28 AM	259°7′16″	259°5′49″	−8′52″	8′59″
Conj.	11 Dec. 1:54 PM	259°13′28″	259°13′28″	−9′7″	9′6″
End	12 Dec. 9:51 AM	260°4′15″	260°16′14″	−11′6″	16′20″

Because of the sensitivity of the computations, the times can be considered secure to within a few minutes and the longitudes to within a few seconds.

3. Sources of error

The principal errors in Magini's *Ephemerides*, based upon Erasmus Reinhold's *Prutenic Tables* (1551), and from that in Scheiner's calculation are: (1) the time of conjunction, 11 December 11:40 PM, is late by about 10h 30m; (2) the daily motion of elongation, 15′1″, leads to a motion of 37.5″ per hour, an error of +1.4″ per hour; (3) the latitude of Venus at conjunction, +9′, should be −9′, a serious error placing Venus to the north when it should be to the south. The reason is that the latitude theory in the *Prutenic Tables*, following Copernicus, is very inaccurate; the first accurate latitude theory is in Johannes Kepler's *Rudolphine Tables*, which were not published until 1627. In addition, Scheiner's radius of the Sun, 17′, is too large by 45″ and the radius of Venus, 1′, is too large by 55″. These increase the duration of the transit and are only slightly compensated for by the excess motion of +1.4″ per hour.

Appendix 5

Galileo's demonstrations

5.1 Second sunspot letter, pp. 113–116

The relation of the versine: vers $\theta = 1 - \cos\theta$.
The abbreviation of chord is crd and crd $\theta = 2 \sin \frac{1}{2}\theta$.

$$\text{CH} = 4°, \text{CG} = 10{,}000,$$
$$\text{CF} = \text{CG vers CH} = 10{,}000 \text{ vers } 4° = 10{,}000\,(1 - \cos 4°) = 24,$$
$$\text{crd CH} = \text{CG} \cdot 2 \sin \frac{1}{2}\,\text{CH} = 10{,}000 \cdot 2 \sin 2° = 698.$$

Galileo gives crd CH = 419, which is an error; it is not clear how he got it. He then says that CH = 17 CF, but in fact CH = 698 = 29 CF.
Now, where CM = 100 and GM = 10,000 + 100 = 10,100,

$$\text{vers MR} = 1 - \cos \text{MR} = \frac{\text{CM}}{\text{GM}}, \cos \text{MR} = 1 - \frac{\text{CM}}{\text{GM}} =$$
$$1 - \frac{100}{10100} = 0.990099,$$
$$\text{arc MR} = \arccos(0.990099) = 8°4',$$
$$\text{MF} = \text{CM} + \text{CF} = 100 + 24 = 124,$$
$$\text{vers MRN} = 1 - \cos \text{MRN} = \frac{\text{MF}}{\text{GM}}, \cos \text{MRN} = 1 - \frac{\text{MF}}{\text{GM}} =$$

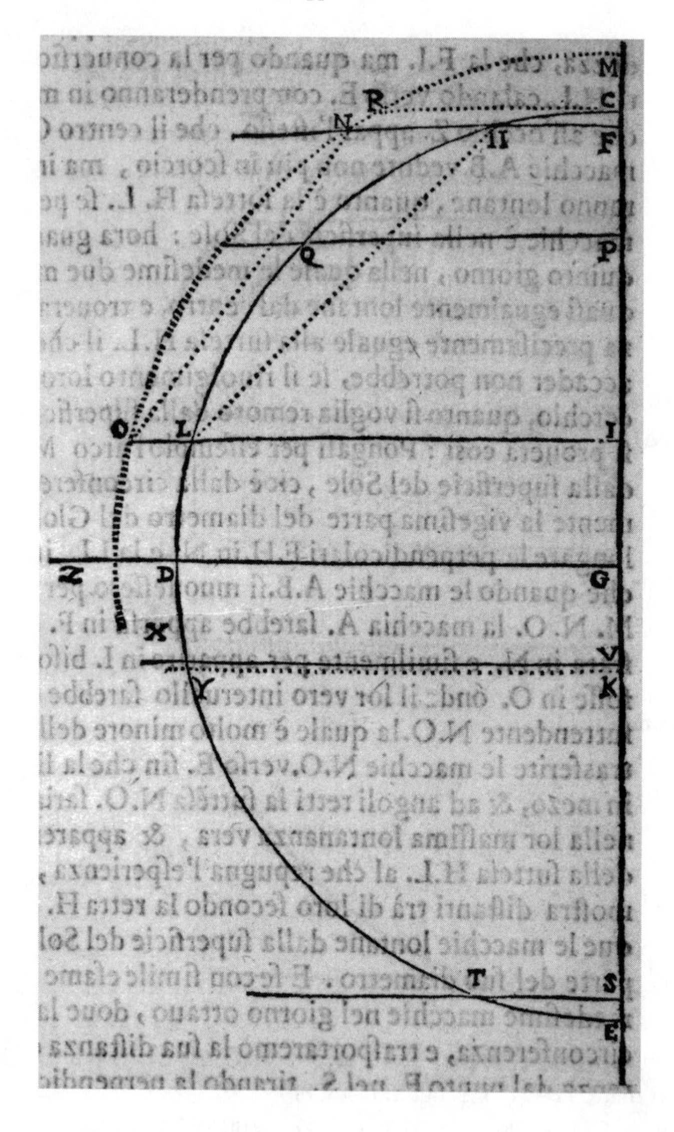

$$1 - \frac{124}{10100} = 0.987722,$$
$$\text{arc MRN} = 8°59'.$$

Galileo has 8°58′, which is not an error, but is just due to rounding.

$$\text{arc RN} = \text{arc MRN} - \text{arc MR} = 8°58' - 8°4' = 54'.$$

Galileo then says that crd RN = 94, but this is an error. Again, it is not clear how he got it. For taking RN = 54′,

$$\text{crd RN} = \text{crd } 54' = 2 \sin \frac{1}{2} \text{RN} = 2 \sin 27' = 0.0157078,$$

and where GM = 10100,

$$\text{crd RN} = 10100 \cdot 0.0157078 = 158.6 \approx 159.$$

The result is about the same using the subtraction formula for crd (8°58′ − 8°4′), which is strictly a more correct way of computing it. Hence RN ≈ 6.6 CF, that is, 159 ≈ 6.6 × 24. But Galileo says that RN is less than four times greater than CF, that is, 94 × 4 · 24. And earlier he said erroneously that CH = 17 CF, when correctly CH = 29 CF.

Still, his main point is correct, that if the sunspots were located even a small distance above the surface of the Sun, by 100 parts where the radius of the Sun is 10,000, that is, by one-hundredth the radius of the Sun, the relation between the true separation of the spots, CH on the surface of the Sun or RN above the surface, and the apparent separation CF would be much smaller above the surface than on the surface. Since the true separation can be seen when the spots are about symmetrical to the center of the Sun, and since it is many times larger than the apparent separation near the limb, the spots must be on the surface of the Sun or very close to it, and not even one-hundredth the radius of the Sun above the surface.

5.2. Third letter, pp. 269–270

The steps in the last sentence are:
Because triangles EMG and FOI are similar,

$$\frac{\text{EG}}{\text{GM}} = \frac{\text{FI}}{\text{IO}},$$

and since

$$\text{EG} = \text{CG and FI} = \text{DI},$$

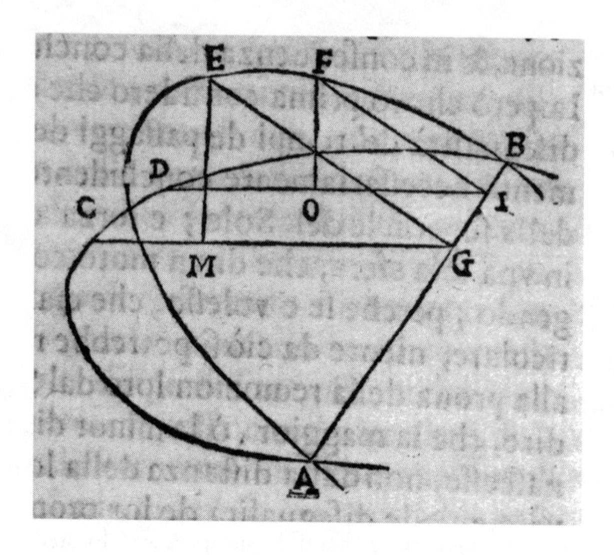

therefore

$$\frac{CG}{GM} = \frac{FI}{IO},$$

and by subtraction:

$$\frac{CG - MG}{MG} = \frac{DI - OI}{OI}, \text{ that is, } \frac{CM}{MG} = \frac{DO}{OI},$$

5.3. Third sunspot letter, pp. 270–272

Because

$$FD^2 = FM^2 + MD^2 \text{ and } ND^2 = NM^2 + MD^2,$$

so

$$FD^2 - ND^2 = (FM^2 + MD^2) - (NM^2 + MD^2) = FM^2 + MD^2 - NM^2 - MD^2,$$

and subtracting MD^2,

$$FD^2 - ND^2 = FM^2 - NM^2.$$

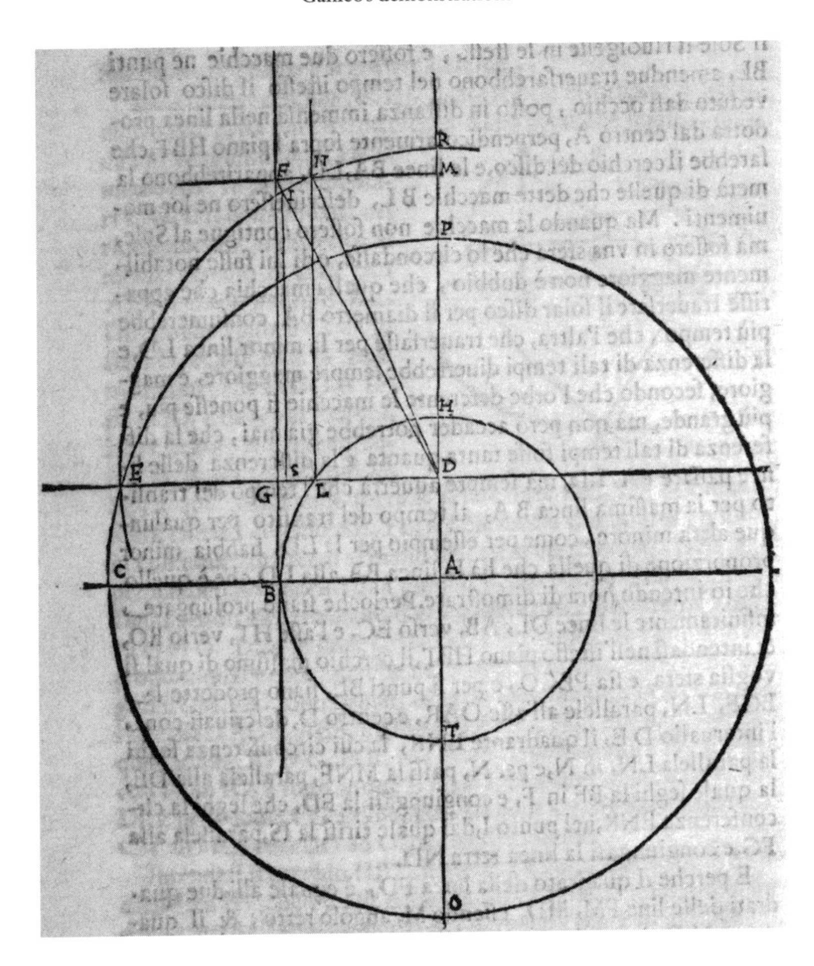

Because

$$FM = BA \text{ and } NM = LD \text{ and } BA^2 - LD^2 = DA^2$$

(if radius AL = radius BA is drawn, this is obvious),
therefore

$$FD^2 - ND^2 = DA^2,$$

because

$$FD^2 - ND^2 = FM^2 - NM^2 = BA^2 - LD^2 = DA^2.$$

But

$$FD^2 = ND^2 + DA^2 = ED^2 + DA^2, \text{ and } CA^2 = ED^2 + DA^2.$$

(If radius AE = radius CA is drawn, this is obvious.)
Therefore

$$FD = CA.$$

Because in triangle FGD, IS is parallel to FG,

$$\frac{FD}{DG} = \frac{CA}{AB} = \frac{ID}{DS} = \frac{ED}{DS},$$

and by subtraction

$$\frac{AC - BA}{BA} = \frac{CD - SD}{SD}, \text{ that is, } \frac{CB}{BA} = \frac{ES}{SD}.$$

5.4. Third sunspot letter, pp. 272–273

In the diagram above, it must be demonstrated that

$$\frac{\text{time of passage along BA}}{\text{time of passage along LD}} < \frac{BA}{LD}.$$

Since it has been established that the time of passage along BA = time of passage along SD, it must be shown that

$$\frac{\text{time of passage along SD}}{\text{time of passage along LD}} < \frac{BA}{LD}.$$

But since

$$\frac{\text{time of passage along SD}}{\text{time of passage along LD}} = \frac{\text{arcIR}}{\text{arcNR}},$$

we must demonstrate that

$$\frac{BA}{DL} = \frac{FM}{NM} > \frac{\text{arcIR}}{\text{arcNR}}.$$

Now, because tri FDN $>$ sec IDN, so

$$\frac{\text{triFDN}}{\text{secNDR}} > \frac{\text{secIND}}{\text{secNDR}}.$$

But ($>>$ means "still greater than")

$$\frac{\text{triFDN}}{\text{triNDM}} >> \frac{\text{triFDN}}{\text{secNDR}},$$

since tri FDN $>$ sec NDR.

Therefore

$$\frac{\text{triFDN}}{\text{triNDM}} >> \frac{\text{secIND}}{\text{secNDR}},$$

and by addition

$$\frac{\text{triFDM}}{\text{triMDN}} > \frac{\text{secIDR}}{\text{secRDN}}.$$

But

$$\frac{\text{triFDM}}{\text{triMDN}} = \frac{\text{FM}}{\text{NM}}.$$

and

$$\frac{\text{sec IDR}}{\text{sec RDN}} = \frac{\text{arcIR}}{\text{arcRN}}.$$

Therefore

$$\frac{\text{FM}}{\text{NM}} = \frac{\text{BA}}{\text{LD}} > \frac{\text{arcIR}}{\text{arcRN}} = \frac{\text{time of passage along BA}}{\text{time of passage along LD}}.$$

5.5. Third sunspot letter, pp. 273–274

Assume with Scheiner that a spot is a body moving in an orbit larger than the body of the Sun and assume that it crosses the Sun 30° from the equator. If we subtract $2 \cdot 30° = 60°$ from a hemisphere of 180°, the path of the spot will be the chord of 120°.

Where the diameter of the circle 2R = 2.0,

$$\text{crd } 120° = 2 \sin 60° = 1.732.$$

Therefore, if the time along the diameter is 16 days,

$$\text{the time along crd } 120° = \frac{1.732}{2.0} \cdot 16\text{d} = 13\text{d } 20\text{h } 33\text{m}.$$

By the preceding demonstration, the *ratio* of the time along the diameter to the time along the chord of 120° cannot be greater than the value computed here, 16d to 13d 20h 33m, about 1.154/1. Hence, it would be impossible for a spot to cross in 13d $20\frac{1}{2}$h, only 3 minutes less, because the ratio of 16d to 13d 20h 30m, about 1.155/1, is greater than 1.154/1. Since Scheiner says that a spot crossed the Sun in 14 days, the ratio is 16d to 14d, about 1.142/1, which is not greater than 1.154/1 although it is only slightly less. Hence, what Scheiner says is not impossible, although it is close to impossible, by less than $+3\frac{1}{2}$h.

5.6. Third sunspot letter, pp. 274–275

After explaining the construction in the figure, it is to be shown that, if the time to traverse BA is $1\frac{1}{7}$ times the time to traverse LD, the semidiameter CA of the sphere carrying the sunspots is more than double the semidiameter BA of the Sun. If it is not greater than double, it will be either double or less; assume first that it is double. (Note that in the figure it looks as though point I coincides with G, which is confusing; but I is actually on line ED between G and L, ever so slightly toward L from G, that is, ID < GD = AB.)
Suppose that

$$\frac{BA}{ID} = \frac{CA}{ED}.$$

Because CA > ED, still more is BA > LD.
Because the time of passage of spot C along AB is equal to the time of passage of spot E along ID, therefore,

$$\frac{\text{time along BA}}{\text{time along LD}} = \frac{\text{time along ID}}{\text{time along LD}} = \frac{\text{arc sin ID}}{\text{arc sin LD}} \text{ in the circle with}$$

semidiameter DE.

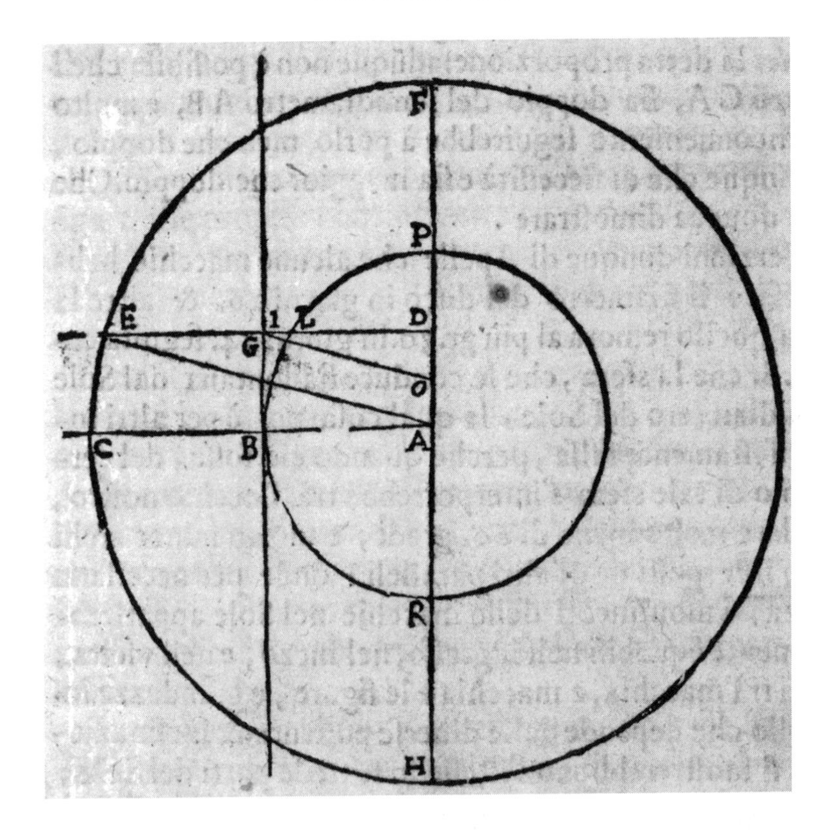

And because IO is parallel to EA and triangle IDO is similar to triangle EDA,

$$\frac{ED}{DI} = \frac{AD}{DO} = \frac{AE}{IO}.$$

But ED = 2 DI because it was assumed that CA = 2 BA. Therefore, AD = 2 DO and AE = 2 IO, and since AE = AC = 2 AB, IO = AB. (Note that the "total sine" is the radius of the circle.)

Now, because it is assumed that arc BL = 30°, the total sine BA = IO = 2 AD = 4 OD.

Letting IO = BA = 1000,

$$OD = BA\,\frac{1}{2}\sin BL = 1000\,\frac{1}{2}\sin 30^\circ = 1000 \cdot \frac{1}{2}\,0.5 = 250,$$
$$DI = (IO^2 - OD^2)^{1/2} = (1000^2 - 250^2)^{1/2} = 968,\ DE = 2\,DI = 1936.$$

The computations underlying the rest are as follows:

$$LD = BA \sin LP = BA \cos BL = 1000 \cos 30° = 1000 \cdot 0.866 = 866.$$

Now, letting ED = 1000 (before it was 1936 where AB = IO = 1000),

$$ID = \frac{1}{2}ED = 500, DL = \frac{1000}{1936} \cdot 866 = 447,$$

from which

$$\text{arc-sin } 500/1000 = 30°, \text{arc-sin } 447/1000 = 26°33'.$$

The proportion of the times of the spots crossing the Sun is the proportion of the length of the chords, or of the sines if only the radius of the Sun is being considered, as here.

$$ID = 1000 \sin 30° = 500.$$

The sine of the arc of which ID $= 1\frac{1}{7}$ is $500/1\frac{1}{7} = 437.4 \approx 437.$

The corresponding arc is thus arc sin $437/1000 = 25°54'$ (more accurately, $25°54'45''$).
It is likely that $25°45'$ in the text is a typographical error for $25°54'$.

5.7. Third sunspot letter, pp. 276–278

Where AM = 100,000, the true width of $\mu = 3°20'$, on the arc below M in the figure, is

$$\text{crd } \mu \approx 100{,}000 \sin 3°20' = 5814.$$

$$\text{arc AB} = 8°, \text{arc BD} = \mu = 3°20', \text{arc ABD} = \text{arc AB} + \text{arc BD} = 11°20'.$$

$$ACO = AM \text{ vers ABD} = 100{,}000\,(1 - \cos 11°20') = 1950,$$

$$AC = AM \text{ vers AB} = 100{,}000\,(1 - \cos 8°) = 973,$$

$$CO = ACO - AC = 977.$$

Hence, when spot μ is at BD, its width is $977/5814 = 1/5.95 \approx 1/6$ its (true) width in the middle of the disc of the Sun; that is, at BD its width is CO = 977 and at μ its width is 5814, about six times as great.
Now, assume a sphere with semidiameter MF in which

$$AM = 100{,}000, AF = 5000, MF = AM + AF = 105{,}000.$$

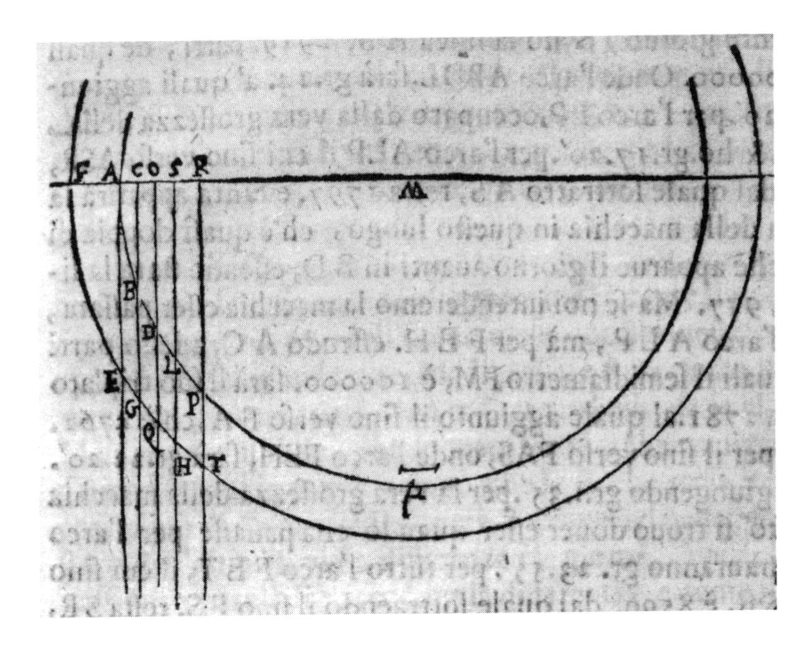

But where MF = 100,000,

$$AF = \frac{100000}{105000} 5000 = 4762, AC = \frac{100000}{105000} 973 = 927,$$
$$CO = \frac{100000}{105000} 977 = 930.$$
$$FAC = AF + AC = 4762 + 927 = 5689,$$
$$FACO = AF + AC + CO = 4762 + 927 + 930 = 6619.$$

Describing circle FEGQ and drawing parallels AE, CG, OQ, we find the arcs of the circle from:

$$1 - \frac{AF}{MF} = \cos FE, 1 - \frac{4762}{100000} = 0.95238, FE = 17°45'.$$

Galileo finds 17°40′.

$$1 - \frac{FAC}{MF} = \cos FEG, 1 - \frac{5689}{100000} = 0.94311, FEG = 19°25'.$$
$$EG = FEG - FE = 19°25' - 17°40' = 1°45'.$$
$$1 - \frac{FACO}{MF} = \cos FEGQ, 1 - \frac{6619}{100000} = 0.93381, FEGQ = 20°57' \approx 21°.$$
$$GQ = FEGQ - FEG = 21° - 19°25' = 1°35',$$

the chord of which, when at μ, would be

$$\text{crd } 1°35' \approx \text{MF} \sin 1°35' = 100{,}000 \sin 1°35' = 2763.$$

Galileo has 2765.

At GQ its width was CO = 930, and $\frac{930}{2765} = \frac{1}{2.97} \approx \frac{1}{3}$, so its chord is less than a third part of 2765. Hence, if the spot moved at a distance from the surface of the Sun of one-twentieth the radius of the Sun, it could never appear more than three times larger than near the limb, contrary to the observations and the previous demonstration, which showed a much greater enlargement of $5814/977 \approx 6/1$, about six times greater.

Now, again placing the spot on the surface of the Sun so that AM = 100,000, let

$$\text{AS} = 3\,\text{AC} = 3 \cdot 973 = 2919 = \text{vers ABDL} = \text{AM}\,(1 - \cos \text{ABDL}),$$

that is,

$$1 - \frac{\text{AS}}{\text{AM}} = \cos \text{ABDL},\ 1 - \frac{2919}{100000} = 0.97081,\ \text{ABDL} = 13°52' \approx 14°.$$
$$\text{ALP} = \text{ABDL} + \text{LP} = 14° + 3°20' = 17°20',$$
$$\text{ASR} = \text{AM vers ALP} = 100{,}000\,(1 - \cos 17°20') = 4541.$$

Galileo finds 4716, which is AM vers 17°40′, so this is an error due to looking up the wrong cosine.

$$\text{SR} = \text{ASR} - \text{AS} = 4716 - 2919 = 1797,$$

which is about twice what it appeared at BD, that is, $1797/977 = 1.8/1 \approx 2/1$.

But if the spot moved above the surface of the Sun through FEH so that FM = 100,000 and AC = 927, now let

$$\text{ACOS} = 3\,\text{AC} = 3 \cdot 927 = 2781,$$
$$\text{FAS} = \text{ACOS} + \text{FA} = 2781 + 4762 = 7543 = \text{vers FEH} = \text{FM}\,(1 - \cos \text{FEH}),$$

that is,

$$1 - \frac{\text{FAS}}{\text{FM}} = \cos \text{FEH},\ 1 - \frac{7543}{100000} = 0.92457,\ \text{FEH} = 22°23' \approx 22°20'.$$

The real size of the spot when moving about the surface of the Sun was GQ = 1°35′. Let it now be at HT so that

$$\text{FET} = \text{FEH} + \text{HT} = 22°20' + 1°35' = 23°55',$$

$$\text{FAR} = \text{FM vers FET} = 100{,}000\,(1 - \cos 23°55') = 8586 \approx 8590.$$

$\text{SR} = \text{FAR} - \text{FAS} = 8590 - 7543 = 1047$, the apparent size of the spot at HT, which is larger than CO = 930 by

$$\frac{1047 - 930}{930} = \frac{117}{930} = 1/7.8 \approx 1/8.$$

Thus, if the spot moved on a circle even one-twentieth larger than the Sun, its apparent size would not have increased by even one-eighth, although in fact it almost doubled in size, from 977 to 1797, that is, exceeded less than $1\frac{1}{8}$ by more than $\frac{7}{8}$. Hence, the spot must be on the surface of the Sun.

Appendix 6

Front matter of *Istoria e dimostrazioni intorno alle macchie solari e loro accidenti*

1. Imprimatur

Let it be printed if it pleases the Most Reverend Father Master of the Sacred Apostolic Palace

Caesar Fidelis Vicegerent

On the order of the Most Reverend Father Master of the Sacred Apostolic Palace, F. Luis Istela of Valencia, I have diligently examined three letters about solar spots written by the Highly Honored and Most Excellent Signor Galileo Galilei to the Most Illustrious Marc Welser, Chief Magistrate of Augsburg. Because I have found that these contain nothing that may be incompatible with the rules of the Sacred Index, but rather extraordinary learning, and new and marvelous observations hitherto unknown and unheard of, explained in a courteous and elegant style, I have declared that they are most worthy of being printed. Written faithfully in my own hand.

I have also examined some letters and disquisitions on the same matter by Apelles, sent to the same Signor Welser, which contain nothing that offends, and judge that these, too, can be published.

Rome, 4 November 1612
Antonio Bucci of Faenza, Citizen of Rome, Doctor of Philosophy and Medicine

Let it be printed.

Fr. Tomaso Pallavicini of Bologna, O.P., associate of the Eminent and Most Reverend Father, F. Luis Istela, Master of the Sacred Apostolic Palace.

2. Dedication [to Filippo Salviati, by Angelo de Filiis]

To the Most Illustrious Gentleman

Filippo Salviati
Member of the Lincean Academy

This gift to the learned community was meant for the judgment of the members of the Lincean Academy, and because I am, by virtue of my particular task,[1] its bearer, I judged it my obligation to choose to whom I should first present it. Thus, considering that this gift derived from the most noble and brilliant fire of the heavens by the philosophical toil and mathematical diligence of the most learned Signor Galilei, who adorns his native country with such celestial labors [*parti*], and bearing in mind the setting [of such labors], the occasion, and Your other qualities, and being ever more struck with Your worthiness and nobility, it seems to me both proper and necessary to offer it to Your Illustrious Lordship, and to the entire republic of philosophers in Your presence. Sublime and celestial objects should be dedicated to persons of eminence and of the most exalted nobility, and who is unfamiliar with the ornaments, the splendor, and the greatness of Your Most Illustrious House, variously dispersed among so many different individuals, but all united and resplendent in You? Works of virtue[2] and erudition are appropriate to those who love and follow the former. In You, virtue itself, fruit of the most exquisite mathematics and the best philosophy, has been allotted to You in such abundance that there being no

1. Angelo de Filiis, the author of this dedication, was the librarian of the Lincean Academy. On the academy's support of Galilean projects, see the recent work of David Freedberg, *The Eye of the Lynx: Galileo, His Friends, and the Beginnings of Modern Natural History* (Chicago: University of Chicago Press, 2002), especially 117–147.

2. The Italian *virtù* implies both prowess and piety.

reason for You to envy others, she gives others many reasons to envy You. The rarer this sort of intelligence, the more You are to be admired and praised by everyone. The Most Illustrious Signor Welser, greatly gifted with every knowledge and virtue, as one who knows and loves You, will be particularly pleased that under Your aegis scholars become acquainted with and enjoy the discoveries that he has obtained for them. And I see Signor Galilei most content that this work of his, sent to seekers of the truth, has reached such a safe haven. And what miracle is it if, in addition to knowledge of Your merits, that bond of friendship with which Galileo loves, admires, and esteems You, You are also bound by the Lincean Academy,[3] by Your native region, and by Your constant companionship? It was with good reason that the noble city of Florence, teeming with gifted minds, and an extraordinary cradle of learning, where all virtues have always flourished and do yet flourish, should first taste and enjoy her own fruits and discoveries. The latter were in fact produced in Your Lordship's own Villa delle Selve, a most agreeable place, while the author stayed there with You and rejoiced with You in the celestial display, over which the villa had thus particular rights. The work thus having been produced by Linceans, it was most fitting that I turn to You, whom they so esteem and regard, and it was to their great satisfaction that I have done so. Having then been sent to the learned community, it arrives most appropriately before You, not only because You are resplendent among lettered men for Your lofty intelligence, assiduous study, and particular knowledge, but also because You favor them, protect them, and exalt them in promoting work of true virtue with heroic munificence. Finally, if because of my role, it is reasonable for me to have any part in this gift, I take great pleasure in being worthy to present it to such a great patron of mine. At last it appears, given and dedicated to You by me, in order to make itself public before Your Illustrious Lordship, certain of being accepted. I pray that the affection with which it is offered pleases You. And I commend myself [to You] in thanks.

From Rome, 13 January 1613
Your Most Illustrious Lordship's
Most devoted Servant
Angelo de Filiis, Lincean.

3. Salviati and Welser were both made members of the Lincean Academy in the fall of 1612; Salviati's name had been proposed in July by Galileo, and Welser's by the German Lincean Johannes Faber. See Giuseppe Gabrieli, *Contribuiti alla storia della Accademia dei Lincei*, 2 vols. (Rome: Accademia nazionale dei Lincei, 1989), 1: 519–522.

3. Preface

[By] Angelo de Filiis, Lincean
To the reader.
If, in this great machine of the universe, celestial bodies are by their own nature the noblest of all, then contemplation of them should without doubt be considered of the very highest rank, and worthy of the most heroic minds, and those who advance and enrich this study deserve no little glory for their part in furthering the innate desire we all have to know[4] in such difficult and remote matters.[5] For this reason if we take such delight in and praise their discoveries so much when investigators of that lower [realm of] nature that lies under our feet reveal the birth of something never before seen—be it plant, animal, or deformed zoophyte[6]—how much more should we not rejoice when we are presented with new lights in that higher realm of the highest heaven and the faces of the noblest discoveries, once veiled, are now revealed to us? How great is our debt to their wise and diligent discoverers, and what praise should we offer them? Here, then, is a great and celestial subject for those minds that search so steadfastly for the truth. And where such searchers once found the heavens bordered and bounded by Herculean pillars,[7] and while from the first astronomers onward nothing new had been seen except for the fixed stars near the South Pole[8]—and this thanks to the new sea

4. "All men naturally desire to know" is the celebrated opening sentence of Aristotle's *Metaphysics*.

5. De Filiis' exalted impression of the study of astronomy was compounded by the fact that his late brother, Anastasio de Filiis, a founding member of the Lincean Academy and its first secretary, had some expertise in the discipline. In the fall of 1603 Anastasio had constructed a planisphere for the academy; it showed "with great accuracy all celestial bodies with their relationships according to the most modern as well as ancient systems of [natural] philosophy." See M. Muccillo, "De Filiis, Anastasio," in *Dizionario biografico degli italiani*, 33: 744–745.

6. On the particular interest which the Linceans had in zoophytes and in other instances of organisms seemingly on the borders of the animal, mineral, and vegetable kingdoms, see Freedberg, *The Eye of the Lynx*, 183, 243–245, 322–345, 408–411.

7. A traditional name for the eastern entrance to the Strait of Gibraltar.

8. A probable reference to Johann Bayer's *Uranometria*, published in Augsburg in 1603 by Christophorus Mang, a part of the printing consortium *Ad insigne pinus*, whose patron was Marc Welser. The *Uranometria* included twelve constellations around the South Pole; on this celestial atlas and its antecedents see N. M. Swerdlow, "A Star Catalogue Used by Johannes Bayer," *Journal for the History of Astronomy* 17.3 (1986): 189–197, and Elly Dekker, "The Light and the Dark: A Reassessment of the Discovery of the Coalsack Nebula, the Magellanic Clouds, and the Southern Cross," *Annals of Science* 47 (1990): 529–560, on 558–559. The *Uranometria*, like Scheiner's *Tres Epistolae* and *Accuratior Disquisitio*, was illustrated by Alexander Mair. Bayer traveled in Italy in 1603 and came

voyages[9]—and some phenomena in the other [stars] perhaps inaccurately observed,[10] now Signor Galilei, going beyond them, reveals to us a new abundance of sparkling bodies and other hidden mysteries of nature. This is taking place under the aegis and happy auspices of the Most Serene Lord Cosimo, Grand Duke of Tuscany, who because of his own virtue and generosity, and in imitation of his forebears, those great Lorenzos and Cosimos and other heroes of the regal family of the Medici, true Mycenas of our native and foreign letters, never ceases to favor the sciences[11] and to assure, for the benefit of the public, their every increase and enlightenment.

And thus Signor Galilei shows us innumerable squadrons of fixed stars, scattered throughout the entire firmament, many, once dim and indistinct, in the Galaxy and in the nebular [stars];[12] he has found Jupiter's royal retinue, and he has discovered that the Moon has a mountainous and varied surface. All of this he has made known and transmitted to everyone in his *Astronomical Announcement*.[13] Astonishment arose quickly, novelty in the heavens being the last of our expectations. Pursuing his enterprise still further, Galilei reveals the

---

into contact then with Federico Cesi; see Giuseppe Gabrieli, *Contribuiti alla storia della Accademia dei Lincei*, 2 vols. (Rome: Accademia nazionale dei Lincei, 1989), 1: 324–325.

9. The information on the southern constellations derived in part from observations made between 1500 and 1517 by Amerigo Vespucci and Andrea Corsali, both of Florence. On these observations, among others, and their garbled appearance in print in the sixteenth and early seventeenth centuries, see Dekker, "The Light and the Dark," and Dekker, "On the Dispersal of Knowledge of the Southern Celestial Sky," *Der Globusfreund* 35–37 (1987): 210–230.

10. It is not clear whether de Filiis had in mind the new stars of 1572 and 1604 or the Coalsack Nebula and Magellanic clouds near the Southern Cross. The new stars were quite troubling to the Linceans in that the interpretation of one early member of the academy, Johannes Eck, differed substantially from that of Federico Cesi, the group's founder, who made radical revisions to Eck's treatise on the nova of 1604 before allowing it to be published as the Linceans' first work in 1605. See in this connection Saverio Ricci, "Federico Cesi e la *Nova* del 1604: La teoria della fluidità del cielo e un opusculo dimenticato di Joannes van Heeck," *Atti dell'Accademia Nazionale dei Lincei: Rendiconti morali*, 8th ser., 48.5–6 (1988): 111–133, and Freedberg, *The Eye of the Lynx*, 91–100. The dark nebulae and Magellanic clouds, on the other hand, were widely but ambiguously reported after the voyages of exploration in the Southern Hemisphere in the early sixteenth century, and would have appeared to be at odds with Galileo's assertions about the nature of stars and clouds. The Coalsack Nebula, described by Vespucci as *canopus niger* and then understood by most readers as a cluster of black stars, would have differed from Galileo's insistence on spherical and luminous bodies; the Magellanic clouds, portrayed as brighter versions of the same phenomenon, would have violated his assertion that large cloud-like objects never strayed far beyond the solar body.

11. To be understood here as referring to learning in general.

12. In the star catalog of his *Almagest*, Ptolemy listed six "nebular stars," while in *De Revolutionibus* Copernicus listed five of these. All can be resolved into individual stars by means of modern telescopes, although there is, of course, true nebular matter in the universe.

13. *Avviso Astronomico*, or *Sidereus Nuncius*.

new triform Venus[14] to be an imitator of the Moon;[15] he moves on to slow and remote Saturn and shows us that it is triple-bodied, accompanied by two stars. He notifies Europe's premier mathematicians of this and he explains everything in prose. In order to dispel doubt with experience—for the former always accompanies the unexpected and the miraculous—he demonstrates to all the way to see and to enjoy for oneself each of these discoveries. Nor does he do this in one place alone, but in Padua, in Florence, and then in Rome itself, where his findings meet with the universal approval of the most learned men, and are expounded with great praise for him from the most public and renowned lecterns. Moreover, he does not leave Rome before demonstrating—and not with words alone—that he has discovered that the Sun is spotted; he proves this with the effect itself, and has us observe the spots in more than one setting, and in particular in the Quirinale Garden of the Most Illustrious Cardinal Bandini, in the company of the cardinal himself, as well as the Most Reverend Monsignors Corsini, Dini, Abbate Cavalcanti, Signor Giulio Strozzi, and other gentlemen.[16]

And as it seemed that not just these celestial discoveries and the means of obtaining them but also the gift of seeing with the mind's eye all that could be inferred from them were reserved for Signor Galilei alone, we waited with widespread interest for his views about these sunspots when at last some Linceans learned that he had written at length about this matter in a few private letters to the Most Illustrious and Learned Signor Welser. Having acquired this correspondence, and seen that the entire undertaking had been brought to the desired conclusion with a long series of observations, they judged it unthinkable that everyone not be permitted to satisfy his own curiosity about both the observations and the solar contemplations [i.e., the letters] themselves. What was private should thus become public, along with Signor Welser's queries.

14. *Triforme Venere. Triforme*, or *Triformis*, was a common way of referring to the Moon—Luna in the heavens, Diana on Earth, and Persephone or Hecate in the Underworld—but it is also possible that some understood the phrase to refer to the three shapes of the lunar body. See for instance Horace, *Carmina* 3.22.4, Virgil, *Aeneid* 4.511, Dante, *Paradiso* 23:25–27, and Ugo Foscolo, "A Diana," 4.

15. Galileo's anagram, published by Kepler in his *Dioptrice*, was *Haec immatura a me iam frustra leguntur o y*; the solution was "Cynthiae figuras aemulatur mater amorum," or "the Mother of Love [Venus] emulates the shapes of Cynthia [the Moon.]" Galileo made this observation in the fall of 1610 and circulated the anagram around Christmas of that year. In his first letter on the sunspots, he expressed surprise that Scheiner was not yet aware of the phases of Venus as proof that the planet goes around the Sun. See pp. 92–94, above.

16. This demonstration must have happened near the end of Galileo's visit to Rome, from late March to late May 1611. It is not to be confused with the famous dinner given by Prince Cesi in Galileo's honor, on 11 April 1611, during which the name "telescopium" was coined for the instrument made famous by Galileo's discoveries. See Rosen, *The Naming of the Telescope*, 29–31.

When I perceived the common desire, in conformity with what my particular office [as the academy's librarian] requires, I gave orders that these letters be published. I judged that they would be well received by all scholars, or, rather, by all those scholars free of an unfortunate fervor about any particular points that would render these letters unacceptable. But scholars who are so troubled—either because of pretensions they may have about the discovery of these spots, or because of their desire to keep their judgment and opinions of the phenomena intact, or even because such novelties and their consequences may perturb the many and sweeping conclusions of their doctrine that have until now been considered as most solid—such scholars will not perhaps receive with openness of mind what originates from Signor Galilei's most sincere affection, and his desire and regard for the truth. But the satisfaction of these people (if any such exist) should be of no great importance, not even to those who because of their private concerns must conceal these genuine and sensate effects that Nature herself makes manifest through the growth of true and real sciences. Then, to those who would claim precedence in observing these spots, the possibility of their having done so without prior notice from Signor Galilei is not denied, as it is also manifest that they have preceded him in publicizing them in print. But it is also more than clear that Signor Galilei gave a private account of them here in Rome to a great many people long before any writings were published, and in particular, as I have said above, in the Quirinale Garden in April of the year 1611, and many months before that to friends of his in Florence, whereas the first writings of other people to have come to light—those of the masked Apelles—have no observations that predate October of that same year, 1611.

Let it however be known to everyone that it is a truly noteworthy conclusion that to a single individual in our age falls not only the celestial use of the telescope, but also the discoveries and observations of such novelties in the stars and superior bodies. Nor let this fact be ascribed to random chance or to luck, as some would have it, perhaps to diminish the glory of the author; these same men convict and condemn themselves because it was they who, for the longest time, denied and ridiculed Signor Galilei's earliest discoveries. But if, after having been informed of the Medicean stars and of his other new observations, they were so slow in ascertaining their existence, how can they not admit that, had it been up to them, these things would have remained forever hidden? Thus, the particular favors that come from above must not be called fortuitous or ascribed to random chance, unless we also wanted to describe thus Galileo's firmness of judgment, his perspicacity of discourse, his integrity of mind, his nobility of spirit, and, in sum, all the other qualities granted us by Nature or divine grace. If, because of the extraordinary novelty of his discoveries, Signor

Galilei has received biting criticism from so many quarters for such a long time, as for example the numerous writings attacking him—writings filled, for the most part, with fury rather than with solid learning and sound reasoning—now, while we learn almost daily that he has proposed nothing that was not true, these men should not begrudge him the honors that usually are and should be the glorious reward for such useful and honest labors.

And I am certain that in enjoying thanks to him the heavens thus revealed and enriched with new orbs and splendors, and in contemplating at your pleasure the Sun no less than these other bright bodies, that you, gracious reader, will be most grateful to him, and especially if you consider carefully in what fashion and with what solid reasoning—for here chance can have played no part—the entirety has been treated and determined. And if you encounter such sound demonstrations in private letters that, though directed to persons of lofty learning, were nonetheless written in haste, you should expect all the more to see these same matters and many others explained with even greater perfection below in this author's individual treatises. Now these letters are made public to you for your delight and benefit. Let the envious and the detractors refrain from such reading; it is not intended for them. Indeed, since they were sent by the author privately to a single individual gifted with great intelligence and an open mind, I should not send them, to his disadvantage, to those endowed with entirely different gifts. I do not so much desire your favor and applause as to refute your censures and contradictions in those matters that might appear doubtful and not well established to you; rather, I assure you that corrections no less than compliments, and arguments no less than agreements, will always be precious to Signor Galilei. Indeed, he welcomes the former more than the latter, to the extent that the former can contribute new knowledge, while the latter merely confirm what has already been gathered. May you live happily.

4. Laudatory poems

To Galileo Galilei

Lincean

By Luca Valerio, Lincean
Professor of Mathematics and Civil Philosophy
In the Academy of the Beloved City

When with your staff, Galileo, you saw the sky laid bare,
The startled Earth gave a strange cry,
For her brazen monuments must all yield to Time,[17]
While you found lasting fame with a fragile glass.

By Johannes Faber of Bamberg, Lincean
Pontifical Herbalist, and Public Professor of Botany in that City

You have no need, Galileo, of wings of Dædalian design
For flight and fall from a flaming Sun,[18]
Nor will you, like some weak and unwarlike boy,
Sail above the stars astride Ganymede's eagle.[19]
Great genius gave to you, as to the sharp-sighted *Lynx*,[20]
The eyes that breached high heaven's walls:
You *first* found new stars in the far firmament.

By Francesco Stelluti, Lincean

Your gifts, Galileo, are now grown so great,
That upon the torch of terrible Night[21]
A hundred Olympian peaks sprang up,
Worthy thrones for your glory.
And when you turned to them unwearied with your glass
The highest heavens inclined in your favor:
Splendid stars were born by the thousand

17. Horace in *Ode* 3.30.1 and Propertius in *Elegy* 3.2.19 offer classical instances of poetic works meant to outlast the Earth's monuments; Galileo referred to the topos in the proem of his *Sidereal Messenger*, suggesting that celestial phenomena such as the Medici Stars were among the most durable memorials.

18. In Greek mythology, Icarus, son of the architect Daedalus, perished when flying too close to the Sun with wings of his father's invention.

19. The mythological figure of Ganymede, a boy of extraordinary beauty, was said to have been carried off by Jupiter himself, the latter being disguised as an eagle. The largest of Jupiter's satellites was soon named "Ganymede" by Galileo's rival Simon Marius, who related in his *Mundus Iovialis* (1614) that in 1613 Johannes Kepler had suggested naming the four moons after four objects of Jupiter's affection. See A. O. Pritchard, "The 'Mundus Iovialis' of Simon Marius," *The Observatory* 39 (1916): 367–381, 403–412, 443–452, 498–504, on 380.

20. The lynx, symbol of the scientific academy established by Federico Cesi, was said to be the most sharp-sighted of animals; see in this connection Christoph Lüthy, "Atomism, Lynceus, and the Fate of Seventeenth-Century Microscopy," *Early Science and Medicine* 1.1 (1996): 1–27.

21. The Moon.

Better to show your valor.
So the day-bearing Sun on lending you his light
Learns that you have long since seized it
From the brightest part of his fair face.
Thus you who showed us first
Those spots scattered about his bright mantle
Artfully adorn your name with a deathless splendor.[22]

5. Printer's note about the addition of Scheiner's tracts

Giacomo Mascardi
Typographer

To the Reader

I considered it essential to present to you here the *Epistolæ* and the *Disquisitio* of the hidden Apelles, for very few copies of these works have arrived from Germany, and I hear that not many copies were distributed elsewhere either. Had I not published them here, it would have been more difficult for you to examine and to weigh them with care. It would be, however, necessary to read and consider these works, for there are many allusions to and discussions of them in the first Phœbean text, which is by the most learned Galilei. For this reason, I have referred with numerous marginal notes to the pages and particular passages in question in the *Epistolæ* and the *Disquistio*; I have done so in two different fonts, the first corresponding to the [original] Augsburg edition, and the second to mine. They refer to the same argument that was sent to the Most Illustrious Welser, for it was my aim to satisfy you, so that here in one volume you will have everything that has been said about the sunspots, and perhaps everything that can ever be said or thought about them. It will be up to you to profit from these things and to use them in contemplation of the Sun, for you will be able to do this, if you like, through the labors of others and, as a solar observer, with the telescope. Thus you will learn what was hidden to all Antiquity. Farewell. Rome, 1 February 1613.

22. Stelluti's poem follows one version of the Petrarchan or Italian sonnet, but is notable for the absence of a *volta*, or change in the direction of the argument, normally occurring after the eighth line and in anticipation of a resolution or synthesis by the poem's end. His suggestion that the heavens somehow colluded to present Galileo with phenomena is a hyperbolic version of Angelo de Filiis' assertion that the fact that almost all these discoveries had been made by one individual was not a matter of random chance, but rather a concession of Nature or divine grace. See "Preface to the Reader," page 377.

Bibliography

Published Sources

Acanfora, Elisa. "Sigismondo Coccapani disegnatore e trattatista." *Paragone* 40 (1989): 71–99.

Aguilon, François. *Opticorum Libri Sex*. Antwerp: Plantijn, 1613.

Alessandrini, Ada. "Originalità dell'Accademia dei Lincei." *Atti dei Convegni Lincei* 78 (1986): 77–177.

Allegri, Alessandro. *Lettere di Ser Poi Pedante*. Bologna: Vittorio Benacci, 1613.

Ambassades du Roy de Siam envoyé à l'Excellence du Prince Maurice, arrivé à la Haye le 10 Septemb. 1608. The Hague, 1608. Facsimile reprint in Stillman Drake. *The Unsung Journalist and the Origin of the Telescope*. Los Angeles: Zeitlin & Verbrugge, 1976.

Andretta, S. "Conti, Carlo." In *Dizionario biografico degli italiani*, 28: 376–378.

Annales Laurissensis et Einhardi. In *Monumenta Germaniae Historica Scriptorum*. Hanover, 1826, vol. 1, pp. 124–218. Issued separately: *Scriptores Rerum Germanicarum in usum scholarum ex Monumentis Germaniae Historicis separatim editi*, vol. 6. Hanover, 1895, 1950.

Annalium et Historiæ Francorum . . . scriptores coetanei. Edited by Pierre Pithou. Paris: Claude Chappelet, 1588.

Annalium de gestis Caroli Magni Imp. Libri V. Opus auctoris quidem incerti, sed Saxonis & historici & poëtæ antiquissimi. Helmstadt: Typis Iacobi Lucij, 1594.

Aratus. *Phaenomena*. In Callimachus, Aratus, Lycophron. *Hymns, Epigrams. Phaenomena. Alexandra*. Translated by A. W. Mair and G. R. Mair. Cambridge, MA: Harvard University Press, 1955, pp. 185–299.

Argoli, Andrea. *Pandosion Sphericum*. Padua: Paolo Frambotti, 1653.

Aricò, Denise. "'In Doctrinis Glorificate Dominum': Alcuni aspetti della ricezione di Clavio nella produzione scientifica di Mario Bettini." In *Christoph Clavius e l'attività scientifica dei Gesuiti nell'età di Galileo*. Edited by Ugo Baldini. Rome: Bulzoni: 1995, pp. 189–207.

Ariew, Roger. "Galileo's Lunar Observations in the Context of Medieval Lunar Theory." *Studies in the History and Philosophy of Science* 15 (1984): 213–226.

———. "The Phases of Venus before 1610." *Studies in the History and Philosophy of Science* 18 (1987): 81–92.

Aristotle. *On the Soul*. In *On the Soul. Parva Naturalia. On Breath*. Translated by W. S. Hett. Cambridge, MA: Harvard University Press, 1935.

Arnold, Christoph. "Vita, genus, et mors auctoris nobilissimi." In Marcus Welser, *Opera historica et philologica, sacra et profana*. Edited by Christoph Arnold. Nuremberg: Wolfgang Moritz, 1682, pp. 3–68.

Ashbrook, Joseph. "Christopher Scheiner's Observations of an Object near Jupiter." *Sky and Telescope*, 42 (1971): 344–345.

Baglione, Giovanni. *Le Vite de' Pittori, scultori, et architetti*. Edited and annotated by Jacob Hess and Herwarth Röttgen. 3 vols. Vatican City: Biblioteca Apostolica Vaticana, 1995.

Baldi, Bernardino. *Le vite de' matematici*. Edited by Elio Nenci. Milan: F. Angeli, 1998.

Baldini, Ugo. "L'Astronomia del Cardinale Bellarmino." In *Novità celesti e crisi del sapere*. Edited by Paolo Galluzzi. Florence: G. Barbèra, 1984, pp. 293–305.

———, ed. *Christoph Clavius: Corrispondenza*. Pisa: Università di Pisa, Dipartimento di Matematica, 1992.

———, ed. *Christoph Clavius e l'attività scientifica dei Gesuiti nell'età di Galileo*. Rome: Bulzoni: 1995.

———. "L'insegnamento della matematica nel collegio di S. Antão a Lisbona (1590–1640)." In *A Companhia de Jesus e a Missionação no Oriente*. Lisbon: Brotéria e Fundação Oriente, 2000, pp. 275–310. Reprinted in *Saggi sulla cultura della Compagnia di Gesù (Secoli XVI–XVIII)*. Padua: CLEUP Editrice, 2000, pp. 129–167.

———. *Legem Impone Subactis: Studi su Filosofia e Scienza dei Gesuiti in Italia, 1540–1632*. Rome: Bulzoni, 1992.

———. "The Portuguese Assistancy of the Society of Jesus and Scientific Activities in Its Asian Missions until 1640." In *Proceedings of the First Meeting History of Mathematical Sciences: Portugal and East Asia*. Camarante: Fundação Oriente, 2000, pp. 49–104.

Baldini, Ugo, and George V. Coyne, eds. *The Louvain Lectures (Lectiones Lovanienses) of Bellarmine and the Autograph Copy of His 1616 Declaration to Galileo*. Vatican City: Specola Vaticana, 1984.

Barbero, G., M. Bucciantini, and M. Camerota. "Uno scritto inedito di Federico Borromeo: *L'Occhiale Celeste*." *Galilaeana, Journal of Galilean Studies* 4 (2007): 309–341.

Barker, Nicolas. *Hortus Eystettensis: the Bishop's Garden and Besler's Magnificent Book.* New York: H. N. Abrams, 1994.
Bartolini, Simone. *I fori Gnomonici di Egnazio Danti in Santa Maria Novella*. Florence: Edizioni Polistampa, 2006.
Bayer, Johann. *Uranometria*. Augsburg: Ad Insigne Pinus, 1603.
Bayle, Pierre. *Dictionnaire historique et critique*. 16 vols. Paris: Desoer, 1820.
Berthold, G. *Der Magister Johann Fabricius und die Sonnenflecken, nebst einem Excurse über David Fabricius*. Leipzig, 1894.
Bettini, Mario, S.J. *Apiaria*. Bologna: Typis Io. Baptistae Ferronij, 1642.
Bevilacqua, A. "Dotti, Vincenzo." In *Dizionario biografico degli italiani*, 41: 548–549.
Biagioli, Mario. "Filippo Salviati: A Baroque Virtuoso." *Nuncius* 7, no. 2 (1992): 81–96.
———. *Galileo Courtier: The Practice of Science in the Culture of Absolutism*. Chicago: University of Chicago Press, 1993.
———. *Galileo's Instruments of Credit: Telescopes, Images, Secrecy*. Chicago: University of Chicago Press, 2006.
———. "Picturing Objects in the Making: Scheiner, Galileo and the Discovery of Sunspots. In *Wissensideale und Wissenskulturen in der frühen Neuzeit: Ideals and Cultures of Knowledge in Early Modern Europe*. Edited by Wolfgang Detel and Claus Zittel. Berlin: Akademie Verlag, 2002, pp. 39–96.
———. *Scientists' Names and Science's Claims*. Forthcoming.
Bianchi, Luca. "Galileo fra Aristotele, Clavio e Scheiner." *Rivista di storia della filosofia* 2 (1999): 189–227.
Bibliorum sacrorum iuxta vulgatam Clementinam nova editio. Edited by Luigi Grammatica. Vatican City: Typis Polyglottis Vaticanis, 1951.
Birch, Thomas. *The History of the Royal Society of London*. 4 vols. London: A. Millar, 1760.
Blackwell, Richard J. *Behind the Scenes at Galileo's Trial*. Notre Dame, IN: University of Notre Dame Press, 2006.
———. *Galileo, Bellarmine, and the Bible*. Notre Dame, IN: University of Notre Dame Press, 1991.
Blumenthal, Arthur. *Theater Art of the Medici*. Hanover, NH, and London: University Press of New England, 1980.
Bosscha, J. "Simon Marius, réhabilitation d'un astronome calomnié." *Archives Néerlandaises des sciences exactes et naturelles, publiées par la Société Hollandaise des Sciences*, 2nd ser., 12 (1907): 258–307, 490–527.
Brahe, Tycho. *Tychonis Brahe Dani Opera Omnia*. 15 vols. in 17. Edited by J. L. E. Dreyer. Copenhagen: Libraria Gyldendaliana, 1913–29; reprint Amsterdam: Swets & Zeitlinger, 1972.
Brant, V. "L'Économie politique et sociale dans les écrits de L. Lessius (1554–1623)." *Revue d'histoire ecclésiastique*, 13 (1912): 73–89, 302–318.
Braunmühl, Anton von. *Christoph Scheiner als Mathematiker, Physiker und Astronom*. Bayrische Bibliothek, vol. 24. Bamberg: Buchnersche Verlagsbuchhandlung, 1891.

Bredekamp, Horst. *Galileo Galilei der Künstler: Der Mond, die Sonne, die Hand.* Berlin: Akademie-Verlag, 2007.

Brótons, Victor Navarro. "Contribución a la historia del copernicanismo en España." *Cuadernos Hispanoamericanos* 283 (1974): 3–23.

Bucciantini, Massimo. "Dopo il *Sidereus Nuncius*: Il Copernicanesimo in Italia tra Galileo e Keplero." *Nuncius* 9 (1994): 15–35.

———. *Galileo e Keplero: Filosofia cosmologia e teologia nell'Età della Controriforma.* Turin: Einaudi, 2003.

———. See Barbero, G., M. Bucciantini, and M. Camerota.

Bury, Michael. *Print in Italy*. London: British Museum, 2001.

Calitti, Floriana. "Fatica o ingegno: Lelio Capilupi e la pratica del centone." In *Scritture di Scritture: Testi, generi, modelli nel rinascimento*. Edited by Giancarlo Mazzacurati and Michel Plaisance. Rome: Bulzoni, 1987, pp. 497–507.

Camerota, Filippo. *La prospettiva del Rinascimento: Arte, architettura, scienza*. Milan: Mondadori, 2006.

Camerota, Michele. "Adattar la volgar lingua ai filosofici discorsi: una inedita orazione di Niccolo Aggiunti contro Aristotele e per l'uso della lingua italiana nelle dissertazioni scientifiche." *Nuncius* 13:2 (1998): 595–623.

———. *Galileo Galilei e la cultura scientifica nell'età della Controriforma*. Rome: Salerno, 2004.

———. See Barbero, G., M. Bucciantini, and M. Camerota.

———. "Flaminio Papazzoni: un aristotelico bolognese maestro di Federico Borromeo e corrispondente di Galileo." In *Method and Order in Renaissance Philosophy of Nature*. Edited by Daniel A. Di Liscia, Eckhard Kessler, and Charlotte Methuen. Aldershot: Ashgate, 1997, pp. 271–300.

———. *Galileo Galilei e la cultura scientifica nell'Età della Controriforma*. Rome: Salerno Editrice, 2004.

Cardi da Cigoli, Giovanni Battista. "Vita di Lodovico Cigoli." In *Lodovico Cardi-Cigoli pittore e architetto*. Edited by Anna Matteoli. Pisa: Giardini Editore, 1980.

Cardi da Cigoli, Lodovico. See Cigoli, Lodovico Cardi da.

Carrara, Bellino, S.J. "L''Unicuique Suum' nella scoperta delle macchie solari." *Memorie della Pontificia Accademia dei Nuovi Lincei* 23 (1905): 191–287; 24 (1906): 47–127.

Casanovas, J. "Early Observations of Sunspots: Scheiner and Galileo." 1st Advances in Solar Physics Euroconference: Advances in the Physics of Sunspots. *ASP Conference Series*, vol. 118. Edited by B. Schmieder, J. C. del Toro Iniesta, and M. Vásquez. 1997.

Chappell, Miles. "Cigoli, Galileo, and *Invidia*." *Art Bulletin* 57 (1975): 91–98.

Christmann, Jacob. *Nodus gordius ex doctrina sinuum explicata. Accedit appendix observationum, quae per radium artificiosum habitae sunt circa Saturnum, Iovem et lucidiores stellas affixas*. Heidelberg, 1612.

Cigoli, Lodovico Cardi da. *Prospettiva Pratica*. Edited by Filippo Camerota. Florence: Leo S. Olschki, 2010.

———. *Macchie di sole e pittura: carteggio L. Cigol–G. Galilei (1609–1613).* San Miniato: Accademia degli Eutelèti, 1959.

Cicero. *Commentariolum petitionis.* In *Letters to Brutus and Quintus. Letter Fragments. Letter to Octavian. Invectives. Handbook of Electioneering.* Edited and translated by D. R. Shackleton Bailey. Cambridge, MA: Harvard University Press, 2002.

Clark, David H., and F. Richard Stephenson. "An Interpretation of the Pre-telescopic Sunspot Records from the Orient." *Quarterly Journal of the Royal Astronomical Society* 19 (1978): 387–410.

Clavius, Christoph. *Christoph Clavius: Corrispondenza.* Edited by U. Baldini and P. D. Napolitani. 6 vols. Pisa: Università di Pisa, Dipartimento di matematica, 1992.

———. *Commentarius in Sphaeram Ioannis de Sacro Bosco.* Vol. 3 in *Opera Mathematica.* 5 vols. Mainz: Antonius Hierat, 1612.

———. *In Sphaeram Ioannis de Sacro Bosco Commentarius.* Rome, 1570.

———. *In Sphaeram Ioannis de Sacro Bosco Commentarius.* Final edition. Mainz: Reinhard Eltz, 1611. Reprinted in *Opera Mathematica.*

———. *Opera Mathematica.* 5 vols. Mainz: Antonius Hierat, 1612.

Colombe, Lodovico delle. *Contro il Moto della Terra.* In *Le Opere di Galileo Galilei,* 3: 253–290.

Colomero, C. "Chiocco, Andrea." In *Dizionario biografico degli italiani,* 25: 11–12.

Copernicus, Nicholas. *On the Revolutions.* Translated with commentary by Edward Rosen. Baltimore: Johns Hopkins University Press, 1978.

Coyne, George V. See Baldini, Ugo, and George V. Coyne.

Cysat, Johann Baptist. *De Loco, motu, magnitudine, et causis cometæ.* Ingolstadt: Elisabeth Angermaria, 1619.

Dame, Bernard. "Galilée et les Taches Solaires (1610–1613)." *Revue d'Histoire des Sciences,* 19 (1966): 306–70. Reprinted as pp. 186–251 in *Galilée: Aspects de sa Vie et de son Oeuvre.* Paris: Centre Internationale de Synthèse, 1968.

Daxecker, Franz. *Briefe des Naturwissenschaftlers Christoph Scheiner SJ an Erzherzog Leopold V. von Österreich-Tirol, 1620–1632.* Veröffentlichungen der Universität Innsbruck, vol. 207. Innsbruck: Publikationsstelle der Universität Innsbruck, 1995.

———. "Christoph Scheiners Lebensjahre zwischen 1633 und 1650." *Beiträge zur Astronomiegeschichte* 5 (2002): 40–46.

———. *Das Hauptwerk des Astronomen P. Christoph Scheiner SJ,* Rosa Ursina sive Sol*—eine Zusammenfassung.* Supplement 13 of *Berichte des Naturwissenschaftlich-Medizinischen Vereins in Innsbruck.* 1996.

———. *The Physicist and Astronomer Christoph Scheiner: Biography, Letters, Works.* Innsbruck: Leopold Franzens University of Innsbruck, 2004.

———. "'Über das Fernrohr' und weitere Mitschriften von Vorlesungen Christoph Scheiners." *Beiträge zur Astronomiegeschichte* 4 (2001): 19–32.

Daxecker, Franz, and Florian Schaffenrath. "Ein Nachruf auf den Astronomen Christoph Scheiner (1673–1650)." *Berichte des Naturwissenschaftlich-Medizinischen Vereins in Innsbruck* 88 (2001): 373–382.

———. "Neue Dokumenten zu Christoph Scheiner: These theologicae, Vorlesungsmitschriften und ein Nachruf aus dem Jahr 1650." *Sammelblatt des Historischen Vereins Ingolstadt* 110 (2001):143–147.

Daxecker, Franz, Florian Schaffenrath, and Lav Subaric. "Briefe Christoph Scheiners von 1600 bis 1634." *Sammelblatt des Historischen Vereins Ingolstadt* 110 (2001): 117–141.

Daxecker, Franz, and Lav Subaric. "Briefe der Generaloberen P. Claudio Aquaviva SJ, P. Mutio Vitelleschi SJ und P. Vincenzo Carafa SJ an den Astronomen P. Christoph Scheiner SJ von 1614 bis 1649." *Sammelblatt des Historischen Vereins Ingolstadt* 111 (2002): 101–148.

———. *Christoph Scheiners* Sol Ellipticus. Innsbruck: Veröffentlichungen der Universität Innsbruck, vol. 226, 1998.

Dear, Peter. *Discipline and Experience: The Mathematical Way in the Scientific Revolution*. Chicago: University of Chicago Press, 1995.

Débarbat, Suzanne, and Wilson, Curtis. "The Galilean Satellites of Jupiter from Galileo to Cassini, Römer and Bradley." In *The General History of Astronomy*. 4 vols. Cambridge: Cambridge University Press, 1984–, IIA:144–157.

DeCeglia, Francesco Paolo. "Giorgio Coresio: Note in merito a un difensore dell'opinione d'Aristotele." *Physis* 37.2 (2000): 393–437.

Deiss, Bruno M., and Volker Nebel. "On a Pretended Observation of Saturn by Galileo." *Journal for the History of Astronomy* 29: 3 (1998): 215–220.

Dekker, Elly. "The Light and the Dark: A Reassessment of the Discovery of the Coalsack Nebula, the Magellanic Clouds and the Southern Cross." *Annals of Science* 47 (1990): 529–560.

Dick, Steven J. *Plurality of Worlds: The Origins of the Extraterrestrial Life Debate from Democritus to Kant*. Cambridge: Cambridge University Press, 1982.

Dizionario biografico degli Italiani. Alberto M. Ghisalberti, general editor. 65 vols. to date. Rome: Istituto della Enciclopedia Italiana, 1960–.

Dollo, Corrado. "Le ragioni del geocentrismo nel Collegio Romano (1562–1612)." In *La diffusione del Copernicanesimo in Italia*. Edited by Massimo Bucciantini and Maurizio Torrini. Florence: Leo S. Olschki, 1997, pp. 99–166.

———. "Tanquam nodi in tabula—tanquam pisces in aqua: Le innovazioni della cosmologia nella *Rosa Ursina* di Christoph Scheiner." In *Christoph Clavius e l'attività scientifica dei Gesuiti nell'età di Galileo*. Edited by Ugo Baldini. Rome: Bulzoni, 1996, pp. 133–158.

Donahue, William H. *The Dissolution of the Celestial Spheres*. New York: Arno Press, 1981.

Drake, Stillman. *Discoveries and Opinions of Galileo*. New York: Doubleday, 1957.

———. "The Dispute over Bodies in Water." In *Galileo Studies: Personality, Tradition, and Revolution*. Edited by Stillman Drake. Ann Arbor: University of Michigan Press, 1970, pp. 159–176.

———. *Galileo at Work: His Scientific Biography*. Chicago: University of Chicago Press, 1978.

———. "Sunspots, Sizzi, and Scheiner." In *Galileo Studies: Personality, Tradition, and*

Revolution. Edited by Stillman Drake. Ann Arbor: University of Michigan Press, 1970, pp. 177–199.

———. *Telescopes, Tides, and Tactics: A Galilean Dialogue about the* Starry Messenger *and Systems of the Worlds*. Chicago: University of Chicago Press, 1983.

———. *The Unsung Journalist and the Origin of the Telescope*. Los Angeles: Zeitlin & Verbrugge, 1976.

———. See *Ambassades du Roi de Siam*.

Dralli, Tommaso. *Theses ex universa philosophia illustrissimo Pietro Card. Aldobrandino . . . inscriptas . . . in collegio Romano Soc. Iesu publicè defendit*. Rome: Bartolomeo. Zannetti, 1613.

Duhr, Bernard. *Geschichte der Jesuiten in den Ländern deutscher Zunge*. 4 vols. Freiburg: Herder, 1907–1913.

Eastwood, Bruce S. "'The Chaster Path of Venus,' Orbis Veneris castior, in the Astronomy of Martianus Capella." *Archives Internationales d'Histoire des Sciences*, 32 (1982): 145–158.

———. "Plato and Circumsolar Planetary motion in the Middle Ages." *Archives d'histoire doctrinale et littéraire du Moyen Age*, 68 (1993): 7–26.

Eddy, John A. "The Case of the Missing Sunspots." *Scientific American* 236, no. 5 (1977): 80–95.

———. "Climate and the Role of the Sun." *Journal of Interdisciplinary History* 10, no. 4 (1980): 725–747.

Einhard. *Vita Caroli Imperatori*. In *Patrologia Latina*. Edited by Jacques-Paul Migne. 221 vols. Paris: Garnier & J.-P. Migne, 1844–1865, vol. 97.

———. *Vita Karoli Magni*. In *Annalium et Historiæ Francorum*. Edited by Pierre Pithou. Paris: Claude Chappelet, 1588.

Elia, Pasquale d'. *Galileo in Cina: relazioni attraverso il Collegio Romano tra Galileo e i Gesuiti scienziati missionari in Cina, 1610–1640*. Rome: Gregorian University, 1947.

Embassies of the King of Siam Sent to his Excellency Prince Maurits, Arrived in The Hague on 10 September 1608. Translated from the French into English and Dutch by Henk Zoomers, edited by Huib Zuidervaart. Wassenaar, Netherlands: Louwman Collection of Historic Telescopes, 2008.

Emler, J., ed. *Cosmae chronicon boemorum cum continuatoribus*, in *Fontes rerum bohemicarum*, 7 vols. (Prague: Nákl. N. F. Palackého, 1873–1932); vol. 2 (1874).

Euclid. *The Thirteen Books of Euclid's Elements*. Translated with commentary by T. L. Heath. 2nd ed. Cambridge: Cambridge University Press, 1956. Reprint, New York: Dover, 1975, 1995.

Evans, R. J. W. "Rantzau and Welser: Aspects of Later German Humanism." *History of European Ideas* 5, no. 3 (1984): 257–272.

Fabricius, Johannes. *Joh. Fabricii Phrysii De Maculis in Sole observatis, et apparente earum cum Sole conversione narratio*. Wittenberg: Impensis Johan Borneri Senioris & Eliae Rehefeldij, 1611.

Favaro, Antonio. "Dal Carteggio di Marco Velser con Giovanni Faber." *Memorie dell'Istituto Veneto* 24 (1891): 79–100.

———. "Iconografia galileiana." *Atti del Reale Istituto Veneto di Scienze Lettere ed Arti* 72.2 (1912–1913): 995–1055.

———. "Oppositori di Galileo. III. Christoforo Scheiner." *Atti del Reale Istituto Veneto di Scienze, Lettere ed Arti,* 78 (1918–1919): 1–107.

———. "Sulla Morte di Marco Velsero e sopra alcuni Particolari della Vita di Galileo." *Bullettino di Bibliografia e di Storia delle Scienze Matematiche e Fisiche* 17 (1884): 252- 270.

———. "Sulla priorità della scoperta e della osservazione delle macchie solari." *Memorie del Reale Istituto Veneto di Scienze, Lettere ed Arti* 22 (1887): 729–790.

Feldhay, Rivka. *Galileo and the Church: Political Inquisition or Critical Dialogue?* Cambridge: Cambridge University Press, 1995.

———. "Producing Sunspots on an Iron Pan." In *Science, Reason, and Rhetoric.* Edited by Henry Krip, J. E. McGuire, and Trevor Melia. Pittsburgh: University of Pittsburgh Press, 1998, pp. 119–143.

Findlen, Paula. "The Economy of Scientific Exchange in Early Modern Italy."In *Patronage and Institutions: Science, Technology, and Medicine at the European Court 1500–1750.* Edited by Bruce T. Moran. Rochester, NY: Boydell Press, 1991, pp. 5–24.

Finocchiaro, Maurice A. *Galileo on the World Systems: A New Abridged Translation and Guide.* Berkeley: University of California Press, 1997.

Folkerts, Menso. "Der Astronom David Fabricius (1564–1617): Leben und Wirken." *Berichte zur Wissenschaftsgeschichte* 23 (2000): 127–142.

Freedberg, David. *The Eye of the Lynx: Galileo, His Friends, and the Beginnings of Modern Natural History.* Chicago: University of Chicago Press, 2002.

Gabrieli, Giuseppe. *Contribuiti alla storia della Accademia dei Lincei.* 2 vols. Rome: Accademia nazionale dei Lincei, 1989.

———. "Marco Welser Linceo Augustano." *Rendiconti della Reale Accademia Nazionale dei Lincei,* 6th ser., 13 (1937): 74–99.

Galilei, Galileo. *Dialogues Concerning Two New Sciences.* Translated by Henry Crew and Alfonso de Salvio. New York: MacMillan, 1914.

———. *Dialogo sopra i due Massimi Sistemi del Mondo Tolemaico e Copernicano.* Edited and annotated by Ottavio Besomi and Mario Helbing. 2 vols. Padua: Antenore, 1998.

———. *Dialogue Concerning the Two Chief World Systems, Ptolemaic & Copernican.* 2nd rev. ed., trans. with revised notes by Stillman Drake; foreword by Albert Einstein. Berkeley: University of California Press, 1967.

———. *Difesa contro alle calunnie et imposture di Baldessar Capra.* In *Le Opere di Galileo Galilei,* 2:513–599.

———. *Discourse presented to the Most Serene Don Cosimo II. Great Duke of Tuscany, concerning the Natation of Bodies Upon And Submersion In, the Water. By Galileus Galilei: Philosopher and Mathematician unto His most Serene Highnesse. Englished from the Second Edition of the Italian compared with the Manuscript Copies, and reduced into Propositions: By Thomas Salusbury, Esquire.* London, 1663.

———. *Discourse on Bodies in Water*. Translated by Thomas Salusbury and edited by Stillman Drake. Urbana: University of Illinois Press, 1960.

———. *Le Messager Céleste*. Translation and commentary by Isabelle Pantin. Paris: Les Belles Lettres, 1992.

———. *Le Opere di Galileo Galilei*. 20 vols. Edited by Antonio Favaro. Florence: G. Barbèra, 1968.

———. *Sidereus Nuncius, or, The Sidereal Messenger*. Translated with introduction, conclusion, and notes by Albert Van Helden. Chicago: University of Chicago Press, 1989.

Gallo, Italo. "Filosofia e scienza agli albori del seicento: Giulio Cesare Lagalla tra Aristotele e Galilei." *Rassegna Storica Salernitana*, N.S., 6 (1986): 55–75; reprinted in *Res Publica Litterarum: Studies in the Classical Tradition* 10 (1987): 111–125.

Gatto, Romano. *Tra scienza e immaginazione: le matematiche presso il collegio gesuitico napoletano (1552–1670 ca.)*. Florence: Leo S. Olschki, 1994.

Gines, Juan Vernet. "Copernicus in Spain." *Colloquia Copernicana* 1 (1972): 271–292.

Gingerich, Owen. "The Galileo Sunspot Controversy: Proof and Persuasion." *Journal for the History of Astronomy* 34 (2003): 77–78.

———. "The Mysterious Nebulae, 1610–1924." *Journal of the Royal Astronomical Society of Canada*, 81 (1987): 113–127.

[Goes, Emmanuel de]. *Commentarii Collegii Conimbricensis Societatis Iesu in quatuor libros De coelo, Meteorologicos et Parva Naturalia*. Coimbra, 1592.

Goldstein, Bernard R. *The Arabic Version of Ptolemy's "Planetary Hypotheses."* American Philosophical Society, *Transactions* 57 (part 4), 1967.

———. "The Pre-telescopic Treatment of the Phases and Apparent Size of Venus." *Journal for the History of Astronomy* 27 (1996): 1–12.

———. "Some Medieval Reports of Venus and Mercury Transits." *Centaurus* 14 (1969): 49–59. Reprinted in *Theory and Observation in Ancient and Medieval Astronomy*. London: Variorum Reprints, 1985, no. 15.

Goldstein, Bernard, and Giora Hon. "*Symmetry* in Copernicus and Galileo." *Journal for the History of Astronomy* 35 (2004): 1–20.

Gorman, Michael John. "Mathematics and Modesty in the Society of Jesus: The Problems of Christopher Grienberger (1564–1636)." In *The New Science and Jesuit Science: Seventeenth-Century Perspectives*. Edited by Mordechai Feingold. *Archimedes: New Studies in the History and Philosophy of Science and Technology*, vol. 6, pp. 1–120. Dordrecht: Kluwer, 2003.

———. *The Scientific Counter-Revolution: Mathematics, Natural Philosophy, and Experimentalism in Jesuit Culture, 1580–c. 1670*. Rome: European University Institute, 1998.

Grant, Edward. "Cosmology." In *Science in the Middle Ages*. Edited by David C. Lindberg. Chicago: University of Chicago Press, 1978, pp. 265–302.

———. *Planets, Stars, and Orbs: The Medieval Cosmos, 1200–1687*. Cambridge: Cambridge University Press, 1994.

[Grassi, Orazio, S.J.] *De tribus Cometis anni MDCXVIII Disputatio Astronomica*. Rome, 1619.

Greco, Vincenzo, Giuseppe Molesini, and Franco Quercioli. "Optical Tests of Galileo's Lenses." *Nature*, 358, no. 9 (July, 1992): 101.

Grillo, Angelo. *Delle lettere . . . raccolte dal Sig. Pietro Petracci*. 3 vols. Venice: Evangelista Deuchino, 1616.

Grotefend, H. *Zeitrechnung des Deutschen Mittelalters und der Neuzeit*. 2 vols. Hanover: Hahn, 1891–1898.

Gualterotti, Raffaello. *Discorso sopra l'apparizione de la nuova stella*. Florence: Cosimo Giunti, 1605.

Guaragnella, Pasquale. "Icone e linguaggio verbale in Galilei: Sulla questione delle macchie solari." *Scienze umane* 6 (1980): 267–278.

Guerrini, Luigi. "Le *Stanze sopra le stelle e macchie solari scoperte col nuovo occhiale* di Vicenzo Figliucci." *Lettere italiane* 50:3 (1998): 387–415.

———. "Nel Dedalo del Cielo: Gassendi e le macchie solari in un inedito del 1633." *Nuncius* 16, no. 2 (2001): 643–672.

———. "Tradizione astronomica e cultura matematica in un orazione gesuitica del 1619." *Giornale critico della filosofia italiana* 79 (2000): 209–235.

Hafenreffer, Samuel. See Maestlin, Michael.

Hakluyt, Richard. *The Principal Navigations Voyages Traffiques & Discoveries of the English Nation*. 12 vols. Glasgow: James MacLehose and Sons, 1903–1905.

Harrison, Thomas G. "The Orion Nebula: Where in History Is It?" *Quarterly Journal of the Royal Astronomical Society* 25 (1984): 65–79.

Haub, Rita. "'Ich bin eines armen Bäckers Sohn!' Matthäus Rader S.J. (1561–1634)." *Archivum Historicum Societatis Iesu* 139 (2001): 173–180.

Hearnshaw, J. B. *The Analysis of Starlight: One Hundred Years of Spectroscopy*. Cambridge: Cambridge University Press, 1986.

Heath, T. L. See Euclid.

Hellman, C. Doris. *The Comet of 1577: Its Place in the History of Astronomy*. New York: AMS Press, 1944.

Hellyer, Marcus. *Catholic Physics: Jesuit Natural Philosophy in Early Modern Germany*. Notre Dame: Notre Dame University Press, 2005.

Herr, Richard B. "Solar Rotation Determined from Thomas Harriot's Sunspot Observations of 1611 to 1613." *Science* 202 (1978): 1079–1081.

Ho, P.-Y. "Natural Phenomena Recorded in the 'Dai-viet Su'ky Toanthu.'" *American Oriental Society Journal* 84 (1964): 127–149.

Hollstein, F. W. H. *German Engravings, Etchings and Woodcuts*. Vol. 12. Edited by Tilman Falk. Amsterdam: Van Gendt & Co, 1983.

Hon, Giora. See Goldstein, Bernard R., and Giora Hon.

Huggins, William, and W. A. Miller. "On the Spectra of Some Nebulae." *Philosophical Transactions of the Royal Society of London* 154 (1864): 437–444.

Humbert, Pierre. *Un Amateur: Peiresc, 1580–1637*. Paris: Desclée de Brouwer et Cie, 1933.

Hutchison, Keith. "Sunspots, Galileo, and the Orbit of the Earth." *Isis* 81 (1990): 68–74.

Huygens, Christiaan. *Oeuvres Complètes de Christiaan Huygens*. 22 vols. The Hague: Martinus Nijhoff, 1888–1950.

Ineichen, G. "Carrara (Alberti), Giovanni Michele Alberto." In *Dizionario biografico degli Italiani*, 20:684–686.

Ingaliso, Luigi. *Filosofia e cosmologia in Christoph Scheiner*. Soveria Mannelli: Rubbettino Editore, 2005.

Jervis, Jane L. *Cometary Theory in Fifteenth-Century Europe*. Dordrecht: Reidel, 1985.

Jiang, Yaotiao. See Xu, Zhentau, David W. Pankenier, and Yaotiao Jiang.

John of Worcester. *The Chronicle of John of Worcester*. Edited by J. R. H. Weaver. Oxford: Clarendon Press, 1908. Reprint, New York: AMS Press, 1989.

Kemp, Martin. *The Science of Art: Optical Themes in Western Art from Brunelleschi to Seurat*. New Haven: Yale University Press, 1990.

Kepler, Johannes. *Admonitio ad astronomos*. Leipzig: J. A. Minzelius, 1629; Frankfurt: Godefridum Tampachium, 1630. In *Johannes Kepler Gesammelte Werke*, 11.1: 475–482.

———. *Ad Vitellionem paralipomena, quibus astronomiae pars optica traditur* (1604). In *Johannes Kepler Gesammelte Werke*, vol. 2.

———. *Dissertatio cum Nuncio Sidereo* (1610). In *Johannes Kepler Gesammelte Werke* 4: 281–311.

———. *Johannes Kepler Gesammelte Werke*. 21 vols. to date. Munich: C. H. Beck, 1937–.

———. *Kepler's Conversation with Galileo's Sidereal Messenger*. Translated with commentary by Edward Rosen. New York: Johnson Reprint Corp., 1965. In *Johannes Kepler Gesammelte Werke* 11.1: 315–438.

———. *Kepler's Somnium*. Translated with commentary by Edward Rosen. Madison: University of Wisconsin Press, 1967.

———. *De stella nova* (1606). In *Johannes Kepler Gesammelte Werke* 1: 147–390.

———. *Phaenomenon singulare seu Mercurius in Sole* (1609). In *Johannes Kepler Gesammelte Werke*. 4: 77–98.

Klug, Joseph. "Simon Marius aus Gunzenhausen und Galileo Galilei." *Abhandlungen der II. Klasse der Königliche Akademie der Wissenschaften* 22 (1906): 385–526.

Krayer, Albert. *Mathematik im Studienplan der Jesuiten: Die Vorlesung von Otto Cattenius an der Universität Mainz (1610/11)*. Stuttgart: Franz Steiner Verlag, 1991.

Krivsky, L. "Naked Eye Observations of Sunspots in Bohemia in the Year 1139." *Bulletin of the Astronomical Institute of Czechoslovakia* 36 (1985): 60–61.

Lagalla, Giulio Cesare. *De phaenomenis in orbe* lunae (1612). In *Le Opere di Galileo Galilei*, vol. 3, part 1, pp. 311–393.

Lattis, James M. *Between Copernicus and Galileo: Christoph Clavius and the Collapse of Ptolemaic Cosmology*. Chicago: University of Chicago Press, 1994.

Leitão, Henrique. "Galileo's Telescopic Observations in Portugal." In *Largo Campo di Filosofare. Eurosymposium Galileo 2001*. Edited by José Montesinos and Carlos Solís. Fundación Canaria Orotava de Historia de la Ciencia, 2001, pp. 903–913.

———. "A Periphery Between Two Centres? Portugal on the Scientific Route from Europe to China (Sixteenth and Seventeenth Centuries)." In *Travels of Learning: A Geography of Science in Europe*. Edited by Ana Simões, Ana Carneiro, and Maria Paula Diogo. Dordrecht: Kluwer Academic Publishers, 2003, pp. 19–46.

Lenk, Leonhard. *Augsburger Bürgertum im Späthumanismus und Frühbarock (1580–1700)*. Augsburg: H. Mühlberger, 1968.

Leopardi, Giacomo. "Dialogo della Terra e della Luna." In *Operette Morali*. Edited by Giorgio Ficara. Milan: Arnaldo Mondadori, 1988, pp. 81–88.

———. *Storia dell'astronomia dalla sua origine fino all'anno MDCCCXIII*. Edited by Armando Massarenti. Milan: Edizioni La Vita Felice, 1997.

Lessius, Leonard. *De Providentia Numinis et Animi Immortalitate*. Antwerp: Plantin, 1613.

Lindberg, David C. *Theories of Vision from Al-Kindi to Kepler*. Chicago: University of Chicago Press, 1976.

Locher, Georg Joachim. See Scheiner, Christoph, S.J., and Georg Joachim Locher.

Lohr, Charles H. "Latin Aristotle Commentaries: Authors C." *Renaissance Quarterly* 28 (1975): 689–741.

Lorini, Giovanni. *Commentarii in Ecclesiasten*. Lyon, 1606.

Lotter, Tobias Conrad. *Germaniae aliorumque quorundam Locorum Europae Poliometria*. Augsburg, ca. 1770.

Ludendorff, H. "Über die erste Verbindung des Fernrohres mit astronomischen Meßinstrumenten." *Astronomische Nachrichten* 213, no. 5112 (1921): 385–390.

Lüthy, Christoph. "Atomism, Lynceus, and the Fate of Seventeenth-Century Microscopy." *Early Science and Medicine* 1, no. 1 (1996): 1–27.

Maestlin, Michael. *Disputatio de Multivariis Motuum Planetarum in Coelo Apparentibus Irregularitatibus, seu Regularibus Inaequalitatibus, earumque Causis Astronomicis. Quam Praeside Michaele Mästlino . . . Defendere Conabitur die 21. et 22. Februarii . . . M. Samuel Hafenreffer*. Tübingen, 1606.

Magini, Giovanni Antonio. *Ephemeridum Coelestium Motuum, ad Annos XL*. Venice: Apud Damianum Zenarium, 1582.

———. *Tabula Tetragonica*. Venice, 1612.

———. *Tabulae Secundorum Mobilium Coelestium*. Venice: D. Zenarus, 1585.

Malapert, Charles. *Oratio habita Duaci*. Douai: Typis Baltazaris Belleri sub Circino Aureo, 1620.

Marius, Simon. *Mundus Iovialis*. Nuremberg, 1614.

———. *Simon Marius, Mundus Iovialis—Die Welt des Jupiter*. Translated by Joachim Schlör, with commentary by Alois Wilder. Gunzenhausen: Schenk Verlag, 1988.

Metius, Adriaan. *Institutiones Astronomicae & Geographicae, Fondamentale ende Grondelijcke Onderwysinghe van de Sterrekonst, ende Beschryvinghe der Aerden, door het Ghebruyck van de Hemelsche ende Aerdtsche Globen*. Franeker, 1614.

Miller, W. A. See Huggins, William, and W. A. Miller.

Mitchell, Walter M. "The History of the Discovery of the Solar Spots." *Popular*

Astronomy 24 (1916): 22–31, 82–96, 149–162, 206–218, 290–303, 341–354, 428–440, 488–499, 562–570.

Molesini, Giuseppe. See Greco, Vincenzo, Giuseppe Molesini, and Franco Quercioli.

Moss, Jean Dietz. *Novelties in the Heavens: Rhetoric and Science in the Copernican Controversy*. Chicago: University of Chicago Press, 1993.

Mossman, J. E. "A Comprehensive Search for Sunspots Without the Aid of a Telescope, 1981–1982." *Quarterly Journal of the Royal Astronomical Society* 30 (1989): 59–64.

Muccillo, M. "De Filiis, Anastasio." In *Dizionario biografico degli italiani*, 33: 744–745.

Mueller, Paul R. "Unblemished Success: Galileo's Sunspot Argument in the *Dialogue*." *Journal for the History of Astronomy* 31 (2000): 279–300.

Müller, Johannes. "Der Zusammenbruch des Welserischen Handelshauses im Jahre 1614." *Vierteljahrschrift für Social- und Wirtschaftsgeschichte* 1 (1903): 196–234.

Nebel, Volker. See Deiss, Bruno M., and Volker Nebel.

Neumeister, Sebastian, and Conrad Wiedemann, eds. *Res Publica Litteraria: Die Institutionen der Gelehrsamkeit in der frühen Neuzeit*. Wiesbaden: Otto Harrassowitz, 1987.

North, John D. "Thomas Harriot and the First Telescopic Observations of Sunspots." In *Thomas Harriot: Renaissance Scientist*. Edited by John W. Shirley. Oxford: Clarendon Press, 1974, pp.129–165.

Odontius, Johann Caspar. *Maculae Solares ex selectis Observationibus Petri Saxonii Holsati* (Nuremberg, 1616).

Olbers, Wilhelm. "Materiellen zu einer Lebensbeschreibung der beiden Astronomen David und Johannes Fabricius." *Astronomische Nachrichten* 31 (1850): 129–142. Reprinted in *Wilhelm Olbers, Sein Leben und seine Werke*. 2 vols. Berlin: Springer Verlag, 1894–1909, 1: 200–211.

Oudemans, J. A. C., and J. Bosscha. "Galilée et Marius." *Archives Néerlandaises des sciences exactes et naturelles, publiées par la Société Hollandaise des Sciences*, 2nd ser., 8 (1903): 115–189.

[Paitoni, Iacopo Maria]. *Lettere d'uomini illustri che fiorirono nel principio del secolo decimosettimo*. Venice: Baglioni, 1744.

Pankenier, David W. See Xu, Zhentau, David W. Pankenier, and Yaotiao Jiang.

Panofsky, Erwin. *Galileo as a Critic of the Arts*. The Hague: Martinus Nijhoff, 1954.

———. "Galileo as a Critic of the Arts: Aesthetic Attitude and Scientific Thought." *Isis* 47 (1956): 3–15.

Papy, Jean. "Lipsius and Marcus Welser: The Antiquarian's Life as *via media*." *Bulletin de l'Institut Historique Belge de Rome* 68 (1998): 173–190.

Peiresc, Nicholas Claude Fabri de. *Lettres de Peiresc*. Edited by Philippe Tamizey de Larroque. 6 vols. Paris: Imprimerie Nationale, 1893.

Pico, Giovanni della Mirandola. *Disputationes in astrologiam*. In *Opera Omnia*. 2 vols. Basel, 1572. Reprint with foreword by Eugenio Garin, 2 vols., Turin: Bottega d'Erasmo, 1971, 1: 411–732.

Pineda, Juan de. *Commentariorum in Iob libri tredecim*. Hispali: In Collegio D. Ermenegildi Societatis Iesu: Excudebat Ioannis Rene, 1598.

Pithou, Pierre, ed. *Annalium et Historiæ Francorum*. Paris: Claude Chappelet, 1588.

Plato. *Timaeus*. In *Timaeus, Critias, Cleitophon, Menexenus, Epistles*. Translated by R. G. Bury. Cambridge, MA: Harvard University Press, 1929.

Pliny the Elder. *Natural History XXXV* in *Natural History XXXIII–XXXV*. Vol. 9. Translated by H. Rackham. Cambridge, MA: Harvard University Press, 1952.

Poetæ saxonis Annalium de gestis B. Caroli Magni libri quinque. In *Patrologia Latina*. Edited by Jacques-Paul Migne. 221 vols. Paris: Garnier & J.-P. Migne, 1844–1865, vol. 99.

Polis, Gary A., ed. *The Biology of Scorpions*. Stanford: Stanford University Press, 1990.

Prickard, A. O. "The *Mundus Jovialis* of Simon Marius." *Observatory* 39 (1916): 367–381, 403–412, 443–452, 498–503.

Ptolemy, Claudius. *Ptolemy's Almagest*. Translated by G. J. Toomer. London: Duckworth; New York: Springer Verlag, 1984.

Quercioli, Franco. See Greco, Vincenzo, Giuseppe Molesini, and Franco Quercioli.

Rader, Matthäus, S.J. *P. Matthäus Rader S.J., Briefwechsel*, vol. 1: *1595–1612*. Edited by Helmut Zäh, Silvia Strodel, and Alois Schmid. Munich: C. H. Beck'sche Verlagsbuchhandlung, 1995.

Rädle, Fidel. "Die Briefe des Jesuiten Georg Stengel (1584–1651) an seinen Bruder Karl (1581–1663)." In *Res Publica Litteraria: Die Institutionen der Gelehrsamkeit in der frühen Neuzeit*. Edited by Sebastian Neumeister and Conrad Wiedemann. 2 vols. Wiesbaden: Otto Harrassowitz, 1987, 2: 525–534.

Reeves, Eileen. "Faking It: Apelles and Protogenes among the Astronomers." *Bildwelten des Wissens. Kunsthistorisches Jahrbuch für Bildkritik* 5.2 (2007): 65–72.

———. *Painting the Heavens*. Princeton: Princeton University Press, 1997.

Remmert, Volker. "Picturing Jesuit Anti-Copernican Consensus: Astronomy and Biblical Exegesis in the Engraved Title-Page of Clavius's *Opera mathematica* (1612)." In John W. O'Malley, Gauvin Alexander Bailey, Steven J. Harris, T. Frank Kennedy, *The Jesuits II: Cultures, Sciences, and the Arts 1540–1773*. Toronto: University of Toronto Press, 2006, 291–313.

Reuber, Justus, ed. *Veterum Scriptorum, qui Caesarum et imperatorum Germanicorum res per aliquot secula gesta, literis mandarunt, Tomus unus*. Frankfurt: Andreas Wechel, 1584.

Ricci, Saverio. "Federico Cesi e la *Nova* del 1604: La teoria della fluidità del cielo e un opusculo dimenticato di Joannes van Heeck." *Atti dell'Accademia Nazionale dei Lincei: Rendiconti morali*, 8th ser., 48, no. 5–6 (1988): 111–133.

Rigaud, Stephen P. "Account of Harriot's Astronomical Papers." In *Supplement to Dr. Bradley's Miscellaneous Works: with an Account of Harriot's Astronomical Papers* (Oxford: Oxford University Press, 1833), pp. 1–70.

Roche, Edouard-Albert. *Recherches sur les offuscations du soleil et les météores cosmiques*. Paris: Leiber, 1868.

Roeck, Bernd. "Geschichte, Finsternis und Unkultur: Zu Leben und Werk des Marcus Welser (1558–1614)." *Archiv für Kulturgeschichte* 72 (1990): 115–141.

Ronconi, Giorgio. "Paolo Gualdo e Galileo." In *Galileo e la cultura padovana*. Edited by Giovanni Santinello. Padua: CEDAM, 1992, pp. 359–371.

Rosen, Edward. "Harriot's Science: The Intellectual Background." In *Thomas Harriot: Renaissance Scientist*. Edited by John W. Shirley. Oxford: Clarendon Press, 1974, pp. 1–15.

———, trans. *Kepler's Somnium*. Madison: University of Wisconsin Press, 1967.

———. *The Naming of the Telescope*. New York: Henry Schuman, 1947.

Sakurai, Kunitomo. "The Solar Activity in the Time of Galileo." *Journal for the History of Astronomy* 11 (1980): 164–173.

Saltzer, Walter G. "Kunst und Wissenschaft Komplimentär." In *Die Kunst und das Studium der Natur vom 14. zum 16. Jahrhundert*. Edited by Wolfram Prinz and Andreas Beyer. Weinheim: Acta Humaniora, 1987, pp. 17–40.

Sarton, George. "Early Observations of Sunspots?" *Isis* 37 (1947):69–71.

[Saxon Poet]. *Poetæ Saxonis Annalium de gestis B. Caroli Magni libri quinque*. In *Patrologia Latina*. Edited by Jacques-Paul Migne. 221 vols. Paris: Garnier & J.-P. Migne, 1844–1865, vol. 99.

———. *Annalium de gestis Caroli Magni Imp. Libri V. Opus auctoris quidem incerti, sed Saxonis & historici & poëtæ antiquissimi*. Helmstadt, Typis Iacobi Lucij, 1594.

Scaliger, Julius Caesar. *Exotericarum Exercitationum Liber XV. De Subtilitate ad Hieronymum Cardanum*. Hanover: Wechel, 1620.

Scheiner, Christoph, S.J. *De Maculis Solaribus et de Stellis circum Jovem errantibus Accuratior Disquisitio*. Augsburg: Ad insigne pinus, 1612.

———. *Pantographice seu Ars Delineandi Res quaslibet per Parallelogrammum lineare seu Cavum Mechanicum mobile*. Rome: Lodovico Grignani, 1631.

———. *Prodromus pro Sole Mobili*. Prague, 1651.

———. *Rosa Ursina*. Bracciano: Apud Andream Phaeum, 1626–1630.

———. *Sol Ellipticus: hoc est novum et perpetuum Solis contrahi soliti Phaenomenon, quod noviter inventum Strenae loco*. Augsburg: Christopher Mangius, 1615.

———. *Tres Epistolae de maculis solaribus*. Augsburg, 1612.

Scheiner, Christoph, S.J., and Johann Georg Locher. *Disquisitiones mathematicae de controversiis et novitatibus astronomicis*. Ingolstadt: Elisabeth Angermaria, 1614.

Schönberger, Georg. *Sol Illustratus*. Innsbruck, 1626.

Schove, D. Justin. *Sunspot Cycles*. Benchmark Papers in Geology, no. 68. Stroudsburg, PA: Hutchinson Ross, 1983.

Seneca. *On Benefits*. Translated by Aubrey Stewart. London: George Bell and Sons, 1900.

Settle, Thomas B. "Danti, Gualterotti, Galileo: Their Telescopes?" *Atti della Fondazione Giorgio Ronchi* 61 (2006): 625–638.

———. "Egnazio Danti: le meridiane in Santa Maria Novella a Firenze e gli strumenti collegati." In *Atti del Convegno Il Sole nella Chiesa: Cassini e le grandi meridiane come strumenti di indagine scientifica: Bologna, Archiginnasio, 22–23 settembre 2005*. Edited by Fabrizio Bònoli, Gianluigi Parmeggiani, and Francesco Poppi. *Giornale di astronomia* 32 (2006): 91–98.

Shapin, Steven. "The Invisible Technician." *American Scientist* 77 (1989): 554–563.

———. "Pump and Circumstance: Robert Boyle's Literary Technology." *Social Studies of Science* 14 (1984): 481–520.

———. *A Social History of Truth*. Chicago: University of Chicago Press, 1995.

Shapin, Steven, and Simon Schaffer. *Leviathan and the Air-Pump*. Princeton: Princeton University Press, 1985.

Shea, William R. *Galileo's Intellectual Revolution*. New York: Science History Publications, 1972.

———. "Galileo, Scheiner, and the Interpretation of Sunspots." *Isis* 61 (1970): 498–519.

Sirtori, Girolamo. *Telescopium: sive ars perficiendi*. Frankfurt: Paul Jacob, 1618.

Smith, A. Mark. "Galileo's Proof for the Earth's Motion from the Movement of Sunspots." *Isis* 76 (1985): 543–551.

[Snellius (Snel), Willebrord.] *De Maculis in Sole Animadversis . . . Batavi Dissertatiuncula*. Leiden: Plantijn, 1612.

———. *Eratosthenes Batavus*. Leiden: Iodocum à Colster, 1617.

Sommervogel, Carlos. *Bibliothèque de la compagnie de Jésus*. 12 vols. Brussels: Schepens, 1890–1932; reprint Brussels: Gregg, 1960.

Southern, R. W. *Medieval Humanism*. New York: Harper & Row, 1970.

Spear, R. E. "Scrambling for *Scudi*: Notes on Painters' Earnings in Early Baroque Rome." *Art Bulletin* 85, no. 2 (2003): 310–320.

Stabile, Giorgio. "Linguaggio della Natura e Linguaggio della Scrittura in Galilei: Dalla *Istoria* sulle Macchie Solari alle Lettere Copernicane." *Nuncius* 9 (1994): 37–64.

Stadler, Ulrich. "Moving Objects, Moved Observers: On the Treatment of the Problem of Relativity in Poetic Texts and Scientific Prose." *Science in Context* 18, no. 4 (2005): 607–627.

Steborius, Christophorus, S.J. *Universa Aristotelis Stagiritae philosophia ab illustrissimo Principe, ac Domino D. Georgio, Illustrissimi Principis Ianussij Ducis ab Ostrog . . . in Catholica et celebri Ingolstadiensium Academia die [XIII] Septemb. publicè propugnata, praeside R.P. Christophoro Steborio*. Ingolstadt: Andreas Angermaria, 1613.

Stephenson, F. R. See Clark, David H., and F. Richard Stephenson.

———. See Willis, D. M., and F. R. Stephenson.

———. See Yau, K. C. C., and F. R. Stephenson.

Swerdlow, Noel M. "A Star Catalogue Used by Johannes Bayer." *Journal for the History of Astronomy* 17 (1986): 189–197.

———. *The Renaissance of Astronomy in the Age of Humanism: Regiomontanus, Copernicus, Tycho, Kepler, Galileo.* Forthcoming.

Tanner, Adam, S.J. *De Creatione Mundi*. In *Universa Theologia Scholastica*, 4 volumes. Ingolstadt: 1626–1627, vol. 1.

———. *Dissertatio peripatetico-theologica de coelis*. Ingolstadt: Gregor Haenlin, 1621.

Tasso, Torquato. "Il Gonzaga secondo overo del giuoco." In *Dialoghi*. Edited by Ezio Raimondi. 2 vols. Florence: G. C. Sansoni, 1958.

Terence. *Phormio*. In *Phormio. The Mother-in-Law. The Brothers*. Edited and translated by John Barsby. Cambridge, MA: Harvard University Press, 2001.

Theophrastus. "De signis pluviarum, ventorum, tempestatis et serenitatis." In *Theophrasti Eresii opera, quæ supersunt*. Edited and translated by Friedrich Wimmer. Paris: A. Firmin Didot, 1866.

———. "Concerning Weather Signs." In *Enquiry into Plants and Minor Works on Odours and Weather Signs*. Edited and translated by Sir Arthur Hort. Cambridge, MA: Harvard University Press, 1916.

Thomason, Neil. "1543: The Year that Copernicus Didn't Predict the Phases of Venus." In *1543 and All That: Image and Word, Change and Continuity in the Proto-Scientific Revolution*. Dordrecht: Kluwer, 2000, pp. 291–332.

Toomer, G. J. See Ptolemy, Claudius.

Topper, David. "Colluding with Galileo: On Mueller's Critique of My Analysis of Galileo's Sunspots Argument." *Journal for the History of Astronomy* 34 (2003): 75–77.

Tuckerman, Bryant. *Planetary, Lunar, and Solar Motions: A.D. 2 to A.D. 1649 at Five-Day and Ten-Day Intervals*. Philadelphia: American Philosophical Society, 1964.

Tufte, Edward R. *Envisioning Information*. Cheshire, CT: Graphics Press, 1990.

[Turpin.] *Historia Karoli Magni et Rotholandi*. In *Veterum Scriptorum, qui Caesarum et imperatorum Germanicorum res per aliquot secula gesta, literis mandarunt, Tomus unus*. Edited by Justus Reuber. Frankfurt: Andreas Wechel, 1584.

Van Helden, Albert. "The 'Astronomical Telescope,' 1611–1650." *Annali dell'Istituto e Museo di Storia della Scienza di Firenze* 1 (1976): 13–35.

———. "Galileo and Scheiner on Sunspots: A Case Study in the Visual Language of Astronomy." *Proceedings of the American Philosophical Society* 140 (1996): 357–395.

———. "The Importance of the Transit of Mercury of 1631." *Journal for the History of Astronomy* 7 (1976):1–10.

———. "Longitude and the Satellites of Jupiter." In *The Quest for Longitude*. Edited by William J. H. Andrewes. Cambridge, MA: Harvard University Collection of Historical Scientific Instruments, 1996, pp. 85–100.

———. *Measuring the Universe: Cosmic Dimensions from Aristarchus to Halley*. Chicago: University of Chicago Press, 1985.

———. "Saturn and His Anses." *Journal for the History of Astronomy* 5 (1974): 105–121.

Van Maelcote, Otto. *Nuntius Sidereus Collegii Romani*. In *Le Opere di Galileo Galilei*, vol. 3, part 1, pp. 291–298.

Van Nouhuys, Tabitta. *The Age of Two-Faced Janus: The Comets of 1577 and 1618 and the Decline of the Aristotelian World View in the Netherlands*. Leiden: Brill, 1998.

Vermij, Rienk. *The Calvinist Copernicans: The Reception of the New Astronomy in the Dutch Republic, 1575–1750*. Amsterdam: Edita, Royal Dutch Academy of Arts and Sciences, 2002.

Verne, Jules. *De la terre à la lune*. Paris: Hachette, 1979.

Veterum Scriptorum, qui Caesarum et imperatorum Germanicorum res per aliquot secula gesta, literis mandarunt, Tomus unus. Edited by Justus Reuber. Frankfurt: Andreas Wechel, 1584.

Virgil, Publius Vergilius Maro. *Eclogues, Georgics, Aeneid I–VI*. Translated by H. Rushton Fairclough and revised by G. P. Goold. Cambridge, MA: Harvard University Press, 1916.

———. *Georgics*. Translated with an introduction and notes by L. P. Wilkinson. London: Penguin, 1982.

Vita Karoli Magni. See Einhard.

Vitali, Girolamo. *Lexicon Mathematicum Astronomicum Geometricum*. Paris: Louis Billaine, 1668.

Vliegenthart, Adriaan W. "Galileo's Sunspots: Their Role in 17th-Century Allegorical Thinking." *Physis* 7 (1965): 273–280.

Vyssotsky, A. N. "Astronomical Records in the Russian Chronicles from 1000–1600." *Meddelande fran Lunds Astronomiska Observatorium* Scr.2. Nr. 126, Historical Papers Nr. 22, Lund, 1949.

Waard, Cornelis de. "Alleaume." In *Nieuw Nederlandsch Biografisch Woordenboek*. Edited by P. C. Molhuysen and P. J. Blok. 10 vols. Leiden: A. W. Sijthoff's Uitgevers-Maatschappij, 1912, vol. 2, cols. 17–19.

Wade, Peter. "Naked Eye Sunspots, 1980–1992." *Journal of the British Astronomical Association* 104 (1994): 86–87.

Wattenberg, Diedrich. *David Fabricius: Der Astronom Ostfrieslands (1564–1617)*. Berlin-Treptow: Archenhold Sternwarte, 1964.

Welser, Marcus. *Opera Historica et Philologica, Sacra et Profana*. Edited by Christoph Arnold. Nuremberg: Wolfgang Moritz, 1682.

Westman, Robert S. "The Astronomer's Role in the Sixteenth Century: A Preliminary Study." *History of Science* 18 (1980): 105–147.

Wiedemann, Conrad. See Neumeister, Sebastian, and Conrad Wiedemann.

Willis, D. M., and F. R. Stephenson. "Solar and Auroral Evidence for an Intense Recurrent Geomagnetic Storm during December in AD 1128." *Annales Geophysicae* 19 (2001): 289–302.

Wohlwill, Emil. "Zur Geschichte der Entdeckung der Sonnenflecken." *Archiv für die Geschichte der Naturwissenschaften und der Technik* 1 (1908–1909): 443–454.

Wootton, David. "New Light on the Composition and Publication of the *Sidereus Nuncius*." *Galilaeana* 6 (2009): 61–78.

Xu, Zhentau, David W. Pankenier, and Yaotiao Jing. *East Asian Archaeoastronomy: Historical Records of Astronomical Observations of China, Japan, and Korea*. Earth Space Institute Series, no. 5. Amsterdam: Gordon and Breach Science Publishers, 2000.

Yau, K. K. C., and F. R. Stephenson. "A Revised Catalogue of Far Eastern Observations of Sunspots (165 BC to AD 1918)." *Quarterly Journal of the Royal Astronomical Society* 29 (1988): 175–197.

Zach, F. X. von. "Beobachtungen des Uranus; . . . und Anzeige von den in England aufgefundenen Harriotschen Manuscripten." In *Astronomisches Jahrbuch für das Jahr 1788*. Edited by J. E. Bode. Berlin, 1785, pp. 139–155.

Zapf, Georg Wilhelm. *Annales typographiæ Augustanæ ab ejus origine MCCCCLXVI usque ad annum MDXXX*. Augsburg: A. F. Bartholomæi, 1778.

Ziggelaar, August. "Jesuit Astronomy North of the Alps: Four Unpublished Jesuit Letters, 1611–1620." In *Christoph Clavius e l'attività scientifica dei Gesuiti nell'età di Galileo*. Edited by Ugo Baldini. Rome: Bulzoni, 1995, pp. 101–132.

Zinner, Ernst. *Astronomie: Geschichte ihrer Probleme*. Freiburg and Munich: Verlag Karl Alber, 1951.

———. *Entstehung und Ausbreitung der Coppernicanischen Lehre*. Erlangen: Mencke, 1943.

Ziolkowski, Theodore. "Judge Bridoye's Ursine Litigations." *Modern Philology* 92, no. 3 (1995): 346–350.

Zuñiga, Diego de (Didacus a Stunica). *In Iob Commentaria*. Toledo, 1584.

Manuscripts

Archivium Romanorum Societatis Jesu (ARSI), Germania Superior 4.

Fabricius, David: Extract of a letter to Micheal Maestlin, 1 December 1611 (o.s.). In Johannes Fabricius's Narratio. *Schafhausen, municipal library.*

Guldin, Paul. Correspondence. Graz Universitätsbibliothek, MS. 159.

Harriot, Thomas, East Sussex Record Office, Harriot MSS, HMC 241/8.

Lanz, Johannes. Correspondence. Dillingen Studienbibliothek, MSS 2°, vol. 247.

Peiresc, Nicholas Claude Fabri de. Papers. Bibliothèque Inguimbertine, Carpentras, Ms 1803, fols. 189r–223r.

Vatican Library, Fondo Urb. Lat. 1077, fol. 260r; 1079, fol. 827v–828r, 834v, 849v, 858r, 872r; 1080, fol. 20r, 26r, 51r.

Index

Page numbers in italics refer to illustrations.